Powder Metallurgy

Principles and Applications

Powder Metallurgy
Principles
and
Applications

By
Fritz V. Lenel

METAL POWDER INDUSTRIES FEDERATION
Princeton, New Jersey

Copyright 1980
Metal Powder Industries Federation

ISBN No. 0-918404-48-7
Library of Congress Catalog Card No. 80-81830

PRINTED IN THE UNITED STATES OF AMERICA

IV

Preface

In the ten years, since the publication of the last comprehensive text on powder metallurgy in the English language, both our understanding of the fundamentals and the technology of powder metallurgy have rapidly progressed. A new treatment of these subjects is, therefore, called for. Having taught a course in powder metallurgy for over 30 years, I am well aware of the need of a textbook for such a course. However, in writing this book, I have included not only those basic principles which should be the primary subject of a textbook, but also the applications of powder metallurgy from an engineering point of view. Without an understanding of how powder metallurgy is applied for practical uses, it is difficult to grasp its significance as a discipline. Making applications an integral part of the book should make it of value not only to those who want an introduction to the field, but also to the practitioners in the growing powder metallurgy industry.

Many of my colleagues, those in universities and particularly those in industry, helped me a great deal in writing the book. I should like to mention those who were kind enough to review individual chapters or sections of chapters, those who provided me with material on certain technological facets and those who helped my understanding of the production side by showing me their plants. They include: Mr. H. D. Ambs, Mr. N. A. Arnold, Mr. L. W. Baum, Dr. Friedrich Benesovsky, Dr. Stoyan Boniff, Mr. W. A. Buerkel, Dr. P. C. Eloff, Dr. H. E. Exner, Mr. Mark Gaydos, Dr. R. M. German, Dr. G. H. Gessinger, Dr. K. E. Grambo, Mr. Ralph Handron, Mr. W. T. Haswell, Dr. R. W. Heckel, Mr. H. H. Hirsch, Dr. W. J. Huppmann, Dr. G. C. Kuczynski, Mr. J. O. Kwolek, Mr. L. H. Mott, Dr. H. S. Nayar, Mr. J. W. O'Brien, Dr. L. F. Pease III, Dr. G. Petzow, Mr. K. H. Roll, Mr. F. A. Schaa, Jr., Dr. Erhard Schelzke, Mr. Hermann Silbereisen, Dr. D. W. Smith, Dr. D. N. Yoon, and Dr. Gerhard Zapf.

It goes without saying that omissions and errors which are inevitable in a book of this kind are the sole responsibility of the author. I would be grateful to have my attention called to them.

The help of those organizations which kindly supplied me with illustrations for the book is acknowledged in the captions. My sincere thanks go to my wife who typed the manuscript.

F. V. Lenel

Troy, N.Y., April 1980

V

*To my former students all over
the world whom I had the
pleasure to introduce to
powder metallurgy*

Table of Contents

CHAPTER 19

**Applications of P/M Structural Parts and Powder Forgings
Their Economics and the Energy Requirements to
Produce Them** 469

CHAPTER 20

Cermets 483

CHAPTER 21

Wrought Powder Metallurgy Products 495

CHAPTER 22

**Powder Metallurgy Applications Involving Sintering
with a Liquid Phase** 521

CHAPTER

1

Introduction

Powder metallurgy may best be defined by comparing it with "fusion metallurgy." In fusion metallurgy a metal or alloy is melted and cast into a mold. The mold may have the shape of the desired product, in which case a casting is produced, or the shape of an ingot may be cast. The ingot is then formed into a wrought product by rolling, forging, extrusion, drawing, machining, etc. In powder metallurgy, metal powders, i.e. metals in finely divided form rather than molten metal, are the starting material. The powders are consolidated into products with a given shape. The basic steps in powder metallurgy are therefore powder production and powder consolidation.

The most common sequence in powder consolidation includes pressing the powder in a die into a compact and sintering the compact, which means heating it to a temperature below the melting point of the metal or alloy to give it the desired physical, mechanical and chemical properties. As in fusion metallurgy the consolidation process may lead to a part with the desired final shape corresponding to a casting or to a compact corresponding to an ingot which must be further shaped by forming.

Metal powders are employed for many other purposes besides consolidation into shapes. They are used in paints, varnishes and printing inks, as reagents in the chemical industry, as explosives particularly in military applications and as food additives. In metal working they are employed in cutting and cleaning and in producing metallic coatings, as e.g. for welding electrodes. None of these uses of metal powders are part of powder metallurgy.

Why would one want to produce a metal product with a given shape starting with metal powders rather than molten metal? One important reason has an economic basis. The cost of producing a part of a given shape to the required dimensional tolerances by powder metallurgy may be lower than by casting or as a wrought product. The powder metallurgy product, often called a "P/M structural part," must, of course, have adequate properties, usually adequate mechanical properties, for the application, but its cost rather than any unusual properties is the main reason for fabricating it from powder.

Producing structural parts is, however, only one and by no means the oldest application of powder metallurgy. Other applications are based on the unusual properties which can be obtained by powder metallurgy. They range from fabricating metals with very high melting points to applications where high wear resistance is needed, to porous materials, to products with special frictional, magnetic and electrical properties and many others. They will be briefly discussed in the second section of this chapter.

Before this review of powder metallurgy applications is presented, the individual processing steps in producing and consolidating metal powders will be discussed from both a practical and theoretical point of view. Certain methods of metal powder production are used primarily for non-powder metallurgy applications, e.g. methods for flake metal powders for paints, etc. However, in general, no sharp distinction between methods for producing powders for powder metallurgy and those for other uses is possible. All methods by which powders for powder metallurgy applications may be produced are included in chapter 2 on metal powder production. One important method which is called "atomizing" involves breaking up a stream of molten metal or alloy into droplets which solidify as powder particles. It may be argued that this method belongs into fusion metallurgy, but it is definitely considered part of powder metallurgy. Other methods of producing metal powders are chemical reactions, such as the reduction of compounds, often oxides, with gaseous, liquid or solid reductants at elevated temperatures, but below the melting point of the metal, or by reduction reactions in aqueous solutions including deposition of a metal by an electric current. Thermal dissociation of compounds, such as carbonyls or hydrides may lead to metal powders. Less commonly, powders may be produced by disintegration of solid metal into powder.

After metal powders have been produced, they must be characterized and tested as described in chapter 3. Such characteristics, as particle size and size distribution, particle shape and specific surface are of interest in most technologies concerned with finely divided materials. Many of the methods for testing metal powders have been adapted from those for cement, pigments and other powder materials.

Chapters 4, 5 and 6 will be concerned with the consolidation of metal powders at room temperature by application of pressure, which is a necessary step in the fabrication of the majority of powder metallurgy products.

It is most often done in rigid dies made of tool steel or cemented carbides. Pressures in the range from 70 to 700 MPa (10,000 to 100,000 psi) are used. The compacts so produced, called "green compacts," are strong enough so that they can be ejected from the die and handled. They are porous and have a lower density than cast and wrought parts of the same metal. An introduction to compacting and the theory of compacting are presented in chapter 4. In this theory the relationship between compacting pressure and the density and density distribution is of principal interest. Powder metallurgy parts often have complex shapes. In order to produce such parts economically, systems of automatic compacting presses and tools have been developed which will be described in chapter 5. Methods of compacting other than in rigid dies are the subject of chapter 6. They include isostatic pressing in flexible molds, powder rolling, also called roll compacting, powder extrusion and powder injection molding.

When powders are pressed into compacts at room temperature or are shaped by powder rolling or extrusion, the resultant products have insufficient strength and ductility for most applications. In order to make them useful they have to be sintered. Many powder metallurgy products are sintered in continuous furnaces, in which the compacts are transported through a preheat, a high-heat and a cooling zone using pusher, endless belt, roller hearth or walking beam mechanisms. A protective atmosphere, or sometimes vacuum, must be maintained in the furnaces to prevent undesirable reaction with the compacts during heating and cooling. Details of sintering practice are presented in chapter 7.

In order to understand what happens during sintering of compacts, the changes in density, in dimensions, in metallographic structure and in mechanical properties during sintering must be studied from an experimental point of view, beginning with those in compacts from a single metal powder or from homogeneous solid solution alloy powders. The results of these experiments are given in chapter 8. The theory of sintering of these compacts or of loose powder aggregates of single metal powders is developed in chapter 9. The driving forces, which cause neck growth between powder particles and densification, and the mechanisms of material transport under the influence of these forces is pointed out. The theory of sintering of metal powder is quite similar to that of ceramics and research in this field has been cooperative work of solid state physicists, ceramists and powder metallurgists.

Sintering of compacts from a mixture of metal powders at temperatures where no liquid phase is formed is discussed in chapter 10. The mechanisms involved are the interdiffusion between the components of the powder mixture in addition to the sintering process occurring in single metal powder compacts.

Compacts of mixtures of metal powders may also be sintered at temperatures where a certain amount of liquid phase is formed. The amount must be small enough to be held by capillary forces within the skeleton of the remaining solid phase so that the compacts do not warp or lose shape. This process is technically important and the mechanisms involved in it have been intensively studied. Chapter 11 is concerned with this process of liquid phase sintering.

The sequence of first compacting and subsequently sintering is not always followed in consolidating metal powders. Occasionally loose powder aggregates or slip castings made from a suspension of metal powder in an aqueous medium may be sintered. This type of sintering, together with the process of infiltrating porous compacts with a lower melting point metal either simultaneously with or following sintering is discussed in chapter 12. In recent years, hot consolidation of metal powders combining compacting and sintering in one step has grown in importance. Hot consolidation techniques including hot pressing in rigid dies, hot isostatic pressing, hot extrusion and hot forging are presented in chapter 13. In hot isostatic pressing metal powder confined in flexible envelopes, often made of steel sheet, is subject to gas pressure at sintering temperatures.

In the second part of this chapter a preview of applications of powder metallurgy is presented from a historical point of view starting with the precursors of powder metallurgy technology and proceeding to the development of the technology in the present century.*

In prehistoric and historic times up to the early years of this century, powder metallurgy was a technology for producing wrought products from metals which could not be melted because their melting point was too high. It began with iron in the form of sponge iron produced by reduction with charcoal in charcoal fired furnaces from iron oxide in the form of more or less pure ores. This sponge was then forged into solid iron. Unless the sponge was exceptionally pure the sponge iron would contain large amounts of non-metallic impurities. Since ancient iron and steel were often remarkably free of inclusions, W. D. Jones[1] believes that the process was modified perhaps in the manner described by Gardi[2] as being used by certain African natives. The sponge after reduction was broken up into powder particles, washed and hand picked to remove as much of the gangue and slag as possible and the powder either compacted or loose sintered into a porous material, which was then forged.

---

*A history of powder metallurgy has been presented by the Italian powder metallurgist, Gian Filippo Bocchini, of the Merisinter Company in Naples, Italy, in his book "L'Uomo e i metalli 'difficili'," Man and the "difficult" metals, privately printed in a limited edition in 1978.

Powder metallurgy again came to the fore between 1750 and 1850, when demand arose for platinum in wrought form. Since pure platinum could not be melted, several rather similar processes to produce wrought platinum were developed in Spain, England and Russia. Platinum powder was pressed and the compacts sintered and hot forged. The processes were described in the scientific literature several years after they had been developed and used commercially. The English process was described by Wollaston in 1829[3] and the Russian one by Sobolevskiy in 1834.[4] The powder metallurgy method of producing platinum became obsolete, when suitable furnaces and refractories for melting platinum were developed. Powder metallurgy entered the stream of modern metals technology early in the 20th century. Again a metal which could not be melted and which, in addition, could be made malleable only with great difficulty, i.e. tungsten, was involved. The interest in tungsten arose from the demand for filaments for incandescent lamps more stable than the Edison carbon filament. Numerous powder metallurgy methods were developed between 1900 and 1910 for producing tungsten wire. The only commercially successful method was the one developed by William Coolidge in a process first described in 1910 and patented in 1913.[5,6] Relatively few changes have been made over the years in Coolidge's procedure; it is still the standard method of producing incandescent lamp filaments used all over the world. In this method very fine tungsten oxide powder, WO_3, is reduced by hydrogen. The powder is pressed into compacts, which are presintered at $1200°C$ to make them strong enough so that they can be clamped into contacts. They receive a final sintering treatment near $3000°C$ by passing a low voltage, high current density current through the compacts. During sintering the compacts shrink and reach a density near 90% that of solid tungsten. The sintered compacts can be worked only at temperatures near $2000°C$. When heated to this temperature they can be swaged into rounds. With increasing amounts of warm work tungsten becomes more ductile so that the swaged bars can be drawn into fine wire at lower temperatures.

The use of powder metallurgy for producing wrought products from metals which could not be readily melted was important in the early technology of the other refractory metals, molybdenum, tantalum and niobium, and of the reactive metals, titanium and zirconium. However, starting about 1940 new methods of melting and casting, i.e. vacuum arc and electron beam melting, were developed, by which all metals regardless of their melting point could be cast. Nevertheless, even today, all tungsten products and a large percentage of those from molybdenum are produced by powder metallurgy.

Only compacts used for incandescent lamp filaments are still sintered by passing a current through them. For other applications, large compacts of tungsten, molybdenum and molybdenum alloys are isostatically pressed from powder and then sintered in furnaces with molybdenum or tungsten heating elements. They are then rolled into sheet or forged or extruded into the desired shape.

In addition to pure tungsten and molybdenum, certain alloys of the refractory metals were developed for special applications. They include alloys for heavy duty electrical contact materials, in which the refractory metals, tungsten or molybdenum, which have high hardness and low rates of material transfer during making and breaking contact, are combined with copper or silver with their high electrical and thermal conductivity. A common method of producing these contact materials is by infiltrating a porous sintered skeleton of the refractory metal with liquid copper or silver.

Not all tungsten alloys are produced by infiltration. The so-called "heavy alloys" developed by Price, Smithells and Williams[7] in 1937 are compacted from a mixture of tungsten powder with less than 10% of either nickel and copper or nickel and iron powders. They may be sintered to theoretical density and are used because of their high density.

Early in the 1920s powder metallurgy techniques were developed for two other types of products, neither of which can be produced by fusion metallurgy. They are self-lubricating bearings and cemented carbides.

Self-lubricating bearings were first successfully produced in the United States. Their characteristic feature is their porosity which can be controlled with regard to total amount, generally in the range from 15-30%, and to size distribution of the pores. They are small enough so that they cannot be detected by the naked eye. They are connected with each other and with the surface of the bearing so that, after being impregnated, the bearing can serve as a reservoir for lubricant. When the bearing is assembled with a shaft and the shaft begins to turn, the temperature rise causes the lubricant to come out of the pores and form a lubricant film between shaft and bearing. The lubricant is reabsorbed in the pores of the bearing structure when the shaft ceases to turn. Self-lubricating bearings are used in very large quantities, particularly in household appliances, where the lubricant in the bearing structure is the only lubricant available for the life of the appliance.

Self-lubricating bearings are compacted from a mixture of 90% copper and 10% tin powder, often with an addition of graphite powder. The mixture is pressed on automatic presses into compacts close to the desired shape of sleeve bearings. The compacts are sintered under non-oxidizing conditions, which originally meant packing them into boxes with spent carburizing compound. This was soon changed to sintering in atmospheres of partially combusted hydrocarbon gases in continuous muffle furnaces at temperatures near 800° C. During sintering, copper and tin alloy and form bronze. Small dimensional changes occur during sintering. The bearings are, therefore, subsequently "sized" in sizing dies and then impregnated with lubricant.

Powder metallurgy applications related to self-lubricating bearings are porous filters, metal brushes and metallic friction materials. Filters are used to remove solid impurity particles from liquids and gases. In contrast to paper, glass and ceramic filters, metallic filters can be plastically deformed and, therefore, readily fitted into housings. Spherical bronze powder filters are produced by loose powder sintering. Filters from stainless steel, nickel alloy and titanium powders were later developed.

Metal brushes produced by mixing copper and graphite powders, pressing and sintering are used in electric motors and generators when brushes higher in conductivity than those from straight carbon are required.

Metallic friction materials consist of a dispersion of a friction producing material, generally silica or a silicate, in a metallic matrix, usually copper or iron base. They are used for brake linings and clutch facings in heavy duty applications, where their ability to absorb energy at a higher rate and their higher wear resistance justify their higher cost compared with the conventional asbestos linings bonded with an organic resin. Several specialized methods for producing metallic friction materials have been developed.

Cemented carbides were first developed in Germany. Karl Schröter of Fried. Krupp[8] received a basic patent in 1925. Cemented carbides consist of a metal carbide and from 3 to 20% of a metallic binder with tungsten carbide, WC, and cobalt being the ones used first and still used most widely. They have high hardness and wear resistance due to the carbide phase and yet are tough enough so that they can be used for many applications for which the carbide alone is too brittle. First developed as a die material for drawing tungsten wire, cemented carbides have their most important use in cutting tools. The tungsten carbide is intimately mixed with cobalt powder by ball milling, pressed and sintered in hydrogen near 1400°C. At this temperature, a liquid phase, a eutectic of carbide and binder metal, is formed. During sintering, cemented carbide compacts shrink considerably and reach near theoretical density without losing their shape. The microstructure of cemented carbides consists of a dispersion of carbide particles in a binder matrix. Such a structure cannot be produced by fusion metallurgy, particularly since WC upon heating below its melting point decomposes into W_2C and carbon.

Many new compositions for cemented carbides have been developed, particularly those for cutting steels; they contain titanium and tantalum carbides. Besides their use as cutting tools in metal cutting and mining, cemented carbides have found other applications where high wear resistance is needed, such as dies and rolls for rolling mills.

A powder metallurgy application related to cemented carbides, which in its origin goes back to the 1930s, is metal bound diamond tool materials, in which diamonds of controlled particle size are dispersed in a metallic matrix. The tools are used for grinding wheels, saws and cut-off wheels for highly abrasive materials which cannot be ground with conventional alumina or silicon carbide wheels.

As was pointed out earlier in this chapter, structural parts are produced by powder metallurgy not because they have unique properties but for economic reasons. The structural parts technology, which started in the late 1930s, grew out of that for self-lubricating bearings. Like bearings, structural parts are pressed on automatic presses and sintered in continuous furnaces. In contrast to bearings, most structural parts are produced from steel compositions and must be sintered at temperatures above 1100° C (2000° F). Continuous furnaces with protective atmospheres operating at these temperatures had been developed for assembling steel parts by copper brazing and were therefore available for structural parts. The first commercially successful structural parts were oil pump gears[9] which replaced gears machined from cast iron slugs. They were easy to compact, since they had a uniform cross section along their length. They were made from a mixture of iron and graphite powders which, after sintering, gave them a structure consisting of a eutectoid steel matrix with about 25% porosity. Because of the porosity, the gears had mechanical properties similar to cast iron rather than steel. The technology of structural parts from steel has developed along two major directions, parts of more complex shape and larger size, and parts with improved mechanical properties approaching those of heat-treated low-alloy steels. To produce parts with more complex shapes, in particular parts with several thickness levels in the direction of pressing, automatic presses incorporating movements which compact each of these levels to constant density have been developed. Also automatic presses of larger tonnage have become available for producing larger and heavier structural parts.

Improved strength and hardness in steel powder metallurgy parts are often achieved by alloying additions—copper and nickel additions are widely used—and by heat-treating the parts after sintering by quenching and tempering, by carburizing or by carbonitriding. The most effective way to improve mechanical properties, in particular to achieve better ductility and toughness, is to increase their relative density compared with that of wrought products of the same composition. Powders with greater compressibility, which give a higher density at a given compacting pressure, were developed. After sintering, structural parts may be repressed and resintered. Iron base parts may be infiltrated with a lower melting metal, such as copper or copper alloy, to fill the pores and increase density. The most effective way to obtain densities near or equal to theoretical density is to hot forge a structural part "preform" after compacting and sintering. The use of this technique is rapidly growing.

The market for structural parts by powder metallurgy has greatly expanded in the last 30 years. Structural parts are the largest field of application for powder metallurgy products, not only as to the amount of powder consumed, but also dollar volume. Parts from iron and low alloy steel constitute the bulk of structural parts production, but parts are also produced from copper, bronze, brass, nickel silver, stainless steel, aluminum and titanium alloys.

Production of structural parts by powder metallurgy has important advantages compared with competing methods. It conserves material because parts do not need subsequent machining. Parts are produced on automatic compacting presses and sintered in continuous furnaces, so that the powder metallurgy approach saves labor and insures high productivity. Fabrication by powder metallurgy requires less energy than other methods and is ecologically clean. Not only structural parts which must have adequate mechanical properties but also certain magnetic parts with satisfactory magnetic properties are produced by powder metallurgy for economic reasons. Soft magnetic parts such as pole pieces, relay cores and computer components with complex shapes are pressed and sintered from metal powders. Alnico permanent magnets are produced from powder as an alternative to being cast and finished by grinding because they cannot be machined.

The applications of powder metallurgy conceived in the first half of this century have undergone further extensive development in the last thirty years, but entirely new applications have also been introduced. The development of cemented carbides, of metal-diamond tool materials, of metallic brushes and metallic friction materials had shown that products in which metallic and non-metallic powders are combined in intimate mixtures could be made by powder metallurgy. These combinations are now called "cermets," because they contain ceramics, i.e. high melting point inorganic non-metallic materials, and metals. A new type of cermets are nuclear fuel elements which consist of a dispersion of a fissionable compound, such as UO_2, UC or UN in a metallic matrix, e.g. aluminum or stainless steel. By dispersing the fissionable compound, radiation damage in the fuel element can be reduced, the total burn-up (percentage of uranium undergoing fission) increased and the life of the element prolonged. To optimize the life of dispersion type fuel elements, the size and the distribution of the fissionable compound particles in the matrix must be carefully controlled.

The new applications of powder metallurgy of greatest significance for the future of the technology are the ones based on hot-consolidation of powders. In several of these applications powder metallurgy has made it possible to produce metallurgical structures and achieve properties which cannot be obtained by fusion metallurgy. Cast beryllium is coarse-grained and completely brittle, while the fine-grained beryllium produced by hot pressing or hot isostatic pressing of beryllium powder has improved, albeit limited ductility. Tool steels, superalloys and titanium alloys may be produced by atomizing the alloys into spherical powders and hot isostatically pressing the powder into completely dense compacts. In powder compacts segregation can be avoided and microstructural features, such as the size and distribution of carbides and intermetallic compounds controlled, leading to improved properties. Fabrication by hot isostatic pressing will also make it possible to produce shapes near those of the final components, e.g. turbine disks in the case of superalloy powders, and thereby save material.

A final group of materials produced by hot consolidation of powders are dispersion strengthened alloys. They consist of a metallic matrix in which particles of a second phase, usually an oxide, are very finely dispersed. The size of the particles is less than 0.1 μm and the dispersed particles must be insoluble in the matrix at temperatures near the melting point of the matrix metal or alloy. For this reason dispersion strengthened alloys have better strength and resistance to creep and recrystallization than other alloys. Dispersion strengthened alloys, hot consolidated from powders, are under intensive development. Certain lead, copper and nickel base alloys are produced commercially.

The discussion of powder metallurgy applications in chapters 14 to 24 includes those in commercial production as well as those in the development stage. For each of these applications, methods of fabrication and important physical, mechanical and chemical properties are presented. On the other hand, no attempt has been made to include all the published properties of powder metallurgy materials which were obtained because they were of scientific interest, but not related to any actual or potential application.

References

1. W. D. Jones, Fundamental Principles of Powder Metallurgy, London, p. 593, (1960).
2. R. Gardi, Der schwarze Hephästus, (1934).
3. W. H. Wollaston, Phil. Trans. Roy. Soc., Vol. 119, p. 1-8, (1829).
4. P. G. Sobolevsky, Ann. der Physik und Chemie, Vol. 33 (7), p. 99-109, (1834).
5. W. D. Coolidge, Trans. AIEE, Vol. 29, Part II, p. 961-965, (May 1910).
6. W. D. Coolidge, U. S. Patent No. 1,082,938, (Dec. 30, 1913).
7. G. H. S. Price, C. J. Smithells and S. V. Williams, J. Inst. of Metals, Vol. 62, p. 239-254, (1938).

8. K. Schröter, U.S. Patent 1,549,615, (1925).
9. F. V. Lenel, Oil Pump Gears, in "Powder Metallurgy," ed. by J. Wulff, Cleveland, p. 502-511, (1942).

CHAPTER
2

Metal Powder Production

In this chapter methods for producing metal powders are treated. Emphasis is put on methods for those types of metal powders which are commercially important in powder metallurgy fabrication, i.e. for producing shaped objects or massive materials from metal powders. There are many other uses for metal powders, e.g. as chemical reagents, for explosives, for paints, inks and varnishes, but methods of producing metal powders for these applications are generally not discussed. Even with these restrictions, the list of methods is quite extensive and includes quite different approaches to making metal powders. As a matter of fact, in the case of iron and copper, powders made by very different methods compete with each other for the same type of powder metallurgy applications. In order to compare these methods with each other, the production methods for each individual metal or group of metals and alloys are discussed. Emphasis is put on commercial methods, presently used, but older methods of historical interest and methods in the development stage are also mentioned.

Before these methods are treated, a general classification and review of the approaches to producing metal powders for powder metallurgy applications is presented. They may be divided into these categories:

> Physical methods
> Chemical methods
> Mechanical methods

The most important physical method of powder production is designated by the rather misleading term "atomization." A stream of molten metal is broken up into droplets which freeze into metal powder particles. In most atomizing processes a stream of liquid, usually water, or of gas impinges upon the liquid metal stream to break it into droplets. Powder production by atomizing is inherently inefficient from the point of view of energy utilization, since only a small percentage of the energy, which the stream of atomizing medium imparts to the molten metal stream, is used to create new surfaces of the powder particles. A fundamental treatment of atomizing from a fluid dynamics point of view appears to be difficult. Attempts to provide such a treatment may be found in chapter 1, "Manufacture of Metal Powders" in W. D. Jones' "Fundamental Principles of Powder Metallurgy" and in a series of recent review articles by Lawley,[1] by Shinde and Tendolkar[2] and by Tallmadge.[3] Most atomizing processes for metal powders have been developed more or less empirically. The practical side of metal powder atomization has been reviewed by J. F. Watkinson,[4] P. Ulf Gummerson[5,6] and, in the case of gas atomization, by E. Klar and W. M. Shafer[7] and is briefly discussed here.

The first methods for producing metal powders by atomizing were those for low melting point metals. The molten metal is held in the liquid state in a tank and is raised by the suction produced by the atomizing medium, commonly hot air, through a pipe to the atomizing nozzle.

For higher melting point metals and alloys, i.e. copper, iron and nickel base alloys, a stream of molten metal issuing from an orifice at the bottom of a tundish is broken up by a jet of atomizing fluid, which may be water, air, steam or an inert gas. As an example, applicable to both water and gas atomization, atomizing with flat stream V-jets is illustrated in Figure 2-1.[6] The following properties of the powders:

> Average particle size
> Particle size distribution
> Particle shape
> Particle chemistry and structure

are controlled by major variables in the atomizing process which is illustrated again in a schematic diagram in Figure 2-2.[6] Finer particle size is favored by low metal viscosity, low metal surface tension, superheated metal, small nozzle diameter, high atomizing pressure, high volume and high velocity of the atomizing fluid, short metal stream and short jet length. The effect of the apex angle is still controversial. The pressures used in water atomization are 3.5-20 MPa (500-3000 psi), those for gas atomization are generally lower in the range of 0.7-2.8 MPa (100-400 psi).

The particle size distribution of powders produced by atomizing is commonly log normal, which means that plotting the weight fraction of particles in a size range vs the size on a logarithmic scale results in the well-known

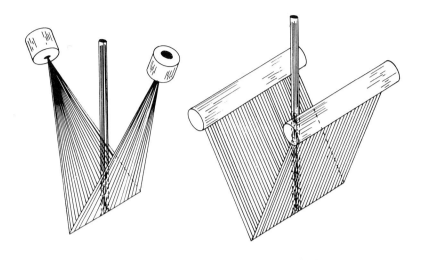

FIGURE 2-1 Principle of atomizing with flatstream V-jets. Two-way plug jet with diverging flatstreams, above. Two way curtain jet, below.

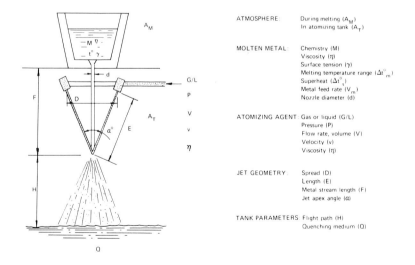

FIGURE 2-2 Major variables in the atomizing process.

15

bell shaped curve. It is difficult to narrow the particle size distribution of an atomized powder by controlling the atomizing variables, e.g. by making the energy transfer from atomizing fluid to liquid metal uniform and as efficient as possible.

The control of particle shape is of major importance when producing metal powders by atomizing. When liquid metal droplets freeze undisturbed, smooth particles with spherical or teardrop shape are formed. Spherical atomized powders are needed when the powders are to be sintered loose or are hot consolidated rather than cold pressed. Powders with irregular particle shape, on the other hand, are required for powder metallurgy applications in which a powder is cold pressed into a compact. Except for soft powders, such as tin and aluminum, compacts from powders with spherical shapes have inadequate green strength. Obtaining irregularly shaped powder particles depends on atomizing conditions. It is generally easier to produce such particles by water than by gas atomization. High water pressures, i.e. higher water velocities, result in more irregular particle shapes owing to greater impact forces and more rapid quench. Superheating the liquid metal, which leads to spheroidization, should be avoided. Particle shape is also influenced by the surface tension of the liquid metal. Small additions of deoxidizers, such as phosphorus, increase the surface tension of copper and copper-base alloys and lead to spherical powder shape. The existence of solid oxide films, such as ZnO, acts in the opposite way and explains why irregular brass powder can be readily produced by air atomization. When austentic stainless steel powders were first produced, it was necessary to add silicon to the melt in order to produce powder with irregular shape; this is no longer necessary. The details of the commercial methods for producing iron, low alloy steel, copper and copper alloy and stainless steel powders by atomizing processes using external nozzles are described in the individual sections for these metals.

Recently a number of special atomizing processes for spherical powders from highly alloyed materials have been developed, which differ significantly from older atomizing processes. They are discussed in a section at the end of this chapter.

Metal powders may be produced by methods related to atomization, such as roller atomization,[8] the vibrating electrode method[9] and the melt drop technique.[10] They are not yet of commercial importance and are only mentioned here.

Chemical methods of powder production are those in which a metal powder is produced by chemical decomposition of a compound of the metal. This includes the large group of reduction reactions. Oxides in the form of finely divided solid powder particles may be reduced, as in the reduction of tungsten oxide to tungsten powder and of copper oxide to copper powder with hydrogen or of iron oxide to iron powder with carbon monoxide. An aqueous solution of a metal compound may be reduced, such as a nickel ammine sulphate solution to nickel powder with hydrogen or an acidified copper sulphate solution to cement copper with scrap iron. Reduction of liquid or gaseous metal halides is less common.

The deposition of a metal by an electric current from an electrolyte containing a metal salt may be treated as a special kind of reduction process. The electro-deposition of copper powder from an aqueous copper sulphate solution is an example.

Metal powders may also be produced by thermal decomposition of certain metal compounds. Examples are the decomposition of titanium hydride into titanium powder and hydrogen or the decomposition of nickel carbonyl into nickel powder and carbon monoxide. The chemical methods are discussed in detail in the sections describing the production of individual powders.

For a review of the equilibria involved in the chemical reactions which are governed by thermodynamics, and of the rates, which are governed by reaction kinetics, reference is made to chapter 1 of W. D. Jones' "Fundamental Principles of Powder Metallurgy." Equilibria and reaction rates of reduction reactions in aqueous solutions are discussed by Burkin and Richardson.[11]

Certain powders may be produced by mechanical comminution processes generally used for brittle materials, such as ores. They are being studied extensively as an important subject in mineral engineering. They are not discussed in detail in this chapter. The reason is that comminution plays only a minor role in metal or alloy powder production because comminuted brittle powders cannot be cold compacted as such. They may, however, be used as minor constituents in powder mixtures. Examples of brittle, mechanically comminuted alloy powders used in powder metallurgy are iron-aluminum alloys used in mixtures for producing alnico permanent magnets, ferro-silicon powders in mixtures for producing soft magnetic components and ferro-phosphorus powders in mixtures for producing structural parts.

Metals or alloys may be embrittled so that they can be comminuted mechanically. This may be done by heat treatment, as in the case of steels, or chemical means may be used for embrittling. Examples are brittle deposits of iron used in electrolytic iron powder production, embrittled permalloy powder for telephone cores and brittle metal hydrides for producing powders of titanium and zirconium and their alloys. Whether these methods should be considered chemical or mechanical methods of powder production is a matter of semantics.

Even ductile metals can be comminuted into powder when they are available as wire and sheet scrap. This wire or sheet scrap is fed into a so-called eddy-mill, in which the metal particles are made to impinge upon each other, thereby work-hardening and embrittling each other so that they finally break up into a powder of dish-shaped particles. The means of making the metal particles impinge upon each other are two air streams produced by impellors at opposite ends of the chamber into which the metal scrap is fed. Comminution by eddy-mill was, at one time, used in Germany for producing powders for powder metallurgy applications, but the method is no longer or commercial importance.[12]

A powder production process, combining atomization and comminution, is the DPG process. It was extensively used for the production of metal powders for powder metallurgy in Germany during the second world war, but is no longer used commercially. Molten metal flows from a tundish through a ceramic nozzle. At the bottom of the nozzle an annular cone of water surrounds the molten metal stream, which impinges upon a rapidly rotating disk upon which a series of knives are mounted.[13]

A recently developed specialized method of producing metal and alloy powders of fine particle size by mechanical comminution is the "cold-stream" process.[14] The size of the feed material for this process should generally not be larger than 2 mm and may, e.g., be coarse spherical atomized powder. It is fed into the blast generator in which it is suspended in a pressurized gas stream at a pressure of 7 MPa or higher. The gas is usually air, but inert gases may be substituted, if oxidation of the powder is to be held to a minimum. The gas stream with the suspended material is expanded through a venturi nozzle into the blast chamber in which the material impinges against a fixed target and size reduction takes place. The comminuted material is separated according to particle size in classifiers with the oversize returned to the blast chamber. When the gas stream is expanded, it is cooled together with the material to be comminuted. The lowered temperature materially aids the comminution process. A principal advantage of the cold stream process compared with other comminution processes is the fact that it can be operated with minimum contamination. The target against which the material impinges may often be made of the same composition as the material to be comminuted, e.g. in the case of tool steel powder. Materials may be comminuted in the cold stream process to powder with particle sizes of 10 μm and finer.

Before individual methods of powder production are presented, another important point which relates to the economics of powder production should be mentioned. Metals in the form of powder may be the product of processes in which a metal is obtained from its ore. In some cases the metal powder so produced may be used directly as the raw material for the powder metallurgy process. This would, of course, give powder metallurgy an economic advantage as a method of powder metallurgy fabrication. This applies particularly to the so-called "continuous P/M processes," such as powder rolling where powder is fabricated into sheet, strip, rod or tubing. On the other hand, the metal powder obtained in the production of a metal from its ore may not be directly usable in powder metallurgy applications. The steps necessary to modify the characteristics of the powder to make it usable may diminish or nullify the economic advantage of obtaining the metals as powders. Examples will be presented in the sections on the production of individual metal powder. These sections are:

I. Iron powder
II. Low alloy steel powder
III. Stainless steel powder
IV. Tool steel powder
V. Copper powder
VI. Copper alloy powder
VII. Silver powder
VIII. Nickel powder
IX. Nickel alloy powder
X. Cobalt powder
XI. Tin powder
XII. Aluminum and aluminum alloy powders
XIII. Magnesium alloy powders
XIV. Tungsten powder
XV. Molybdenum powder
XVI. Tantalum powders
XVII. Powders of titanium, zirconium and their alloys
XVIII. WC, TiC and TaC
XIX. Special atomizing methods
XX. Beryllium powder

I. Iron Powders

The principal use of iron powders is for P/M structural parts, both those processed conventionally, i.e. by cold compacting and sintering without pressure application and those produced as preforms and then hot forged into components. Other uses for iron powder are in magnetic parts, in certain self-lubricating bearings and in friction materials. In their use for structural parts, four properties of iron powders are important, the first two of which also apply to other powders for structural parts. Even though powder characterization and properties are to be discussed in detail in chapter 4, these properties are mentioned here because they determine whether a powder produced by a certain method is suitable for P/M structural applications. The properties are:

1. The powders should have enough flowability so that they readily fill the die cavities in automatic compacting presses. This means they should not contain too high a percentage of sub-sieve (finer than 44 μm) size powder (generally not over 50%).

2. Compacts from the powders should have adequate green strength so that they can be ejected from the press and transferred to the sintering furnace without breaking. This applies not only to compacts from the powder itself, but to those from mixtures of powder with lubricant since they generally have a lower green strength than compacts from the powder itself. How much strength is needed depends upon the configuration of the compact.

3. Compacts from iron powder and from iron powder with additions of graphite, copper, nickel powder, etc. should show minimum dimensional changes (less than $\frac{1}{2}$%) when sintered under the usual commercial sintering conditions ($\frac{1}{2}$-1 hour at temperatures in the range from 1100 to 1300° C). With these small dimensional changes during sintering, it is easier to control variations in shrinkage from part to part or lot to lot and to minimize distortion during sintering.

4. Iron powders should have high compressibility, i.e. they should produce compacts of high green density using commercially practical compacting pressures. In compacts which do not shrink during sintering, high green density also means high sintered density. Since the mechanical properties of structural parts are, to a considerable extent, a function of sintered density, high green and consequently high sintered densities mean good mechanical properties.

The requirements for powders used for preforms which are hot forged differ from those for powders used in conventional processing in several respects. High green and sintered density of the preforms is less important than in conventional parts, since the parts are completely or nearly completely densified during hot forging. Powders for these applications may also be coarser than those for conventional processing with a considerable percentage of particles larger than 150 μm, since good surface smoothness is obtained during hot forging even for preforms from coarse powders. On the other hand, the requirements for cleanliness, i.e. a low content of acid insoluble material in the powder, are more stringent for powders used in hot forged parts.

The commercially important methods for producing iron powders may be divided into three groups:

> Reduced iron powders

Iron powders in which both atomization and reduction are involved
Atomized iron powders

Atomizing furnishes the largest percentage of iron powders produced today, but was the last method being developed commercially. Methods of producing iron powder by reduction are the oldest ones, followed by methods combining atomization and reduction. In addition to these three groups of methods, two other methods of producing iron powder, electrolytic deposition and decomposition of iron carbonyl, will be briefly treated. Finally, the possibility of using iron powder produced from machining chips as raw material for structural parts will be discussed.

A. Reduced Iron Powder

Among the methods of producing iron powder by reduction, the Swedish Sponge Iron process developed in Höganäs, Sweden, is the oldest. It was developed early in the century. The process was originally intended to produce metallic iron in sponge form as the raw material for making crucible or later electric furnace steel. As such it is one of numerous processes for the direct reduction of iron ores at temperatures below the melting point of iron.[15] The sponge iron produced by most of these processes is not sufficiently pure as raw material for iron powder for powder metallurgy. However, the Höganäs process is based on the use of quite pure magnetite (Fe_3O_4) ores found in northern Sweden and has been developed for producing sponge iron powder rather than merely iron sponge both in Sweden and in the United States. Powder by a very similar process is also produced in the Soviet Union.

In the Höganäs process the iron ore is reduced with a carbonaceous material. The flow chart, Figure 2-3, shows the steps in the production of powder. They are illustrated in the schematic drawing, Figure 2-4. The ore is ground to a particle size distribution determined by that of the desired iron powder. The ore powder is placed in the center of cylindrical ceramic containers ("saggers" made of silicon carbide) surrounded on the outside by a concentric layer of a mixture of coke and limestone. The saggers are placed in layers upon cars which are conveyed through a fuel-fired tunnel kiln. The carbon monoxide produced from the coke reduces the ore to iron. Total reduction time is on the order of 24 hours, reduction temperature is 1200° C. The limestone serves to bind any sulfur in the coke and prevents its contaminating the iron. The sponge iron is mechanically removed from the saggers, ground and the resulting powder magnetically separated from impurities. In a final reduction step the powder is conveyed through a continuous furnace in an atmosphere high in hydrogen on a belt made of stainless steel. The cleanliness, i.e. the acid insoluble content of Höganäs

IRON BY CARBON REDUCTION

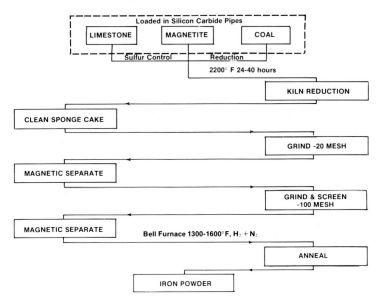

FIGURE 2-3 Flow chart for the production of Hoeganaes Sponge Iron Powder.

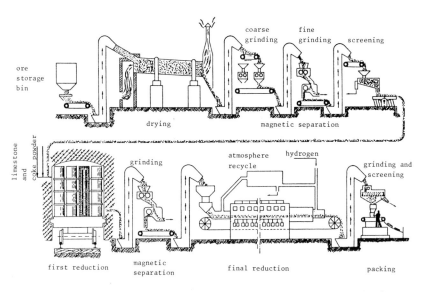

FIGURE 2-4 Schematic of Hoeganaes sponge Iron Powder Production.

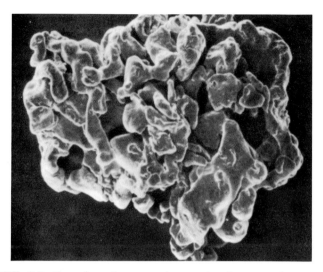

FIGURE 2-5 Scanning electron micrograph of sponge iron powder particle, x400.

powder, will depend on the amount and the size of the particles of gangue in the ground ore. They determine how much of the gangue can be removed in the magnetic separation step. The principal reduction step in the Höganäs process is a batch process in which the ground ore does not move during reduction, but is static in contrast to other direct reduction processes which are continuous. A micrograph of a sponge iron powder particle is shown in Figure 2-5.

One of the continuous direct reduction processes, the so-called H-iron process, was adapted to the production of iron powder for structural parts. Even though the process is no longer used commercially, it is briefly presented because it illustrates the fact that low-cost continuous processes for producing metal powders may not be suitable for powders for powder metallurgy applications. In the H-iron process the ground ore and later the iron powder is conveyed through the process by "dense phase transport," suspended in a stream of gas, i.e. in the manner of a liquid. The reduction of the preheated ore powder takes place in a cylindrical pressure vessel in which it is suspended in a counter current stream of hydrogen, also pre-heated, in three stages of fluid bed reduction at 540° C and a pressure of 2.75 MPa.

At the high pressure, the reduction is reasonably fast in spite of the low reduction temperature, which is necessary to keep the iron powder particles

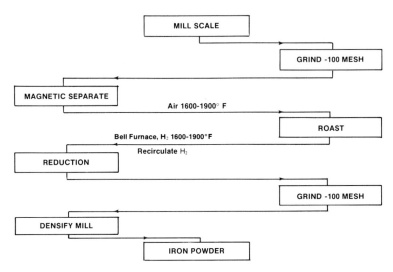

FIGURE 2-6 Flow chart for the production of Pyron iron powder.

from sintering together. The water vapor in the hydrogen is removed by condensation and the gas recirculated. The automatic fluid bed reduction process is low cost, when low cost hydrogen produced from natural gas is available. However, in order to make the powder suitable for compacting and sintering into structural parts it had to be given additional high temperature reduction treatments in which the powder particles sintered together and had to be subsequently ground into powder in a hammer mill. These additional steps made the method non-competitive with other methods of producing iron powder.

Instead of finely ground iron ore as the raw material for reduced iron powders, mill scale, an iron oxide of the approximate composition Fe_3O_4, may also be used. Mill scale is obtained as a by-product of steel mill operations. Iron powder is produced from mill scale by reduction with hydrogen, which is also available as a by-product of chemical manufacturing. This iron powder is available under the trade name "Pyron iron powder." As indicated in the flow chart, Figure 2-6, the mill scale is ground, magnetically separated and first roasted in air to convert the Fe_3O_4 to Fe_2O_3 because the rate of reduction of Fe_2O_3 with hydrogen is faster than that of Fe_3O_4. The oxide is then reduced in a belt furnace at temperatures near 980° C. The reduction product is ground and mechanically densified to make it suitable for production of structural parts. The pores in the interior

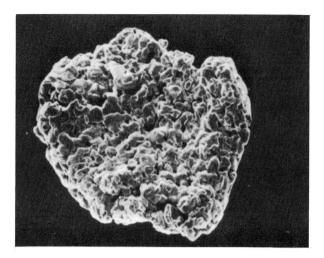

FIGURE 2-7   Scanning electron micrograph of Pyron iron powder particle, x400.

structure of Pyron iron powder particles are considerably finer than those in Swedish sponge iron powder particles, because in the latter the long time reducing treatment at higher temperature coarsens the pores. Because of the fine pore structure, compacts from Pyron powder sinter faster than those from other commercial iron powders. A micrograph of a Pyron powder particle is shown in Figure 2-7.

B.   Iron Powder in Which Both Atomizing and Reduction are Involved

It is easier to produce iron powders which have good green strength by methods which combine atomizing and reduction than by straight atomizing. One of these combination methods is the so-called "RZ process" developed in Germany and presented schematically in the flow chart, Figure 2-8. The iron-carbon alloy is melted in a cupola. The stream of molten alloy is atomized with air using the apparatus schematically shown in Figure 2-9. Many of the particles formed from the freezing droplets are hollow because of the formation of CO through reaction of the carbon in the alloy with the oxygen in the air. At the same time, a thin oxide layer is formed on the surface of the particles. The resultant product is ground to $-150\ \mu$m powder. Suitable batches of the atomized and ground product are selected to provide a feed for the annealing process, which yields, by

IRON POWDER BY ATOMIZATION (RZ)

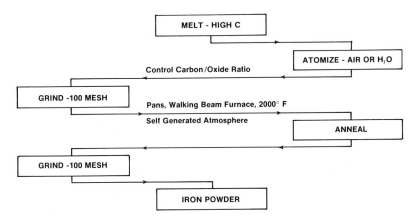

FIGURE 2-8 Flow chart for the production of "RZ" iron powder.

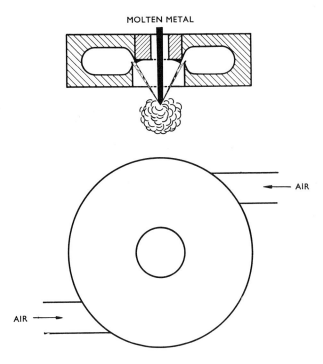

FIGURE 2-9 Schematic of RZ iron powder atomization.

IRON POWDER BY ATOMIZATION (DOMFER)

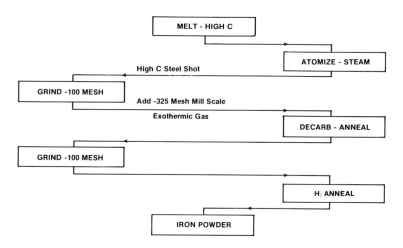

FIGURE 2-10 Schematic of "Domfer" iron powder production.

reaction of the carbon and oxygen which form carbon monoxide, an iron powder low in either carbon or oxygen. The powder is annealed near 1100° C in a walking beam furnace using the CO developed as an annealing atmosphere. The resulting cake is ground to –150 μm powder.

Another combination process ("Domfer" iron powder) was developed in Canada and is shown schematically in the flow chart Figure 2-10. The iron-carbon alloy is melted in an electric furnace, shotted with steam and ground to –150 μm powder. The powder is mixed with –44 μm mill scale. The amount of oxide is adjusted so that during the subsequent decarburizing anneal the oxygen in the oxide reacts with the carbon in the ground iron-carbon alloy forming CO. The product of the annealing treatment is a cake which is ground to –150 μm powder.

Both Domfer and RZ powders have a certain proportion of porous particles. For this reason, compacts from the powder have a better green strength than those from powders with solid particles.

C. Atomized Iron Powder

In the last few years iron powders atomized from molten iron-carbon alloys lower in carbon than those in the RZ and Domfer powders were developed. These are sometimes called "atomized steel powders," since the raw mate-

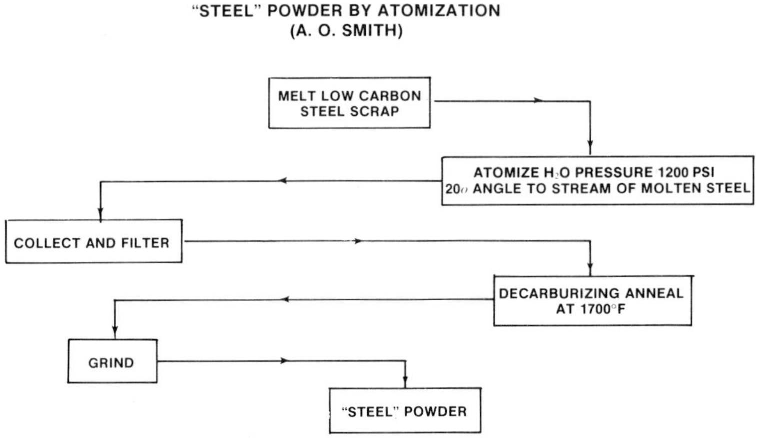

"STEEL" POWDER BY ATOMIZATION
(A. O. SMITH)

```
                    ┌─────────────────────┐
                    │  MELT LOW CARBON    │──────────────────┐
                    │    STEEL SCRAP      │                  │
                    └─────────────────────┘                  │
                                                             │
                              ┌──────────────────────────────────────────────┐
              ┌───────────────│  ATOMIZE H₂O PRESSURE 1200 PSI               │
              │               │  20° ANGLE TO STREAM OF MOLTEN STEEL         │
              │               └──────────────────────────────────────────────┘
    ┌─────────────────────┐
    │  COLLECT AND FILTER │─────────────────────┐
    └─────────────────────┘                     │
              │                    ┌──────────────────────────────────┐
              │               ┌────│  DECARBURIZING ANNEAL            │
              │               │    │        AT 1700°F                 │
        ┌───────────┐         │    └──────────────────────────────────┘
        │   GRIND   │─────────┘
        └───────────┘
              │         ┌──────────────────┐
              └─────────│  "STEEL" POWDER  │
                        └──────────────────┘
```

FIGURE 2-11 Flow chart for the production of A. O. Smith Corp. atomized iron powder.

rial being atomized is molten steel. However, with regard to composition, in particular carbon content, they do not differ markedly from other iron powders. The steps in the first of these processes developed by A. O. Smith Corp. are schematically shown in the flow chart, Figure 2-11. Although the powder as atomized is relatively low in carbon, the rate of quenching the molten droplets is fast enough so that the particles are quite hard. They also contain enough oxygen so that they have to be given a decarburizing anneal at 925° C in order to produce a cake which can be ground into powder suitable for structural parts. In other atomizing processes for producing iron powders, such as the ones used by Hoeganaes Corporation in Sweden and the U.S.A. and by Mannesmann in Germany, the atomized product is higher in carbon than in the A. O. Smith process. It is reduced to a low level during the reducing decarburizing anneal.

The atomized product obtained in the Quebec metal powder process is highest in carbon. It is obtained from atomizing a molten iron-carbon alloy, which is the product of electric furnace processing of ilmenite. The process yields a titania slag, raw material for titanium dioxide, in addition to the alloy. The Quebec metal powder process may be classified as one in which both atomizing and reduction are involved. The green strength of a compact from atomized iron powders whose particles are solid rather than porous is generally lower than that of compacts pressed from powders produced by reduction or by combination atomizing-reduction processes, but is ade-

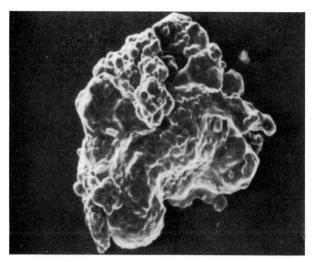

FIGURE 2-12 Scanning electron micrograph of A.O. Smith Corp. atomized iron powder particle, x400.

quate for most structural parts because of the irregular surface configuration of the powder particles. A micrograph of atomized iron powder particles is shown in Figure 2-12. On the other hand, the compressibility of powders with solid, non-porous particles, which also includes electrolytic iron powders, is generally higher than that of powders with porous particles.

D. Electrolytic Iron Powder

Electrolytic iron powder is no longer of major commercial importance for iron and steel components for structural parts. It is still used for some magnetic applications and is also used outside of powder metallurgy. Before the atomized iron powders with solid, non-porous particles had been developed, electrolytic iron powders, in spite of their higher cost compared with other grades, were in demand because of their good compressibility, particularly for small parts, such as those in office machines, where powder cost was only a small proportion of the total part cost. The present grades of atomized powders with solid, non-porous particles have as high a compressibility, are as low in impurities as electrolytic powders and are considerably lower in cost. In producing electrolytic iron powder,[16,17] brittle, lightly adhering solid sheets are deposited on stainless steel cathodes. Armco iron or low carbon steel serves as anode material and either chloride or sulphate electrolytes are used. The deposits are brittle because they contain hydrogen and/or oxygen as impurities and can readily be milled into powder in ball mills. The powder is annealed in hydrogen to purify it and make it soft. The resulting friable cake is broken up into powder in a hammer mill.

E. Carbonyl Iron Powder

Iron powders produced by the decomposition of iron carbonyl, $Fe(CO)_5$, are not used for structural parts, but only in specialized applications, such as Alnico magnets. Carbonyl iron powders are much finer than other iron powders with particle sizes generally in the range of 2-20 μm. Therefore they do not flow, although such flow could be obtained by agglomerating the particles into coarser aggregates. Compacts from carbonyl iron powder shrink when they are sintered, particularly at temperatures below the α-γ transformation temperature of iron. Finally the cost of carbonyl iron powder is higher than that of other iron powders. Even though these characteristics make carbonyl iron powder unsuitable for fabricating parts, its production should be briefly considered here. A number of German studies[18] in the early 1930s are concerned with loose powder sintering of carbonyl iron powder into high density bodies which could be further densified by forging. Also, studies of sintering carbonyl iron powder compacts have made important contributions to an understanding of the mechanisms of sintering.[19]

Iron carbonyl, $Fe(CO)_5$, the raw material from which carbonyl iron powder is produced, is a liquid with a boiling point of 103°C which is prepared by passing carbon monoxide over sponge iron at temperatures of 200-220°C under a pressure of 70-200 atmospheres. The iron carbonyl is then decomposed into powder in the gas phase at temperatures in the range of 200 to 300°C and atmospheric pressure under conditions which prevent decomposition at the wall of the reaction vessel.[20] By controlling the conditions under which decomposition takes place, e.g. by addition of NH_3 to the gas stream or by a subsequent annealing treatment of the powder in hydrogen, grades of powder differing in chemical composition and in particle shape may be obtained. The E-grade of powder containing several tenths of a percent of both carbon and oxygen has an onionlike spherical structure, is quite hard and could, therefore, not be pressed into compacts by conventional methods. It is used for radio frequency cores, in which the powder particles are electrically insulated from each other by etching with phosphoric acid and bonded by a layer of a thermosetting resin, compressed at high pressures and cured. Many radio frequency cores on the basis of carbonyl iron have been replaced by ceramic cores with an iron oxide spinel basis ("ferrite" cores). The H-grade of carbonyl iron powder, on the other hand, is much lower in oxygen and carbon, has an irregular particle shape, can be readily pressed into compacts and is used for research on the mechanism of sintering.

F.   Iron Powder from Steel Turnings

Steel turnings and chips, a by-product obtained in machining steel, would obviously be quite a low cost raw material for structural parts. Early attempts to hammermill steel turnings into a coarse powder, press it into preforms and hot forge the preforms into sleeves were made by Tormyn of Chevrolet in 1941,[21] but did not lead to commercial use. A similar process, called "micromesh process," was investigated at General Motors Research Laboratories starting in 1968.[22] Machining chips are vapor degreased, trash is separated from the chips, the chips are reduced into a −20 +65 mesh product in a hammer mill and are annealed at 700° C. Experiments have shown that the coarse powder so produced can be pressed into compacts and that the compacts after sintering have similiar properties to those from conventional powder. Nevertheless, the process has not been widely used commercially. The principal problem appears to be to make available sources of sufficient quantities of strictly segregated uncontaminated turnings and chips.

II.   Low Alloy Steel Powders

Prealloyed powders, which contain alloying elements, in particular nickel and molybdenum, in solid solution in the iron, are produced by the same methods as atomized iron powder. The alloying elements are added to the molten steel bath before atomizing. After atomizing the powders are annealed and reduced in the same manner as straight iron powders and, therefore, have the same low carbon contents as these powders. Low alloy steel compacts may be produced from mixtures of prealloyed low alloy steel powder and graphite powder or from mixtures of iron powder, graphite powder and the elemental powders of the alloying ingredients. The advantages and disadvantages of producing alloy steels by these two methods are discussed in chapter 11 and in chapters 17 and 18.

III.   Stainless Steel Powders

Stainless steel powders are used for structural parts and for filter applications. The principal commercial method of producing stainless steel powders is water atomization.[23] Austentitic stainless steels of the composi-

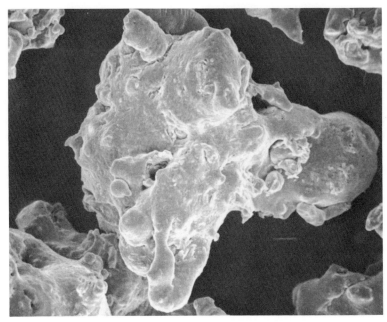

FIGURE 2-13   Scanning electron micrograph of atomized grade 316L stainless steel powder particle x650 (Courtesy: Glidden Metals, Div. of SCM Corp.).

tions 303 L, 304 L and 316 L and martensitic steel of the composition 410 L are the most important grades of stainless steel powder produced by this method. Virgin raw materials, armco iron, electrolytic nickel, low carbon ferrochrome and ferromanganese are melted in high frequency induction furnaces under a protective slag cover. The melt is poured at a temperature of 3000° F (1650° C) into a preheated tundish at the bottom of which is an orifice, ¼″ to ½″ (6 to 12 mm) in diameter which meters the flow of molten metal into the atomizing chamber. The atomizing water streams are directed from the atomizing head in a series of converging jet streams which intersect with the metal stream at an angle of approximately 45°. Pressures up to 2000 psi (14 MPa) are used to produce a powder, all of which passes through a 100 mesh (–150 μm) sieve with 35 to 50% finer than 326 mesh (–44 μm). The atomized molten droplets fall to the bottom of the atomizing chamber filled with water. The slurry of powder and water is dewatered, dried and screened. While austenitic stainless steel powders are used without further treatments, martensitic stainless steel powders must be annealed in a protective atmosphere. A scanning electron micrograph of a water atomized stainless steel powder, grade 316 L, is shown in Figure 2-13.

Spherical stainless steel powders may be produced by argon or nitrogen atomization. These powders are used in fabricating seamless stainless steel tubing by hot extrusion of preforms produced from the powders by cold isostatic pressing in mild carbon steel molds.[24]

Older methods for producing austenitic stainless steel powders involve sensitizing (heating at 500-750° C in order to precipitate chromium carbide at grain boundaries) and disintegrating the alloy. The disintegration may be by attack with a corroding agent (e.g. Strauss reagent with 11% $CuSO_4$, 10% H_2SO_4, balance H_2O)[25] or electro-chemically by making the sensitized steel the anode in an electrolyte of hydrofluoric and sulfuric acid and hydrogen peroxide.[26] These methods are high in cost and no longer used commercially.

IV. Tool Steel Powders

Two processes for producing powders with tool steel compositions by water atomizing melts of the alloys should be briefly mentioned, even though full details of the processes are not yet available. One of these processes, the "Fuldens" process developed by Aerojet General Corp. and improved by Consolidated Metallurgical Industries,[27] involves water atomizing melts of the required tool steel composition to a powder with an average particle size of 100 μm. The hard particles are further pulverized by the cold stream process to a powder with a Fisher subsieve particle number of 6-12 μm and then annealed in hydrogen. The other process is the British "Powdrex process," which also involves water atomizing molten tool steel to a –100 mesh (150 μm) powder, which is vacuum annealed.

V. Copper Powder

The most important powder metallurgy application of copper powder is in self-lubricating bearings having a bronze composition which are produced from mixtures of elemental copper and tin powders. The grades of copper powder used are free flowing granular powders with an apparent density in the range of 2.5-3.2 g/cm³. Copper powders quite similar to those for self-lubricating bearings are used

> as an ingredient in mixes for iron-base structural parts
> for metallic brushes pressed from mixtures of copper and graphite powder
> for structural parts with a bronze composition
> for straight copper powder structural parts for electrical and electronic applications
> for the bond material for bronze base metal bonded diamond tools

For friction materials a copper powder of low apparent density which has unusually high green strength is required.

The three principal methods for producing copper powder for powder metallurgy applications are:

  (1) Electrolytic deposition of copper powder
  (2) Gaseous reduction of copper oxide
  (3) Atomization

Two other methods which are of lesser commercial importance for powder metallurgy applications are:

  (4) Precipitation of copper powder from a copper sulfate solution with iron
  (5) Reduction of an aqueous solution of a copper salt with hydrogen under pressure

A. Electrodeposition of Copper Powder

Copper powder is electrolytically deposited by a process[29] similar to the one which is an important step in copper refining. However, instead of using impure cast copper anodes, electrolytically refined copper anodes are used and the conditions of electrodeposition are modified so that powder, instead of a smooth adherent deposit, is deposited on cast antimonial lead cathodes. The most important modifications are in the current density and the concentration of the electrolyte. Instead of using an electrolyte with 150 g/l of $CuSO_4.5H_2O$ and a current density of 100-200 A/m^2 which is normal for copper refining, the concentration is reduced to 50 g/l and the current density increased to 500 A/m^2. The powder deposited on the cathodes is brushed down by a squeegee which lets it fall to the bottom of the tanks. The powder is washed and filtered and finally given an annealing and reducing treatment at temperatures between 500 and 800° C in a belt furnace with an atmosphere of partially combusted hydrocarbon gas. The properties of the powder, particularly its apparent density, are primarily controlled by the reducing treatment after electrolytic deposition. The reduced powder forms a cake which must be broken up into powder by a hammermill. A schematic flowsheet of the process is shown in Figure 2-14.

The powder particles as deposited on the cathode have dendritic shapes. The reducing treatment causes sufficient sintering so that the dendritic appearance is lost. An electrolytic powder, merely rapidly vacuum dried after deposition but not reduced, which therefore has a dendritic particle shape, as shown in the scanning electron micrograph of Figure 2-15 is being commercially produced in Germany.

COPPER BY ELECTROLYTIC DEPOSITION

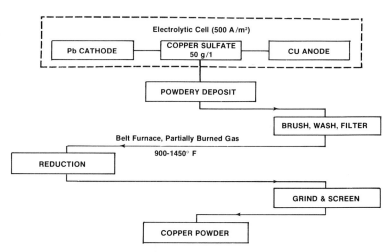

FIGURE 2-14 Flow chart for production of electrolytic copper powder.

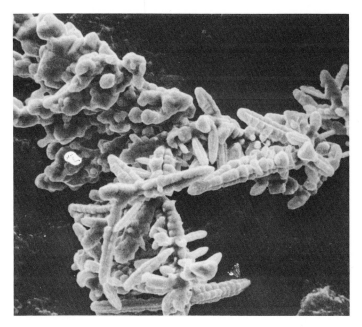

FIGURE 2-15 Scanning electron micrograph of electrolytic copper powder particle, x2800 (Courtesy: Norddeutsche Affinerie, Hamburg, Germany).

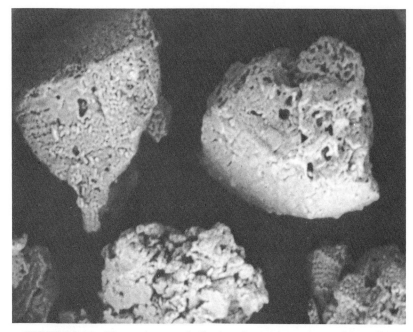

FIGURE 2-16   Photomicrograph of reduced copper powder particles, x525 (Courtesy: Glidden Metals, Div. of SCM Corp.)

B.   Gaseous Reduction of Copper Oxide

The amount of copper oxide in the form of scale available as by-product of copper fabrication is quite limited. Copper oxide is therefore produced from metallic copper, primarily scrap. The metal is melted, atomized into fairly coarse, more or less spherical particles and then converted into copper oxide by an oxidizing roast in a fluid bed. The oxide is ground, screened and reduced in a belt furnace using hydrogen as the reducing agent. The resulting sinter cake is broken up into copper powder. A micrograph of reduced copper powder is shown in Figure 2-16.

C.   Atomized Copper Powder

Atomizing was not widely used as a method of producing copper powder for self-lubricating bearings and similar applications because the powders so produced had spherical and spheroidal shape and, therefore, compacts pressed from them had poor green strength. These drawbacks of atomized powders have been overcome by suitable control of the conditions for

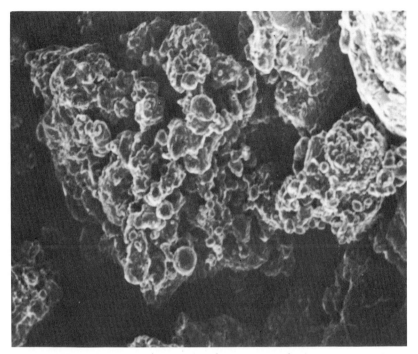

FIGURE 2-17 Photomicrograph of water atomized copper powder particles, x450 (Courtesy: Glidden Metals, Div. of SCM Corp).

atomizing, e.g. producing a very fine powder and partially agglomerating the powder during annealing. Atomizing of copper powder is therefore of increasing commercial importance. A photomicrograph of atomized powder particles is shown in Figure 2-17.

Coarse spherical copper powders by atomizing for metallic filters will be discussed under copper alloy powders.

D. Precipitation of Copper Powder from a Sulfate Solution with Iron

Copper powder precipitated from an acidified solution of copper sulfate with iron in the form of steel scrap, e.g. detinned tin can scrap, is called "cement copper." It is an important product in the hydrometallurgy of copper and for recovering copper from solutions, which are by-products of other copper refining processes. Most cement copper is melted and cast instead of being used as powder. However, the low apparent density of cement copper powder and the high green strength of compacts pressed from it make this material particularly suitable for use in friction materials. Low apparent density copper powders may also be produced by modification of the electrolytic deposition process.

E. Copper Powder by the Reduction of Aqueous Solutions of Copper Salts with Hydrogen under Pressure

A process similar to the one used for nickel powder, in which the powder is precipitated in an autoclave by reacting a solution of the nickel ammine sulfate with hydrogen under pressure at elevated temperature, has also been developed for copper powder.[30] It is no longer used commercially.

VI. Copper-Alloy Powders

The raw materials for self-lubricating bronze bearings are elemental copper and tin powders. For bronze structural parts either elemental powders or a prealloyed atomized bronze powder is used. On the other hand, structural parts from brasses (copper-zinc alloys) and nickel silvers (copper-nickel-zinc alloys) are always produced from prealloyed powders. The presence of zinc in the molten droplets of the alloys, which is easily oxidized, affects the surface tension of the droplets so that they can be readily air atomized into irregular shape particles. A photograph of brass powder is shown in Figure 2-18.

Copper-aluminum alloy powder produced by atomizing the molten alloy with a stream of nitrogen forms the basis for the fabrication of recently developed copper, dispersion strengthened with alumina.[31]

Copper-lead powders are used in the production of bearings which consist of a bearing alloy lining on a steel shell primarily for main and connecting rod bearings. Since copper and lead are not completely miscible even in the liquid state, the alloys from which the powder is produced are melted in electric induction furnaces whose stirring action keeps the metals finely dispersed. The alloys are then water atomized into powder.

Copper-base metallic filter materials have commonly a copper-tin bronze composition and are produced from spherical powder. Loose powder sintering of powders with a fairly narrow particle size distribution results in filters having an optimum combination of properties required in filters: retention of fine impurity particles and reasonable fluid permeability. Spherical bronze powders for this application may be readily produced by atomizing. More commonly, however, spherical particles of pure copper are coated with a fine dispersion of tin in water. The coated powders are only superficially alloyed before being sintered. Sintering the particles, which have a tin-rich surface coating, produces first an excellent interparti-

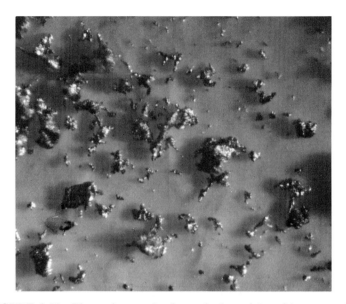

FIGURE 2-18 Photomicrograph of atomized particles of brass powder (78.5% Cu, 20% Zn, 1.5% Pb) x100 (Courtesy: G & W Natural Resources Group).

cle bond and eventually a homogeneous alloy. Narrow particle size distributions cannot be produced by atomizing. Therefore, for use in filters from closely sized coarse powder fractions, copper wire of the required diameter is chopped into cylindrical pieces, the pieces are tumbled to give them a near spherical shape and they are then coated with tin.

VII. Silver Powder

The principal use of silver powder is in electrical contacts, such as silver-tungsten, silver-tungsten carbide, silver-molybdenum, silver-cadmium oxide, silver-graphite and silver-nickel contacts. Several of the methods used for copper powder may also be applied for producing silver powder, i.e. electrolysis of aqueous solutions, precipitation from a salt solution by a less noble metal, which may be copper in the case of silver powder, and by atomizing. In addition, silver powder may be precipitated from the solution of a silver salt with an organic reducing agent. Silver powder produced by this method is amorphous and spongy.

Schreiner[32] has described the details of producing high purity silver powder by electrolysis using an acidic solution ($p_H = 1$ to 1.5) of silver nitrate (10-50g Ag/l) and sodium nitrate (100-150g $NaNO_3$), anodes and cathodes of pure silver and a current density of at least 400 A/m^2. As in the case of copper, the deposit of spongy silver powder is stripped from the cathode at regular intervals. The electrolytic silver powder is crystalline and dendritic and has particles generally finer than 40 μm.

VIII.   Nickel Powder

Much research and development work has been done on producing strip and sheet from metal powder by powder rolling. At present only nickel and nickel alloy sheet and strip are produced commercially by this process. This is directly related to the character of the two principal processes for producing nickel powder:

Decomposition of nickel carbonyl
Reduction of aqueous solution of a nickel salt with hydrogen under pressure.

Both of these processes are used to produce metallic nickel from its ores. In the case of nickel from nickel carbonyl, the metal may be produced as pellets or as powder. The nickel powder produced by reduction of an aqueous solution may be pressed into briquettes. Nickel may be introduced into melts of nickel containing alloys as pellets or briquettes. However, the fact that the metal is obtained as powder at a cost not higher than nickel in ingot form is an incentive for powder metallurgical uses.

In addition to nickel and nickel alloy sheet and strip, powder metallurgy applications for nickel powder include the preparation of porous electrodes for batteries and fuel cells and uses as alloying ingredients in low alloy steel structural parts and in Alnico magnets both of which are produced from mixtures of nickel powder and other powders. A potential use is in the nickel base super alloys prepared by mechanical alloying.

Compared with the two principal processes of producing nickel powder, other methods, such as atomizing, reduction of the oxide and mechanical comminution are of minor importance for powder metallurgy applications.

A.   Nickel Powder by Decomposition of Nickel Carbonyl

The production of nickel in the form of pellets or powder by decomposition of nickel carbonyl, Ni $(CO)_4$, goes back to a process developed early in the century based on an invention of Mond et al.[33] In this process, which in modified form is still used at the refinery of International Nickel Company at Clydach, Wales, nickel oxide produced from nickel sulfide is reduced to nickel sponge by hydrogen, activated by sulfiding and volatilized as car-

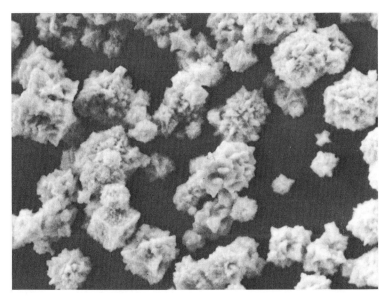

FIGURE 2-19 Scanning electron micrograph of carbonyl nickel powder particles, discreet particle type, x3000 (Courtesy: INCO Res. and Dev. Center).

bonyl. The rate of the carbonylization reaction can be increased by increasing pressure and temperature. The nickel carbonyl so produced is purified by fractional distillation.

While Inco's Clydach process is based on refined nickel oxide as starting material, a new process, the INCO Pressure Carbonyl (IPC) process, has been developed which is now of primary commercial importance. Matte produced pyrometallurgically from complex ores using the Kaldo oxygen top-blown converter is subjected to pressure carbonylation in 150 ton capacity pressure reactors at 7000 kPa and 180° C. Over 95% of the nickel in the matte is extracted in this process.[34] The carbonyl is condensed and the liquid carbonyl separated into nickel carbonyl vapor and iron-rich liquid carbonyl. The nickel carbonyl vapor may be used for producing nickel powder in a decomposer chamber in which it is rapidly heated or for producing nickel pellets. Both powder and pellets are of high purity containing more than 99.99% nickel. By control of process variables, two types of carbonyl nickel powder may be produced, type 123 and the type 200 series which includes types 255 and 287. The two types of powder have different morphologies. Type 123, shown in the micrograph Figure 2-19, consists of compact discreet particles of a characteristic spiky nature and equiaxed

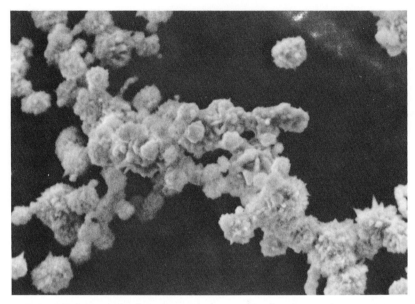

FIGURE 2-20   Scanning electron micrograph of carbonyl nickel powder particles, filamentary grade, x3000 (Courtesy: INCO Res. and Dev. Center).

shape. Type 255, shown in Figure 2-20, and type 287 have an open filamentary structure with the particles having a spiky irregular surface. The filaments consist of chains 10 or more μm in length with a linear cross section of about 1 μm.

B.   Reduction of an Aqueous Solution of a Nickel Salt With Hydrogen Under Pressure

This process[35] was developed by Sheritt Gordon Company in Fort Saskatchewan, Alberta, Canada. A flowsheet of the process is shown in Figure 2-21. The nickel salt solution is obtained by leaching complex copper-nickel-cobalt ores. Before the nickel is precipitated as a metallic powder, the copper is removed by precipitation as sulfide. The nickel remaining in solution is oxidized to nickel ammine sulfate. For the precipitation of the first nickel powder nuclei a catalyst (ferrous sulfate) is added to the solution. The nickel powder nuclei are formed through reaction of the solution with hydrogen at 1400 kPa (200 psi) pressure at 200° C. The nuclei are allowed to settle in the autoclave, the barren solution is decanted and a new batch of solution is introduced in the autoclave. The nickel powder nuclei are

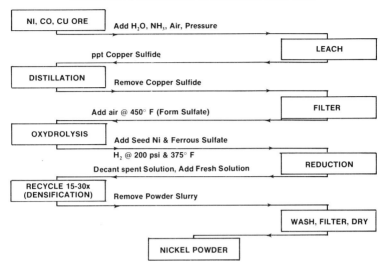

NICKEL BY H₂ REDUCTION IN AQUEOUS SOLUTION

FIGURE 2-21 Flow chart for the production of nickel powder by precipitation with hydrogen from ammine sulfate—ammonium sulfate aqueous solution.

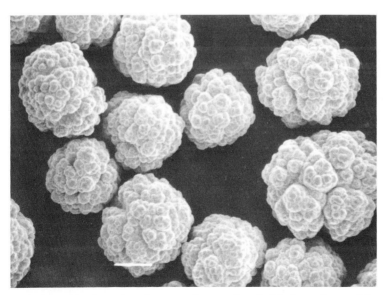

FIGURE 2-22 Scanning electron micrographs of nickel powder particles by precipitation with hydrogen from solution, x200 (Courtesy: Sherritt Gordon Mines, Ltd.).

suspended in the solution by agitation and the nickel produced by reduction of the nickel salt in the solution is precipitated on the existing nuclei. This process, called "densification," is repeated many times (15-30 times). Finally the nickel powder is removed from the autoclave, washed and dried. A micrograph of powder particles is shown in Figure 2-22. This nickel powder is the raw material for the Sherritt-Gordon process of rolling nickel sheet from powder.

IX.   Nickel Base Alloy Powders

Several nickel base alloys are produced in powder form by atomizing. Of commercial interest is a 50% nickel, 50% iron powder which is compacted and sintered into magnetic components with high permeability and low coercive force and monel powders, approximately 70% Ni - 30% Cu, for corrosion resistant structural parts.

Another group of nickel base alloys for magnetic applications are the so-called permalloys with approximately 80% nickel. A special production method[36,37] was developed for powder of one of these permalloys, containing 81% Ni, 2% Mo, balance Fe, which is used in cores for induction coils for audio frequencies. A few thousandths of one percent sulfur is added to the alloy. The ingots cast from the molten alloy are carefully hot rolled at elevated temperatures where the alloy is sufficiently ductile, until a strip is achieved of a grain size which corresponds to the particle size of the alloy powder to be produced. At room temperature the alloy is completely brittle because of the intergranular sulfide film at the grain boundaries. The hot rolled strip is quenched and breaks up into fragments which are pulverized first in a gyratory crusher and then in an attrition mill to equiaxed −120 mesh powder particles. The 81% nickel, 2% molybdenum, balance iron powder is now also produced by atomizing.

X.   Cobalt Powder

Cobalt powder is used primarily in the production of cemented carbides. Most of this powder is produced by the reduction of cobalt oxide with hydrogen at temperatures near 800° C in order to obtain the −44 μm reactive powder needed for cemented carbide manufacture.

XI.   Tin Powder

Tin powders are used in self-lubricating bearings. The powder is produced by atomizing. A schematic of an apparatus for atomizing tin powder is shown in Figure 2-23. The stream of preheated air flowing through the hot

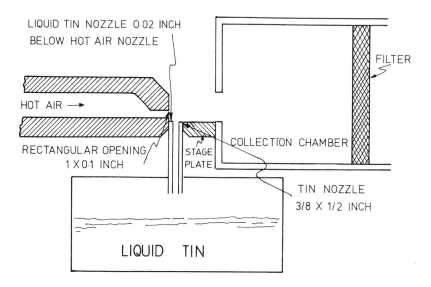

LIQUID TIN NOZZLE 0·02 INCH
BELOW HOT AIR NOZZLE

FILTER

HOT AIR →

RECTANGULAR OPENING
1 X 0·1 INCH

STAGE
PLATE

COLLECTION CHAMBER

TIN NOZZLE
3/8 X 1/2 INCH

LIQUID TIN

FIGURE 2-23 Schematic of apparatus for atomizing tin powder.

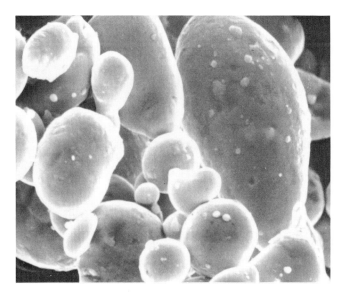

FIGURE 2-24 Scanning electron micrograph of atomized tin powder particles, x3000 (Courtesy: Alcan Metal Powders).

air nozzle aspirates the liquid tin from its reservoir, breaks it up into fine droplets, which freeze into powder particles, and carries the particles into the settling chamber. A suction fan maintains the settling chamber under negative pressure. The finest powder which has not settled in the settling chamber is retained by filters in the bag house. The variables which control the particle size and size distribution of the powder are design of the atomizer nozzles, pressure and temperature of the air and temperature of the molten tin. Figure 2-24 shows a scanning electron micrograph of atomized tin powder particles with their typical rounded shapes.

XII. Aluminum and Aluminum Alloy Powders

The most important powder metallurgy use of elemental aluminum powder is in structural parts from aluminum alloys which are produced by compacting and sintering mixtures of elemental aluminum powder and the alloying ingredients in powder form. Much larger quantities of aluminum powder are used for non-powder metallurgy applications.

Special atomizing nozzles have been designed for producing atomized aluminum powder.

Aluminum and aluminum alloy powders are being produced for applications which are at present in the development stage. These include prealloyed powders atomized from molten aluminum alloys which are the raw material for fully dense "wrought powder metallurgy products" to be discussed in chapter 21. Also discussed in this chapter are the SAP (sintered aluminum powder) alloys which consist of a fine dispersion of aluminum oxide in aluminum and are produced from a special flake type aluminum powder. A coarse aluminum alloy powder of acicular particle shape was developed as raw material for hot rolling aluminum alloy strip from powder, as described in chapter 6. It was produced by pouring molten aluminum alloy into a spinning perforated cylinder.

XIII. Magnesium Alloy Powders

In their extensive work on magnesium powder metallurgy, described in chapter 21, the Dow Chemical Company developed a specialized atomizing process in which magnesium alloy powders having coarse particles, called by Dow magnesium alloy pellets, are produced. A vertical stream of molten alloy is delivered to the center of a rapidly rotating disk. The disk spreads the metal into a thin sheet which breaks up into droplets. The droplets freeze into spherical particles. A natural gas atmosphere is used in the atomizing process.

XIV.   Tungsten Powder

Tungsten powder is used for incandescent lamp filaments, for electrodes for inert arc and atomic hydrogen welding, for massive tungsten and tungsten alloy forgings and extrusions, for producing tungsten carbide for cemented carbides and for tungsten alloys, such as "heavy metal" (W-Cu-Ni and W-Ni-Fe alloys) and "compound metals" (W-Cu and W-Ag alloys). Most of the tungsten powder for these applications is produced by reducing tungsten oxides or ammonium paratungstate with hydrogen. Since this method of producing tungsten powder is an essential part in the fabrication of tungsten products it will be discussed in chapter 14, in which the powder metallurgy of the refractory and reactive metals is described.

Two methods for producing tungsten powder based on reduction of tungsten halides have had only limited commercial use. One of these methods developed by Allied Chemical Company[38] is based on vapor deposition from WF_6. The powder is rather coarse (40-650 μm) and the particles roughly spherical. The powder cannot be cold compacted but only hot consolidated by hot isostatic pressing. The other method[39] is based on hydrogen reduction of tungsten chlorides. Powders of very closely controlled particle shape, particle size and particle size distribution are available by this method. The process has been developed primarily for producing very fine tungsten powder for carburization to WC and production of special grades of cemented carbides.

XV.   Molybdenum Powder

Molybdenum powder is used for fabricating molybdenum wire, sheet and strip, employed, e.g., for the support of the tungsten filaments in incandescent lamps and for other electronic applications, for molybdenum wire for heating elements for electric furnaces, for molybdenum alloys, such as TZM (containing titanium, zirconium and carbon) and for compound materials (Mo-Ag for electrical contacts, $Mo-ZrO_2$ for thermocouple protection tubes). Molybdenum powders are prepared from purified MoO_3 by reduction with hydrogen, very similar to the preparation of tungsten powder. As in the case of tungsten powders, details are given in chapter 14 on the powder metallurgy of refractory and reactive metals.

XVI.   Tantalum Powder

Tantalum powder for fabricating tantalum capacitors is produced by two methods:[40,41]

Reduction of tantalum fluoride with metallic sodium

> Deposition of tantalum powder by electrolysis from molten baths of K_2TaF_7, KF and TaO_2 at 900° C.

The particle size of the powder used for capacitors should be in the range between 5 and 40 μm in order to get the best combination of high capacitance and low leakage losses in the capacitors. One method of obtaining the desired particle size distribution is by the coldstream process discussed above in which coarser powder particles are broken up by propelling them by high velocity air against a target lined with tantalum.

XVII.   Titanium and Zirconium Powders

Titanium is one of the metals which was first produced in massive form by powder metallurgy, starting with powder obtained by the Kroll process. In this process titanium chloride vapor is reacted with liquid magnesium at temperatures between 800 and 900° C forming titanium metal and magnesium chloride. When the reaction product is leached with dilute hydrochloric acid a coarse titanium powder (–30 mesh) is obtained. To produce solid titanium the powder was compacted and vacuum sintered. The Kroll process is still the most important process for producing metallic titanium, but instead of leaching, the reaction product of the process is now subjected to vacuum sublimation. In this way titanium sponge, rather than titanium powder, is obtained. The sponge is generally vacuum arc melted.

The commercial applications for titanium powder metallurgy products are still fairly limited and include titanium filters and titanium alloy structural parts produced by compacting and sintering. Titanium powder for these applications may be sponge fines, by-products of titanium sponge production by reacting titanium chloride with magnesium or sodium. These fines may be leached to modify their composition, flow or particle size. Nevertheless, residual chloride is generally present in sponge fines which may foul sintering furnaces and produce porosity in welds. Powder obtained by disintegration of sponge is an alternate source of titanium powder.[42] Structural parts from titanium alloys, e.g. the well-known 6Al, 4V alloy, may be compacted and sintered from mixtures of elemental titanium and the alloying ingredients as elemental or masteralloy powders.

The hydride-dehydride process

Not only titanium, but also zirconium, hafnium, vanadium, niobium, tantalum, thorium and uranium and their alloys can be converted to their hydrides by heating the metal or alloy in the form of chips or turnings or even compact metal in hydrogen. The hydrides are quite brittle and can be readily attrited into powder of the desired fineness. In the case of uranium, mechanical disintegration is not necessary since during hydriding a fine powder of UH_3 is immediately obtained; it may be used as an intermediary for producing uranium carbide or nitride. The hydride powders may be dehydrided by heating them in a good vacuum in the temperature range of hydride formation or somewhat higher. Titanium and titanium alloys are hydrided at 450° C, cooled, attrited, the hydride powder dehydrided at 700-750° C and the resulting cake attrited into powder. Because the refractory metals are excellent getters for oxygen, nitrogen and carbon, contamination during hydriding and dehydriding must be carefully avoided. The zirconium alloy, Zircalloy 2 (1.5% Sn, 0.12% Fe, 0.10% Cr, 0.05% Ni), which was developed for its low neutron capture cross section and its resistance to attack by super-heated steam, may be converted into powder by the hydriding-dehydriding process. The powder was used for hot-pressing it in graphite molds into components for nuclear reactors.

A final method for producing titanium alloy powder is the rotating electrode process (REP) which will be discussed in connection with the special atomizing processes for powders from highly alloyed materials. The powder so produced is spherical and must be consolidated by hot-pressing, hot isostatic pressing, hot forging or hot extrusion.

XVIII. Tungsten Carbide, Titanium Carbide and Tantalum Carbide

Tungsten carbide powder is the principal raw material for producing cemented carbides. Steel cutting grades of cemented carbides contain titanium and tantalum carbides in addition to tungsten carbide. There are also types of cemented carbides in which titanium carbide is the principal ingredient.

Tungsten carbide powder is usually manufactured by heating an intimate mixture of tungsten powder, whose production will be described in chapter 14, and of carbon powder in the form of lamp black to a temperature of approximately 1500° C. The amount of carbon in the mixture is slightly higher than the stoichiometric proportion in WC. The powder mixture may be lightly compacted into briquettes which are fitted into a carbon boat. The boats are heated in high frequency induction or resistance furnaces. The reaction product is broken in jaw-crushers and ball milled into powder in carbide lined ball mills. The particle size of the tungsten carbide depends on that of the tungsten powder and on the carburizing conditions. From very fine tungsten powder (particle size less than 0.5 μm) carburized at relatively low temperatures of 1350-1400° C tungsten carbide powders with particle sizes under 1 μm may be produced. On the other hand, fairly coarse tungsten powder with particle sizes under 6 μm are converted into a coarse

WC powder with particle sizes less than 10 μm by carburizing at 1600° C. Sumitomo Electric[43] have developed a continuous direct carburization process of tungsten carbide from WO_3 and C.

In producing titanium carbide the object is to obtain a product as nearly stoichiometric in composition as possible. A good method[44] to achieve this purpose is to wet mix TiO_2 and carbon black, thoroughly dry the mixture and press it into compacts. The compacts held in graphite crucibles are reacted in a high-frequency heated vacuum furnace at a temperature of 1900-2300° C. Titanium carbide may also be produced by reacting titanium metal powder with carbon. The reaction products must be balled-milled into powder.

In the system Ti-C there exists a compound which is single phase over the range from 35 to somewhat less than 50 atomic % C. The compound with the highest melting point has a composition somewhat below that of stoichiometric TiC. TiC dissolves TiO in solid solution. The carbon and oxygen content of the carbide affect the cutting properties of cemented carbides containing TiC.

Tantalum carbide powder may be produced either from the oxide or from pure tantalum metal or tantalum hydride powder with carbon. When tantalum metal powder is used, the reaction temperature is similar to that used for tungsten carbide (approximately 1500° C). The reaction product is milled into powder.

XIX. Special Atomizing Processes for Spherical Powders from Highly Alloyed Materials, such as Tool Steels, Super Alloys and Titanium Alloys

Several new atomizing processes have been developed in recent years for powders of highly alloyed materials which are consolidated by hot pressing, hot isostatic pressing, hot extruding or hot forging. All of the atomizing processes have in common that they produce powders with spherical or near spherical shapes and that the process variables are carefully controlled to limit the oxygen content of the powders to values of 100 ppm or less. The processes are:

(1) Inert gas atomizing used for tool steel, stainless steel and super alloy powders.
(2) The rotating electrode process of Nuclear Metals Company applied to many highly alloyed materials, in particular super-alloys and titanium alloys.
(3) The vacuum atomizing process of Homogeneous Metals, primarily for superalloy powders.

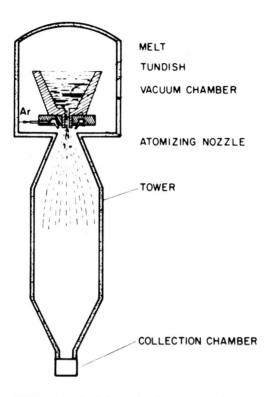

MELT

TUNDISH

VACUUM CHAMBER

ATOMIZING NOZZLE

TOWER

COLLECTION CHAMBER

FIGURE 2-25   Schematic of argon atomizer.

A.   Inert Gas Atomizing

The process of producing relatively coarse more or less spherical powder particles by inert gas atomizing is simple. The alloy is melted, in the case of the superalloys generally from a prealloyed vacuum cast ingot. The melt is tapped into a tundish on top of an atomizing chamber equipped with one or more nozzles. Several designs of nozzles have been developed. The molten alloy is atomized by high velocity gas jets. For superalloys argon is used, which is recirculated. Argon and nitrogen are both used for tool steel and stainless steel atomization. The drops solidify as they fall through the chamber, which is, therefore, quite high. A schematic drawing of the inert gas atomizing apparatus is shown in Figure 2-25. The particle size and size distribution of the powder is controlled by the atomizing conditions. Superalloy powders are often finer than 150 μm; considerably coarser high

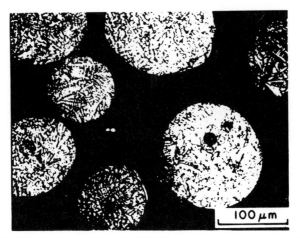

FIGURE 2-26   Cross section of IN-100 alloy powder particles produced by argon atomization, x200.

speed steel atomized powders have been consolidated. The oxygen content of the powder is below 100 ppm and often below 50 ppm. The powder may, however, contain trapped argon and require a vacuum degassing treatment. A micrograph of a cross section of inert gas atomized IN-100 powder is shown in Figure 2-26.

B.   Rotating Electrode Process

In the rotating electrode process (REP) of Nuclear Metals Company[45] a cast rod of the alloy to be produced in powder form is rotated. The front end of the rod is heated by an electric arc or plasma. Molten droplets are thrown by centrifugal force and collected in a chamber filled with helium. The powder is even more spherical than inert gas atomized powder and has a relatively narrow particle size distribution. A schematic drawing of the apparatus for the rotating electrode process is shown in Figure 2-27 and a micrograph of a cross section of IN-100 powder atomized by the process in Figure 2-28.

Variations of the rotating electrode process have been developed in Europe. A French process uses an electron beam to melt the rotating electrode. A process being developed in Germany[46] is a combination of a vertical drip melting process and an atomizing process with a rotating disk. An electron beam is used to drip melt the slowly rotating electrode. The drops are dripping into the center of a rapidly rotating disk, from the edge of which the metal is atomized.

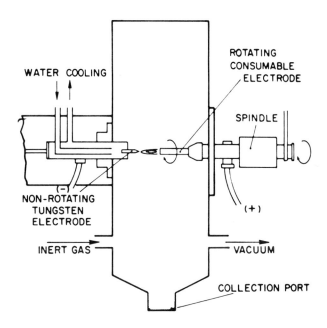

FIGURE 2-27   Schematic of rotating electrode process.

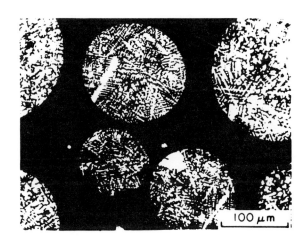

FIGURE 2-28   Cross section of IN-100 alloy powder particles produced by the rotating electrode process, x200.

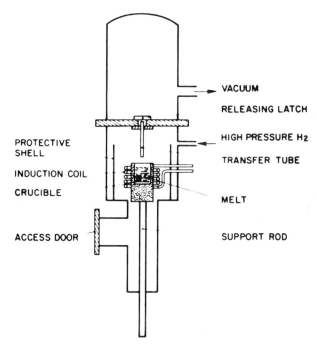

FIGURE 2-29 Schematic of vacuum atomization technique using hydrogen.

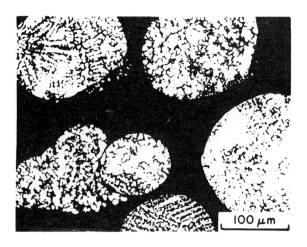

FIGURE 2-30 Cross section of IN-100 alloy powder particles produced by vacuum atomization, x200.

C. Vacuum Atomizing

In the vacuum atomizing process of Homogeneous Metals the alloy is melted under vacuum. The charge is superheated before the melting chamber is pressurized with hydrogen, which dissolves in the molten metal. The melt is then released upwards through a ceramic tube into the atomizing chamber held under vacuum. The molten alloy explodes into fine spherical droplets as it emerges from the tube. The particle size distribution of vacuum atomized powder is relatively flat, compared with the inert gas and the rotating electrode processes. Figures 2-29 and 2-30 show a schematic drawing of the apparatus for vacuum atomizing and a micrograph of a cross section of IN-100 powder produced by the process.

In addition to the three processes described, several other atomizing processes have been or are being developed, but have not yet found commercial applications. One of these is called the Rapid Solidification Rate (RSR) powder process, since its aim is to quench the liquid metal droplets at very high rates. The process resembles the Dow process for magnesium alloy pellets. A stream of molten superalloy falls on a disk, coated with a high temperature ceramic, rotating at a speed of about 400 revolutions/sec. As the molten particles are ejected from the periphery of the disk, a blast of helium solidifies the particles having diameters of 50 μm or less and cools them at rates of $10^{6\circ}$/sec.

Other centrifugal atomization processes are briefly discussed by Lawley, who also gives further references (see references 1).

XX. Beryllium

Beryllium technology is based on powder metallurgy because this technique is the only one by which fine-grained beryllium can be produced commercially. A fine-grained structure in beryllium is necessary for satisfactory mechanical properties, in particular adequate ductility. Beryllium powder is produced from cast beryllium ingots. The ingots are melted from pebbles obtained by reacting beryllium fluoride with magnesium or by vacuum melting beryllium flakes electrolytically deposited from a molten electrolyte. The steps in producing powder from the ingots are: machining into chips and impact-attriting the chips.

References

1. A. Lawley, An overview of atomization processes and fundamentals, Int. J. of Powder Metallurgy, vol. 3, p. 169-188, (1977).
2. L. Shinde and G. S. Tendolkar, Analyses of atomization, a review, Powder Metallurgy Int., vol. 9, p. 180-184, (1977).

3. J. A. Tallmadge, Powder production by gas and water atomization of liquid metals, in "Powder Metallurgy Processing" ed. by H. A. Kuhn and A. Lawley, N.Y., p. 1-32, (1978).

4. J. F. Watkinson, Atomization of metal and alloy powders, Powder Metallurgy, No. 1/2, p. 13-23, (1958).

5. P. U. Gummeson, Modern atomizing techniques, Powder Metallurgy, vol. 15, No. 29, p. 67-94, (1972).

6. P. U. Gummeson, High pressure water atomization, in "Powder Metallurgy for High Performance Applications" ed. J. J. Burke and V. Weiss, Syracuse, p. 27-55, (1972).

7. E. Klar and W. M. Shafer, High pressure gas atomization of metals, ibid., p. 57-68.

8. A. R. E. Singer and A. D. Roche, Roller atomization of molten metals for the production of powder, in "Mod. Dev. in Powder Metallurgy," vol. 9, ed. by H. H. Hausner and P. W. Taubenblat, p. 127-140, (1977).

9. G. Matei, E. Bicsak, W. J. Huppmann, N. Claussen, Atomization of metal powders using the vibrating electrode method, ibid., p. 153-160.

10. F. Aldinger, E. Linch and N. Claussen, A melt-drop technique for the production of high-purity metal powder, ibid., p. 141-152.

11. A. R. Burkin and F. D. Richardson, The production of metal powder from aqueous solutions, Powder Metallurgy, vol. 10, No. 19, p. 33-57, (1967).

12. C. G. Goetzel, "Treatise on Powder Metallurgy," vol. 1, p. 40, N.Y., (1949).

13. ibid., p. 43.

14. J. Walraedt, The coldstream process, a new powder production equipment, Powder Metallurgy Int., vol. 2, No. 3, p. 77-80, (1970).

15. M. Wiberg, Jernkontorets Annaler, vol. 142 (6) p. 289-355, (1958). In this reference the Höganäs process is also described in detail.

16. I. Ljungberg, Electrolytic production of straight and alloyed metal powder, Powder Metallurgy, No. 1/2, p. 24-32, (1958).

17. W. M. Shafer and C. R. Harr, Electrolytic iron powders-production and properties, J. of the Electrochem. Soc., vol. 105, p. 413-416, (1958).

18. See, e.g., L. Schlecht, W. Schubardt and F. Duftschmid, Z. Elektrochemie, vol. 37, p. 485, (1931).

19. See, e.g., G. Cizeron and P. Lacomb, Revue de Metallurgie, vol. 52, p. 771-783, (1955).

20. C. G. Goetzel, "Treatise on Powder Metallurgy," vol. 1, p. 173, N.Y. (1949).

21. A. H. Allen, Steel, p. 74, (May 26, 1941).

22. P. Vernia, Macromesh-metal powder from machining chips, Res. Pub. GMR 1171, (Feb. 1972).

23. H. D. Ambs and A. Stosuy, chapter 29, The powder metallurgy of stainless steel, in "Handbook of Stainless Steel", ed. by D. Peckner and I. M. Bernstein, N.Y., (1977).

24. C. Aslund, A new method of producing stainless seamless tubes from powder, Preprints, vol. 1, p. 278-283, 5th European Powder Metallurgy Symposium, Stockholm, (1978).

25. J. Wulff, Stainless steel powders, in "Powder Metallurgy", ed. by J. Wulff, Chapter 11, p. 137-144, Cleveland, (1942).

26. P. P. Marshall, Porous stainless steel, Part II, Production of 18/8 austenitic stainless steel powder in "Symposium on Powder Metallurgy", Spec. Rep. No. 58, The Iron and Steel Institute, p. 181-184, London, (1956).

27. T. Levin and R. P. Harvey, P/M alternate to conventional processing of high speed steel, Metal Progress, vol. 115, No. 6, p. 31-34, (1979).

28. Anonymous, The Powdrex Process, Metals and Materials, p. 36-37, (Sept. 1977).

29. F. Willis and E. J. Klugston, Production of electrolytic copper powder, J. Electrochemical Soc., vol. 106, p. 362-366, (1959).

30. W. Kunda and D. J. I. Evans, Hydrogen reduction of copper to powder from the ammine carbonate system, "Mod. Dev. in Powder Metallurgy," vol. 4, ed. by H. H. Hausner, p. 49-61, (1971).

31. A. V. Nadkarni, E. Klar and W. M. Shafer, A new dispersion strengthened copper, Metals Engineering Quarterly, vol. 16, No. 3, p. 10-15, (1976).

32. H. Schreiner, "Pulvermetallurgie elektrischer Kontakte," Berlin, 1964, p. 24.

33. L. Mond, C. Langer and F. Quincke, Proc. Chem. Soc., vol. 86, p. 112, (1890).

34. P. Queneau, E. D. O'Neill, A. Illis and J. S. Warner, Some novel aspects of the pyrometallurgy and vapometallurgy of nickel, Part II - The Inco pressure carbonyl (IPC) process, Journal of Metals, vol. 21, No. 7, p. 41-45, (1969).

35. K. O. Cockburn, R. J. Loree and J. B. Haworth, The production and characteristics of chemically precipitated nickel powder, Proc. of the 13th Annual Meeting of the Metal Powder Association, p. 10-24, (1957).

36. E. E. Schuhmacher, Magnetic powder and production of cores for induction coils, in "Powder Metallurgy" ed. by J. Wulff, Cleveland, p. 166-172, (1942).

37. V. J. Pringel, Heat treatment of 2-81 molybdenum permalloy powder cores, The Western Electric Engineer, vol. 14, No. 2, p. 2-10, (1970).

38. Allied Chemical Company, Properties of Allied Chemical Tungsten, Morristown, N.J., (1963).

39. L. Ramqvist, A new tungsten powder for producing carbides, "Mod. Dev. in Powder Metallurgy," vol. 4, ed. by H. H. Hausner, p., 75-84, (1971).

40. Clifford A. Hempel, Tantalum, Ch. 25 in "Rare Metals Handbook," ed by C. A. Hempel, 2nd ed., N.Y., p. 469-481, (1961).

41. R. Titterington and A. G. Simpson, The production and fabrication of tantalum powder, in "Symposium on Powder Metallurgy," The Iron and Steel Institute, London, Special Report No. 58, p. 11-18, (1956).

42. R. E. Garriott and E. L. Thellmann, Titanium powder metallurgy, a commercial reality, in "Mod. Dev. in Powder Metallurgy," vol. 11, ed. by H. H. Hausner and P. W. Taubenblat, p. 63-78, (1977).

43. M. Miyake, A. Hara, T. Sho, and Y. Kawabata, Continuous direct carburization process of WC from WO_3, Preprints vol. 2, p. 93-98, 5th European Powder Metallurgy Symposium, Stockholm, (1978).

44. E. M. Trent and A. Carter, Sintered titanium carbide, in "Symposium on Powder Metallurgy", The Iron and Steel Institute, London, Spec. Rep. No. 58,

p. 272-276, (1956). See also: F. Benesovsky, Carbide, in "Ullmanns Enzyklopädie der Technischen Chemie" 4th ed., vol. 9, p. 122-136, (1975).

45. A. S. Bufferd and P. U. Gummeson, Applications outlook for superalloy P/M parts, Metal Progress, vol. 99, No. 4, p. 68-71, (1971).

46. H. Stephan and J. K. Fischhof, Production of high purity metal powders by EBRD-electron beam rotating disk process, "Mod. Dev. in Powder Metallurgy," vol. 9, ed. by H. H. Hausner and P. W. Taubenblat, p. 183-190, (1977).

3

Powder Characterization and Testing

Before a lot of metal powder can be tested, a sample has to be taken from the lot. Methods of taking a representative sample are described in the first section of this chapter. The second section is concerned with chemical methods for testing metal powders. In general the methods for chemical analysis of metal powders are the same as for metals, but certain empirical tests have been developed which are specifically applicable to metal powder. Particle size, particle size distribution, particle shape and structure and specific surface are the subjects of the third section of the chapter. They determine to a considerable extent the properties of finished compacts from powder. They also have a strong effect upon the behavior of the powder during processing by compacting and other methods of consolidation and by sintering. Methods of measuring particle size, particle size distribution and specific surface are not specific for metal powders. Most of them have been developed for other materials in powder form, e.g. cement, pigments, ground minerals, etc. Detailed discussions of these methods are available in the books by Orr and Delavalle,[1] by Irani and Collis[2] and by Allen.[3] Only those methods which are in general use for metal powders are treated in this chapter.

The behavior of metal powders during processing largely depends upon the particle size, particle size distribution, particle shape and structure of the powder, but data on these properties can often not be translated directly into values characterizing processing behavior. For this reason a number of specific tests for processing behavior have been developed. They are tests for flowrate, also called flowability of powders, for apparent density, for compactibility, also called compressibility, for compression ratio, for green strength and for dimensional changes during sintering. These tests are discussed in the last section of this chapter.

I. Sampling of Powders

Standard methods for sampling finished lots of metal powder have been developed by ASTM Committee B-9 and the standards committee of MPIF and are found in ASTM Standard B 215 and in MPIF Standard 1. Taking a representative sample of a lot, which may be the quantity of powder which is blended in a large blender, is relatively simple for the powder producer. He combines these samples from the entire cross section of the stream of powder as it flows from the blender, the first taken when the first shipping container is half full, the second when half of the burden of the blender has been discharged and the third when the last shipping container is half full. Portions of these samples are blended and from the resulting blend a sample of the proper size for the desired test is obtained using a sample splitter. It is not as easy for the powder consumer to take representative samples from a shipment consisting of several drums. "Thieves" which are devices to take samples from different layers (top, center, bottom) of drums filled with powder are used.

II. Chemical Tests

The two tests for the chemical analysis of metal powders which have been standardized by ASTM and MPIF are:

> ASTM standard E 159, MPIF standard 2 for the so-called hydrogen loss of copper, tungsten and iron powder and
>
> ASTM standard E 194, MPIF standard 6 for acid-insoluble content of copper and iron powder.

In the hydrogen loss test a sample of powder is heated in a stream of hydrogen for a given length of time and at a given temperature. The loss of weight is determined, which is an approximate measure of the oxygen content of the powder. However, the hydrogen loss value will be lower than the actual oxygen content to the extent that oxygen may be present in the form of oxides not reduced by hydrogen under the test conditions such as SiO_2, Al_2O_3, CaO etc. The hydrogen loss value may be higher than the actual oxygen content in the presence of elements forming volatile compounds with hydrogen, such as sulfur or carbon, or in the presence of metal volatile at the test temperature, such as zinc, cadmium and lead. To avoid measuring the content of carbon, sulfur or volatile metals in the metal powder, a modified hydrogen loss test is being standardized by Technical Committee 119 of the International Standards Organization in which the amount of water vapor produced by heating in a stream of dry hydrogen is determined by titration. To determine the total amount of oxygen in a metal powder including oxygen in refractory oxides, the standard test for oxygen in steels and other alloys may be used, which is based on fusing a sample contained in a small, single-use graphite crucible under a flowing inert atmosphere at a temperature of 2000° C or higher. The oxygen is released as carbon monoxide and measured by infrared absorption or alternatively converted to carbon dioxide and measured by a thermal conductivity difference.

In the test for acid insoluble content samples of iron powder are dissolved in hydrochloric and those of copper powder in nitric acid under specified conditions. The insoluble matter is filtered out, ignited in a furnace and weighed. The insoluble matter consists of silica, insoluble silicates, alumina, clays and other refractory materials which have been introduced into the powders as impurities in the raw material or from furnace linings, fuels, etc. In iron powder the acid insoluble may also include insoluble carbides. Similar to the hydrogen loss test the test for acid insoluble content is an empirical test used as an indication for powder quality. Particularly in iron powder, refractory materials present in the powder as impurities will not only affect the properties of the parts fabricated from the powder, but may also increase abrasive wear of the tools in which compacts from the powder are pressed.

III. Particle Size, Particle Size Distribution, Particle Shape and Structure, and Specific Surface

The following methods of determining particle size and particle size distribution in metal powders are discussed in this section:

- (a) Sieving
- (b) Microscopic sizing
- (c) Methods based on Stokes' Law
    - (c_1) the Roller air analyzer
    - (c_2) the Micromerograph
    - (c_3) Light and X-ray (Sedigraph) turbidimetry
- (d) Coulter Counter and Particle Analysis by Light Obscuration
- (e) Laser Light scattering; the Microtrac particle analyzer

For powders whose particle size distribution includes primarily particle sizes larger than 44 μm sieving is by far the most important method. The Roller Air Analyzer, the Micromerograph, the Coulter Counter, the Light Obscuration Particle Analyzer and the Microtrac Analyzer are used for powders with finer particle sizes, i.e. below 44 μm, most commonly in the range from 1 or 2 to 50 μm. The turbidimetric methods have been developed for the very fine refractory metal and refractory compound powders with particle sizes from less than 1 μm to 10 μm.

The two methods used for the determination of the specific surface of metal powders are:

    (f) the Fisher Subsieve Sizer which is based on permeametry
    (g) the gas adsorption (BET) method

Before details of the methods for particle size, particle size distribution and specific surface are discussed in the last part, part E of this section, a few general remarks on the following subjects are presented:

    A. Particle size. An answer to the question "What is particle size?" is attempted.
    B. Particle size distribution: How are particle size distributions tabulated and plotted?
    C. Particle shape and structure
    D. Specific surface. How are particle size and size distribution of powders related to specific surface and what is measured in specific surface measurements by permeametry and by gas adsorption?

A. Particle Size

The term particle size must be defined. Only for spherical particles is there a unique dimension which defines particle size, i.e. the diameter of the sphere. For irregular particles, "particle size" will depend upon the method by which this size is measured. To illustrate this point, the definition of particle size when it is determined by sieving, by microscopic sizing, by sedimentation and by the Coulter Counter method are discussed.

In sieving the powder is shaken through a woven wire mesh screen with square openings. The particle size is defined by the aperture of the sieve which will just retain a given particle. Needle-like particles with dimensions in the length direction considerably larger than the sieve aperture may pass through the sieve, if the dimensions perpendicular to the length direction are smaller than the sieve aperture.

In sedimentation and elutriation methods the particle size determination is based on Stokes' law which gives the settling velocity v of spherical particles with a diameter x and a density ρ in a fluid medium with density ρ_F and viscosity η

$$v = \frac{g\,(\rho - \rho_F)}{18\,\eta} x^2$$

where g is the gravitational constant. Particles which are not spherical will also settle. Their "Stokesian" size is defined by the diameter of a sphere of the material which has the same settling velocity as the irregular powder particle.

The Coulter Counter method is based on measuring the change in electrical resistance when particles suspended in an electrically conducting liquid pass one by one through a small aperture in the liquid on either side of which electrodes are mounted. The change in resistance produces a voltage pulse which is proportional to the particle volume. Particle sizes determined by this method are therefore defined by the cube root of the particle volume. The fact that these definitions of particle size all give slightly different results for the size of a given particle or for the size distribution of a given powder should be kept in mind in attempting to check one method against another or in using data obtained partly by one method and partly by another.

B. Particle Size Distribution

Since most methods of producing metal powders yield particles of a range of sizes and since these powders are suitable for most powder metallurgy applications, the particle size distribution of metal powders must be characterized. Many base metal powders are relatively coarse with a maximum particle size near 150 μm, but a distribution extending to 20 μm or less. Refractory metal powders, on the other hand, generally have distributions in the range from 1 to 10 μm. Particle size distributions may be presented in the form of tables or graphs. In Table 3-1[4] the size distribution of a sample of glass beads is shown using a linear scale for the size ranges as seen in column 1. In column 3 the percent frequency by number of particles in each range are shown. Distribution by number percent frequency are directly obtained in microscopic sizing, while sieving would yield distributions by weight percent frequency shown in column 2. Instead of giving a frequency distribution for each size range, a cumulative distribution, percent cumulative finer than, or percent cumulative coarser than, again by number or by weight may be tabulated as shown for the same glass bead powder in Table 3-2. The values for distribution by number and by weight are, of course, quite different, but number distributions can be readily converted into weight distributions which are used more frequently for metal powders.

63

Table 3-1   Size Distribution of a Sample of Glass Beads, Frequency Distribution

Size Range	Weight Distribution: Percent Frequency	Number Distribution: Percent Frequency
0-5	0.0	0.0
5-10	0.1	1.0
10-15	1.5	12.8
15-20	8.9	28.2
20-25	18.5	26.0
25-30	21.5	17.0
30-35	17.9	8.0
35-40	12.9	4.2
40-45	10.4	1.6
45-50	2.8	0.7
50-55	3.0	0.35
55-60	1.1	0.06
60-100	1.9	0.09
0-100 Total	100.0	100.00

Table 3-2   Size Distribution of a Sample of Glass Beads, Cumulative Distribution

Size (microns)	Weight Distribution		Number Distribution	
	% Cumulative (Greater than)	% Finer than	% Cumulative (Greater than)	% Finer than
5	100.0	0.0	100.0	0.0
10	99.9	0.1	99.0	1.0
15	98.4	1.6	86.2	13.8
20	89.5	10.5	58.0	42.0
25	71.5	28.5	32.0	68.0
30	50.0	50.0	15.0	85.0
35	32.1	67.9	7.0	93.0
40	19.2	80.8	2.8	97.2
45	10.8	89.2	1.2	98.8
50	6.0	94.0	0.5	99.5
55	3.0	97.0	0.15	99.85
60	1.9	98.1	0.09	99.91
100	0.0	100.0	0.0	100.0

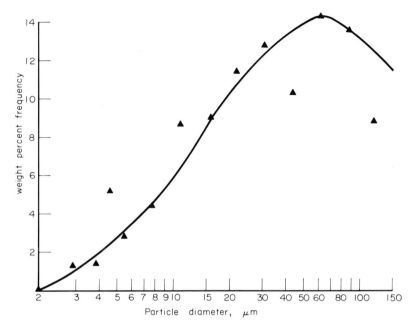

FIGURE 3-1 Weight frequency plot of particle size distribution of tin powder.

In Tables 3-1 and 3-2 linear scales for the size ranges, i.e. 0-5, 5-10, 10-15 μm etc. were used. Linear scales for size distributions are commonly used for narrow size distributions and are frequently found in tabulations for particle size distributions of refractory powders by the turbidimetric method. On the other hand, geometric scales for size distributions are usually preferable, particularly in cases where the size distributions are fairly wide. As an example, the size distribution for an atomized tin powder as determined by the Microtrac instrument is shown in Figures 3-1, 3-2 and 3-3. Figure 3-1 is a linear plot of weight frequency as the ordinate, Figure 3-2 a linear plot of cumulative weight percent frequency as the ordinate and a logarithmic scale of particle diameter with values at size ratios of $\sqrt{2}$ as the abscissa in both figures. In Figure 3-3 the particle diameter on a logarithmic scale as the ordinate is plotted against the cumulative weight percent below a stated size on a probability scale as the abscissa. This last plot is commonly used for size distributions. The weight frequency plot is essentially the derivative or slope of the cumulative plot and small deviations from a smooth curve are accentuated. The weight frequency plot shows a distribution of particle sizes near log normal, which is not uncommon for size

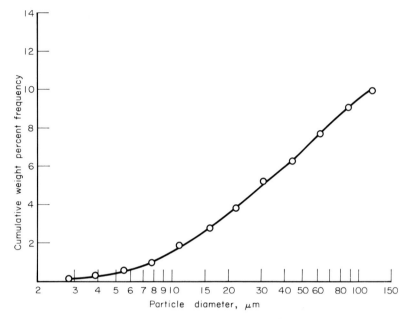

FIGURE 3-2   Cumulative weight percent plot of particle size distribution of tin powder, linear ordinate.

distributions of metal powders. However, skewed distributions, distributions with a flat maximum or bimodal or polymodal distributions with two or more maxima may also occur. On a plot of cumulative weight percent finer than, as in Figure 3-2 and 3-3, the median size, i.e. the size corresponding to 50%, can be read directly from the plot.

C.   Particle Shape and Structure

Although quantitative shape factors for powder particles have been defined, they have been little used in characterizing metal powders. In the vocabulary issued in standard 3252 of the International Standards Organization, the following qualitative descriptions of particle shapes have been defined and illustrated by figures:

acicular	needle shaped - See Figure 3-4
angular	sharply edged or roughly polyhedral - See Figure 3-5
dendritic	of branched shape - See Figure 3-6
fibrous	having the appearance of regularly or irregularly shaped threads - See Figure 3-7

66

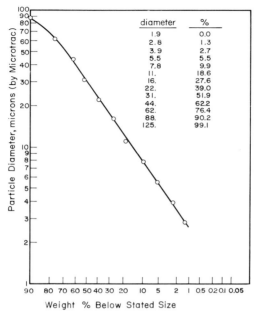

FIGURE 3-3 Cumulative weight percent plot of particle size distribution of tin powder, abscissa: probability scale.

flaky	plate like - See Figure 3-8
granular	approximately equidimensional but of irregular shape - See Figure 3-9
irregular	lacking any symmetry - See Figure 3-10
nodular	of rounded irregular shape - See Figure 3-11
spheroidal	roughly spherical - See Figure 3-12

With regard to particle structure, agglomeration of powder into aggregates is frequently observed, particularly in fine refractory metal powders. In order to determine the particle size distribution of such agglomerated powders, methods of deagglomerating the aggregates in a defined manner have been developed. They will be discussed in connection with the turbidimetric method of determining particle size distribution.

Metal powder particles may be single crystals or may be polycrystalline. The difference between grain, particle and agglomerate is illustrated in Figure 3-13 taken from ISO Standard 3252.

Another structural characteristic of many metal powders is porosity of the individual particles. An at least semiquantitative measure of the porosity of metal powders is possible by using different methods of specific surface determination, as will be discussed in the next subsection.

FIGURE 3-4   Acicular powder particles.

FIGURE 3-5   Angular powder particles.

FIGURE 3-6 Dendritic powder particles.

FIGURE 3-7 Fibrous powder particles.

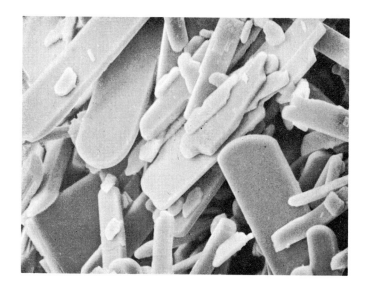

FIGURE 3-8 Flaky powder particles.

FIGURE 3-9 Granular powder particles.

FIGURE 3-10   Irregular powder particles.

FIGURE 3-11   Nodular powder particles

FIGURE 3-12   Spheroidal powder particles.

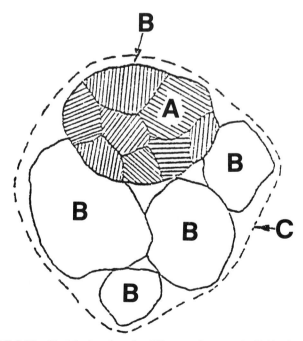

FIGURE 3-13   Sketch showing the difference between individual particles (B) grain (A) and agglomerate of particles (C).

D. Specific Surface

The specific surface of a powder is the surface area in square meters per kg or square centimeters per gram. For spherical powders of size x μm, specific surface in cm^2/gm is $6 \times 10^4/\rho x$, ρ being the density of the particle. For irregular particles a shape factor, θ, which is larger than 1 may be inserted. The specific surface then becomes $6\theta \times 10^4/\rho x$. If the particle size distribution of a powder is known, its specific surface may be calculated. For a powder whose size distribution has been determined on a geometrical scale with a $\sqrt{2}$ ratio between size fractions the relationship between specific surface σ and weight percent retained w to size x in μm is

$$\sigma = \frac{y_1\sigma_0}{100} + \frac{600\,\theta}{\rho} \left[\frac{w_1}{1.2\,x_1} + \frac{w_2}{1.2\,x_2} + ... \frac{w_m}{1.2\,x_m} \right]$$

in which θ is the shape factor, $x_1, x_2....x_m$ are sizes in the $\sqrt{2}$ size scale in μm, $w_1, w_2....w_m$ are the corresponding weight percent retained, y_1 is the cumulative weight percent finer than the finest size measured (x_1) and σ_0 is the specific surface of the fraction finer then the finest size measured (x_1). The factor 1.2 is included because the average size of a size fraction retained on x and passing $x\sqrt{2}$ should be near the geometric average of the sizes, which is $x\sqrt{2} = 1.189$ x or the arithmetic average of the sizes, which is $(1 + \sqrt{2}x)/2 = 1.207x$. Only if y_1, the cumulative weight percent finer than x_1, the finest size measured, is very small, can the specific surface σ_0 of the fraction finer than x_1 be neglected.

If the specific surface of a powder is known, an "average" particle size may be calculated, as is done by the chart of the Fisher Subsieve Sizer.[5] The assumption is made that all particles are of equal size and spherical.

The two principal methods in use for determining the specific surface of metal powders, permeametry and gas adsorption, often given very different values of specific surface. In permeametry, the permeability of a bed of powder, i.e. the volume of air which permeates a certain cross section and thickness of the bed in a given time under a given pressure, is determined. The specific surface of powders measured by permeametry includes primarily the exterior surfaces of the particles, but not the surface of interior pores. The Brunauer-Emmet-Teller[6] method of measuring specific surface is based on the determination of the amount of gas which is adsorbed on the surface of a powder in a monomolecular layer. The specific surface so determined includes the surface of interior pores.

To illustrate the difference in these two methods of specific surface determination, data obtained on the 104-149 μm mesh fraction of three types of iron powder are presented. The powders are water-atomized iron

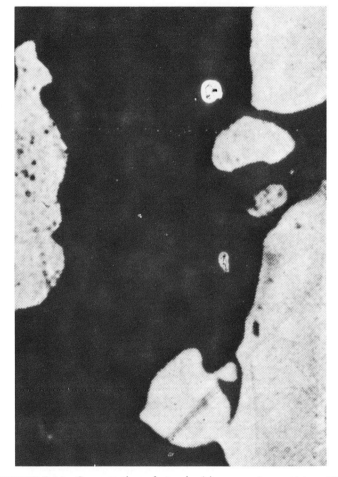

FIGURE 3-14 Cross section of atomized iron powder particle, x600.

powder (A. O. Smith grade 300M), sponge iron powder (Hoeganaes grade MH 100) and iron powder by hydrogen reduction of mill scale (Pyron-100). Scanning electron micrographs of the powders were shown in chapter 2 in Figures 2-12, 2-5 and 2-7. Micrographs of cross sections through these powders are shown in Figures 3-14, 3-15 and 3-16. The atomized powder particles are solid, those of the sponge iron powder have relatively large pores, those of the Pyron powder, produced by hydrogen reduction at low temperature, have fine pores. The specific surfaces in cm^2/g of the 104-149 μm mesh fraction of the three powders as measured by the permeability and

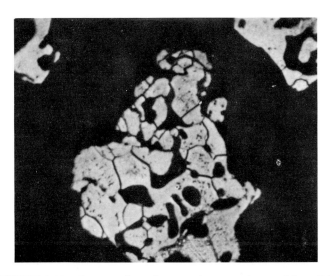

FIGURE 3-15 Cross section of sponge iron powder particle, x600.

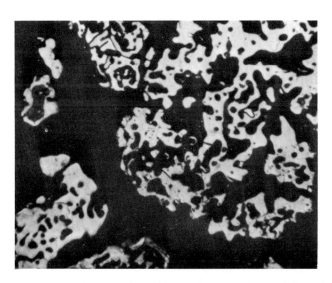

FIGURE 3-16 Cross section of Pyron iron powder particle, x600.

the BET method are shown in Table 3-3.[7] The range of specific surface of the three powders as measured by the permeability method is relatively narrow from 397 to 547 cm^2/g. These values are, however, 6 to 9 times as large as the calculated specific surface of 62 cm^2/g for a powder consisting of spheres of

Table 3-3   Specific Surfaces of the 104-149 μm-Mesh Fractions of Three Powders in cm²/g

Powder	by Permeability Method	by BET Method	Difference
A. O. Smith-Inland, Inc.			
Water-atomized 300M	397	320	−77
Hoeganaes MH-100	501	984	+483
Pyron-100	547	1620	+1073

122 μm in diameter (geometrical mean between 104 and 149 μm). This ratio may be taken as a measure of the pronounced roughness of the outside surface of the particles in all three powders. The roughness is evident from an inspection of the scanning electron micrographs of the particles (Figures 2-12, 2-5 and 2-7).

While the values of specific surface of the three powders as measured by permeametry are within a relatively narrow range, those measured by the BET method differ widely. The larger surface area of the sponge and the Pyron powders is evidently due to their porosity. Depending on their volume and size, the pores add additional area to the specific surface of the powders above that measured by permeametry.

E.   Details of Individual Methods for Determining Particle Size Distribution and Specific Surface

(a) Sieving

ASTM standard B 214 and MPIF standard 5 describe how the sieves analysis of granular metal powders is determined. A set of wire cloth sieves 8″ in diameter and 1 or 2″ in height fitted with bronze cloth are assembled from the finest to the coarsest in ascending order with a collecting pan at the bottom under the finest sieve. A 100 g sample—50 g for powders with an apparent density under 1.5 g/cm³—is placed on the top sieve and a solid cover is added to close the nested assembly. It is fastened in a mechanical sieve shaker and shaken for 15 minutes under conditions detailed in the standards. A widely used sieve shaker is the Tyler Rotap shaker. The quantity of powder retained on each sieve of the set is determined by removing it from the sieve and weighing it as described in the standards.

The sieves chosen in the standards are taken from the series of sieves specified in ASTM standard E 11 or the Tyler sieve series which is close to

Table 3-4   ASTM sieves and equivalent Tyler standard sieves for sieve analysis of metal powders passing a No 100 (150 μm) sieve.

ASTM sieves, size and U.S. standard sieve designation, μm	Tyler sieves, size and Tyler sieve series designation, μm
180 (No 80)	175 (80 mesh)
150 (No 100)	149 (100 mesh)
106 (No 140)	104 (150 mesh)
75 (No 200)	74 (200 mesh)
45 (No 325)	44 (325 mesh)

that in ASTM E-11. Sieves standardized by the International Standards Organization are also practically interchangeable with ASTM and Tyler sieves. Traditionally, sieves are designated by the number of wires per inch in the cloth, e.g. 100 mesh, i.e. 100 wires per running inch. Actually, what is important in a sieve is the square aperture opening in μm which depends on both the number of wires per unit length and the diameter of the wire. The openings are the decisive parameter in the designation of a sieve. The ratio of the openings in consecutive sieves in the ASTM and Tyler series is $\sqrt[4]{2}$, but for sieving metal powders this ratio is too small. The sieves in ASTM standard B 214 and MPIF standard 5 are shown in Table 3-4.

It is seen that the ratio of the openings between the 150 (Tyler 149) and the 106 (Tyler 104) μm and between the 106 (Tyler 104) and 75 (Tyler 74) μm sieves is quite near $\sqrt{2}$, but that the ratio between the 180 (Tyler 175) and the 150 (Tyler 149) sieves is near $\sqrt[4]{2}$ and that between the 75 (Tyler 74) and the 45 (Tyler 44) μm sieve is equal to $\sqrt[4]{8}$. The reasons for this deviation from a strictly geometrical series are:

(1) A 44 μm (325 mesh) sieve was included because it is the finest woven wire mesh sieve readily available and separates "sub-sieve" sizes from coarser sizes.

(2) The 100 mesh, 150 mesh and 200 mesh were traditionally included in determining particle size distribution in metal powders by sieving.

(3) The 80 mesh sieve was considered necessary to determine oversize for powders passing 100 mesh sieves.

As pointed out by Haertlein and Sachse[8] typical specifications for the sieve analysis of many metal powders giving maximum and minimum values for the percentages in each particle size range are often more restrictive than necessary for the use for which the powders are intended. They make it difficult and expensive for the powder producer to supply acceptable powder. The authors give an example of a specification for a powder for which no particles should be coarser than 177 µm and approximately half should pass through a 44 µm sieve:

Retained on sieve	Passing sieve	Min. %	Max. %
No 80 (177 µm)			0
No 100 (149 µm)	No 80 (177 µm)		1.0
	No 325 (44 µm)	40.0	60.0

Powder fulfilling this type of specification can be readily supplied by the manufacturer of metal powder and should be satisfactory to the user for many P/M base metal products.

Data on tolerances for the openings in wire mesh sieves are given in ASTM standard E 11. Haertlein and Sachse[8] present data from a round robin test for the sieve analysis of a nickel powder which allows a statistical analysis of the precision possible in sieving metal powders with wire mesh sieves.

Besides wire woven test sieves, sieves produced by electroforming nickel in meshes of precisely square openings from 10 to 120 µm have been developed and are commercially available as Micromesh sieves.[9] They are much more expensive, but also have closer tolerances than woven wire sieves and are of particular interest for particle size analysis in the region between 40 and 10 µm. The smaller micromesh sieves have quite small open areas. Since the sieves are fragile, they are available only in the three inch frame size and should not be overloaded. In sieving with micromesh sieves and in cleaning them an airjet and ultrasonic vibration is often used.

(b) Microscopic Sizing

Microscopic sizing used to be considered the standard method upon which other methods of determining particle size and particle size distribution of metal powders were based. The powder was dispersed on a microscope slide in a suitable mounting medium. The dimensions of individual particles were determined using a filar eye-piece or a method was used in which the projected are of a particle as seen in the microscope was matched with the size of graticules engraved on the eye-piece reticle. These methods were time consuming and afflicted with observer bias. Automatic methods of sizing and counting particles have been developed in which the scanning spot of a television camera is used for determining size distribution. The number of intercepts of the spot by particles are recorded and the size distribution is derived by successive scans of the field at different discriminatory levels. Although these methods have made microscopic sizing less time consuming and independent of operator bias, they are not widely used for metal powders.

When particle sizes below the limit of resolution of the optical microscope are to be determined, sizing with an electron microscope may be used, in which the sample is deposited on a thin membrane of carbon or plastic. Again automatic sizing by scanning is possible. The particle sizes of metal powders used in powder metallurgy are seldom so small that microscopic sizing with the electron microscope is called for.

Even though microscopic sizing has been abandoned as a routine method for particle size determination, it is still of value in calibrating other methods particularly those based on sedimentation.

(c) Methods Based on Stokes' Law

Among the large number of methods available for the determination of particle size and particle size distribution by elutriation and sedimentation only a few are commonly used for metal powders. Before these methods are presented, certain conditions, which must be observed to obtain valid results, should be mentioned: convection currents in the suspending fluid must be avoided and the relative rate of motion between the fluid and the powder particles must be slow enough to guarantee laminar flow, which means that the Rynolds number, given by x $v\rho_F/\eta$ should be less than 0.2 where x is the particle size, v the settling velocity, ρ_F the density of the fluid and η its viscosity. This generally restricts the particle size which can be determined by Stokes' Law methods to those in the subsieve range. On the other hand, the particles should be large compared with inhomogeneities in the fluid, which for elutriation or sedimentation in air restricts the methods to particles larger than 5 μm, while for sedimentation in liquids particle sizes down to 0.1 μm can be determined.

The particles in the suspension must be perfectly dispersed and the suspension must be dilute enough to guarantee independent motion, which means maximum concentration of about 1% by volume of particles in the suspending medium. Finally, wall effects must be guarded against which means that the inside diameter of the elutriation or sedimentation chamber should be sufficiently large.

(c₁) The Roller Air Analyzer[10]

The Roller air analyzer is an apparatus in which the particle size distribution of metal powders is determined by elutriation in a stream of air. With this apparatus, shown in the diagram Figure 3-17, a powder may be classified into particle size fractions in the range between 5 and 40 μm. The

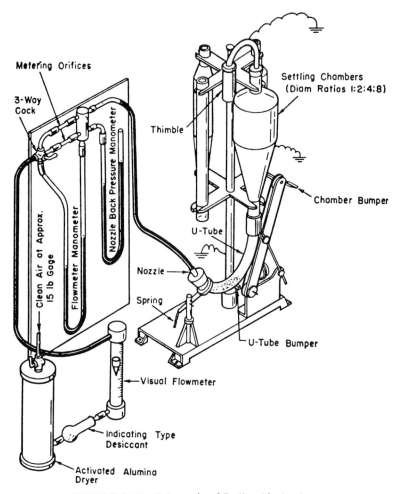

FIGURE 3-17 Schematic of Roller Air Analyzer.

procedure is described in ASTM standard B 283 and in MPIF standard 10 for the subsieve analysis of granular metal powders by air classification. A stream of high velocity air flowing through a nozzle of suitable size impinges upon the powder sample contained in a U tube so that it becomes dispersed in it. The suspension of powder in air rises through a cylindrical settling chamber. The velocity v of the air stream through the chamber in cm/sec which just balances the settling velocity of particles with diameter x in μm and a density ρ in g/cm^3 can be calculated from Stokes law in the form

$$v = 29.9 \times 10^{-4} \, \rho \, x^2$$

In this equation the values for the gravitational constant and the viscosity of air are inserted and the very small density of air compared to that of metal powder particles is neglected. To obtain this velocity v in cm/sec of the air stream through a cylindrical settling shamber of diameter D in cm, the volume rate of air flow F in cm^3/sec must be

$$F = 47.1 \times v \times D^2$$

With this volume rate of flow, particles with a size smaller than x will be carried through the settling chamber into the collecting system which consists of an extraction thimble. Large particles will fall back into the U-tube. By using a series of vertical settling chambers with diameters in the ratio 1:2:4:8 and a constant volumetric rate of flow, the powder may be classified into particle size fractions with the maximum sizes in a ratio of 1:2:4:8 (e.g. 5, 10, 20 and 40 μm). The apparatus is equipped with manometers to regulate the volume rate of air flow. Means to vibrate the U-tube to insure proper suspension of the powder in the air stream and to vibrate the settling chamber to dislodge particles adhering to the side of the chamber are provided.

(c₂) The Micromerograph

The micromerograph[11] is a type of sedimentation balance occasionally used for determining the particle size distribution of subsieve metal powders. The powder is suspended in air by projecting the sample with a burst of nitrogen through a deagglomerating device consisting of a conical annulus into the settling chamber. The chamber consists of a thermally insulated vertical aluminum tube 10 cm inside diameter and 2.5 m high. The pan of an automatic balance weighing the amount of powder settling on it is located at the bottom of the chamber. A recorder records the cumulative weight of powder settled as a function of time, from which the particle size distribution is calculated on the basis of Stokes' law. Particle size distribution in the range between 2 and 100 μm may be determined with the instrument. A drawback of the method is the tendency of the powder to cling to the wall of the column with no assurance that this clinging will be uniform for all particle sizes.

(c₃) Light and X-Ray Turbidimetry

A sedimentation method widely used to determine the particle size distribution of refractory metal powders, such as W and Mo, and of refractory metal compound powders, such as WC, is the turbidimeter[12] standardized

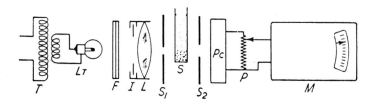

FIGURE 3-18   Schematic of Turbidimeter.

in ASTM standard B 430. A diagrammatic sketch of a turbidimeter is shown in Figure 3-18. The procedure for determining the particle size distribution of refractory metal powders with this instrument is described in ASTM standard B 430. A sample of powder dispersed in a liquid is filled into a glass cell and allowed to settle. A collimated beam of light is passed through the cell at a level having a known vertical distance h from the liquid level. The intensity of the light beam is determined by the current generated in a photocell. The current is passed through a potentiometer where the voltage drop across it is measured by a recording millivoltmeter. The Beer-Lambert law gives a relation between the intensity of the light beam and the projected area of the powder particles suspended in the liquid

$$\ln I_o/I = k.S.c.l/4$$

in which I_o is the intensity of the light beam passing a cell filled with clear liquid, I the intensity of the beam passing through the suspension, S the projected area of the powder particles per unit mass, c the concentration of the suspension in unit mass per unit volume, l the length of the light beam and k the so called extinction factor, which is generally treated as a constant for all particle sizes.

The reading of the millivolt recorder for the intensity of the light beam shining through the clear solution is adjusted to 100% while the concentration of the suspension is adjusted so that the reading for the intensity of the light beam through the suspension before any settling has taken place is in the range between 20 and 40% of that through the clear solution. As the suspension settles the projected area of the particles in the suspension decreases and the intensity of the light beam increases.

At the beginning of settling all particle sizes are uniformly distributed through the volume of the sedimentation cell. As settling proceeds, larger particles settle faster then small ones. The time t_x necessary for a particle of size x to settle through the distance h from the top of the suspension to the level of the light beam may be calculated from Stokes' law:

$$t_x = \frac{18\,\eta h}{(\rho - \rho_F)gx^2}$$

where η is the viscosity of the liquid, ρ and ρ_F are the densities of the particles and the suspending liquid and g the gravitational constant. After the time t_x all particles larger than x have settled below the level of the light beam. The concentration of particles at the light beam level is now equal to the original concentration of particles minus all particles with sizes equal to and larger than x. The projected surface of the particles at time t_x is, therefore, smaller than that of the particles in the original suspension and the intensity of the transmitted light is greater. Referring to the Beer-Lambert law, the percent decrease in the projected area of the particles is proportional to the negative of the logarithm (the logarithm of the reciprocal) of the increase in light transmission measured by the millivolt recorder expressed as percent of the reading through the clear liquid.

In order to obtain information on particle size distribution from the plot of light intensity vs. time, use is made of the relationship between the total weight of n particles of size x, which is proportional to nx^3, and the projected surface area of the n particles, which is proportional to nx^2. Therefore the cumulative weight of particles up to a given particle size x_{lim} which is $\int_0^{x_{lim}} dW$ is proportional to $\int_0^{x_{lim}} x dS$, the integral of the product of particle size and projected surface area integrated from 0 to size x_{lim}.

The weight percent of particles in each of a series of size ranges may be obtained by evaluating the integral $\int_0^{x_{lim}} x dS$ by a summation. The settling times through distance h for particles at the bottom and top of each size range, x_1 to x_2, x_2 to x_3, etc., generally in arithmetic progression, e.g. 0-1 μm, 1-2 μm, etc., are calculated from Stokes' law. The intensities I_{x_1}, I_{x_2}, corresponding to these settling times are read from the light transmission vs. time curve. The relative weight ΔW_{1-2} of particles in the range between x_1 and x_2 may then be obtained from the product of the arithmetic mean of the particle sizes and the difference of the logarithms of the light intensities:

$$\Delta W_{1-2} = \frac{x_1 + x_2}{2}(\log I_{x_1} - \log I_{x_2})$$

Values of ΔW are determined for each of the particle size ranges chosen. The sum of the values is $\Sigma \Delta W$. The weight percent of the particles in the size range x_1 to x_2 is calculated as

$$\text{Weight } \% = (\Delta W_{1-2} / \Sigma \Delta W) \cdot 100$$

Turbidimetric measurements using white light can be fairly low cost and have proven themselves quite reproducible and useful. They are used in research and in routine analyses comparing different lots of refractory metal powders with each other. When particle size distribution is determined with this method, the extinction factor k in the Beer-Lambert law is used as a constant and not a function of particle size. This use is incorrect as the factor decreases slightly with particle size to about 2 μm and then increases rapidly for smaller particles. As a result, greater weight may be given to particles smaller than 2 μm in calculating the particle size distributions from turbidimetric data. This is not too detrimental as the method is generally used for comparative work on the same type of material.

When X-rays rather than white light are used in determining the particle size distribution of a subsieve particle suspension, the attenuation of the X-ray beam intensity is proportional to the mass of the powder particles rather than their projected area. An instrument called the Sedigraph[13] is based on measuring the variation of intensity with time when a finely collimated X-ray beam is transmitted through a settling suspension of metal powder. Particle size analysis with the instrument is similar to that of the turbidimeter, except that a scintillation counter rather than a photocell is used to measure the intensity of the transmitted beam. The instrument has another feature which greatly enhances the speed with which subsieve particle size distributions can be determined. Instead of holding constant the distance h through which particles settle from the top of the suspension to the level of the X-ray beam, an arrangement is provided by which the cell is driven downward relative to the X-ray beam as a function of elapsed time, thereby continually decreasing h. A programmed controller simultaneously establishes the cell position as a function of time from the start of sedimentation and positions the x-axis of the recorder in such a way that the particle size distribution is directly indicated in a plot of cumulative-weight-percent-less-than in terms of the Stokesian equivalent spherical diameter.

As indicated above in subsection C on particle shape and structure, the particles of fine refractory metal powders are often agglomerated by sinter bonds. In order to determine the particle size distribution of these powders by turbidimetry, this fact must be taken into account. In ASTM method B 430, determination of particle size distribution of the powder is described in the as-supplied condition and after the powder has been deagglomerated by rod milling (laboratory milling). In order to get reproducible results for particle size distribution by turbidimetry not only the turbidimetric procedure, but also the deagglomeration procedure must be standardized. Deagglomeration of refractory metal powder samples by milling the powder with tungsten rods in a milling bottle is described in the method. Other methods of conditioning, i.e. deagglomerating the powders, and the reproducibility of these methods have been discussed by Buerkel.[12] In addition to the deagglomeration procedure the dispersion of the powder in the liquid before turbidimetric analysis must be standardized, as shown in ASTM method B 430.

(d) Coulter Counter[14] and Particle size analysis by light obscuration[15]

These two methods of determining particle size and particle size distribution are treated together because they both rely upon suspending particles in a liquid and drawing this suspension through a sensing zone. The suspension is sufficently dilute so that the particles pass through the zone one by one. During this passage their size is determined. The method of measuring the size is quite different in the two methods. In the Coulter Counter, for which a block diagram is shown in Figure 3-19, the suspending liquid must be electrically conducting. The particles pass through an orifice which has immersed electrodes on either side. The passage of the particle changes the resistance of an electric circuit through the liquid between the electrodes causing voltage pulses proportional to the particle volume which are counted. In particle size analysis by light obscuration the sensing zone is a collimated light beam from a high intensity quartz halogen lamp source. After first passing through a square glass window of precisely known dimensions and then through the liquid, the beam impinges upon a photodetector which causes an electric current proportional to the amplitude of the light. As a particle traverses the light beam the amount of light reaching

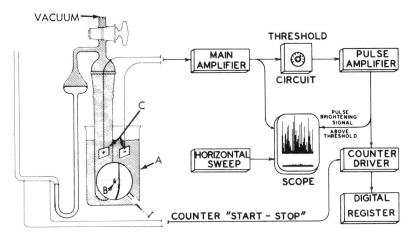

FIGURE 3-19   Schematic of Coulter Counter.

the photodetector is momentarily reduced by an amount proportional to the projected area of the particle and the resulting voltage pulses are counted. In both methods the determination of particle size distribution is, therefore, based on counting voltage pulses for individual particles. Both methods require calibration to correlate the height of the voltage pulses produced by passing particles with particle size. Also in both methods a series of size thresholds or discriminators are employed to count only voltage pulses larger than a given value corresponding to particles larger than a given size. The raw counts are corrected for coincidence counts when two particles pass the sensing zone simultaneously. Counts are repeated for each size threshold until the entire range of particle sizes in the sample is covered. With the aid of microprocessors the number of particles counted is translated into particle volumes which are tabulated or shown as plots of cumulative percent coarser or finer.

An important requirement in these methods of particle size analysis, which is not always easy to fulfill, is that no changes in the concentration of the suspension and particularly no settling out of particles or clogging of the sensing zone occurs while counts for the various particle size thresholds are made. The Coulter Counter has been in use for determining particle size distribution in metal powders for several years. Particle sizes down to less than 1 μm can be analysed. Several hundred thousand particles are counted in each analysis. Data collection requires 10 to 20 minutes. Particle size analysis by the light obscuration principle, as in HIAC Model PA 520 "ASAP" of Hiac Instrument Division, Pacific Scientific, has been introduced only recently for metal powders.

(e) Particle size analysis by laser light scattering, the Microtrac particle size analyzer.[16]

When coherent monochromatic laser light is scattered by small particles the angular distribution of the diffraction pattern depends upon the size of the particles. Different size particles have their pattern of scattered light distributed at different angles from the primary laser beam. The smaller the particle, the larger is the angle. Figure 3-20 shows schematically the Microtrac particle size analyzer which uses laser light scattering. A suspension in a liquid of the particles whose size distribution is to be measured flows through a sample cell which is illuminated by the beam of a helium-neon gas laser. The light scattered by the particles is collected by a lens and focused in the plane of a rotating spatial filter. This filter is so designed that the deconvolution of the transmitted scattered light patterns produces signals proportional to the cube of the particle diameter, i.e. to the particle volume. An integral part of this filter is a rotating disk with openings at 13 selected

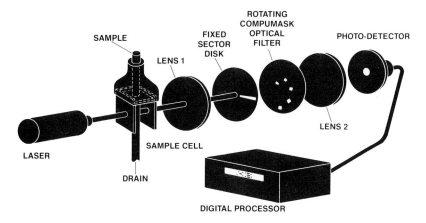

EXPLODED VIEW OF COMPLETE SYSTEM

FIGURE 3-20   Schematic of Microtrac Particle Analyzer (Courtesy: Leeds and Northrup Co.).

angular positions. A second lens collects the transmitted scattered light and directs it onto a photodetector. The output of the detector is fed to a microprocessor where the successive signals for each angular position, i.e. each particle size range, are manipulated to provide a histogram print-out of the particle size distribution. The instrument is capable of measuring the size distribution of powders in 13 particle size ranges on a geometric scale with a $\sqrt{2}$ ratio from 1.9 to 176 μm. A new sample can be analyzed every two minutes. In contrast to the methods of particle size analysis with the Coulter Counter and by light obscuration where provisions must be made to pass particles in the suspension through the sensing zone one by one, the handling of the suspension used in the Microtrac analyzer is much simpler. The suspension is pumped from a mixing chamber provided with a stirrer directly into the sample cell and recirculates to the mixing chamber. When sufficient data are accumulated the mixture is dumped into the drain, debubbled water is pumped into the chamber to clean out the old sample and to prepare the apparatus for the introduction of a new sample.

(f) Specific surface by permeametry; the Fisher Subsieve Sizer[5,17]

The Fisher subsieve sizer is widely used to characterize metal powders with regard to particle size. The determination of "average particle size of powders" by this instrument is covered in ASTM standard B 330 and MPIF standard 32. The test is easily performed, the apparatus is relatively low cost and the reproducibility between laboratories is excellent. In spite of its name the Fisher Subsieve Sizer does not actually measure particle size, but the specific surface of a powder. It is one of several types of apparatus developed to determine specific surface by permeametry.

The basis of the permeametry method is the relationships between the resistance of a packed bed of powder to fluid flow and the specific surface of the powder. For conditions in which the fluid flow is viscous, this relationship is given by the equation

$$S_w = \frac{1}{\rho} \sqrt{\frac{1}{5\alpha} \cdot \frac{\epsilon^3}{(1-\epsilon)^2}}$$

which has been suggested by Carman.[18] S_w is the specific surface of the powder in cm^2/g, ρ is the density of the material, α is the permeability coefficient in cm^2 and ϵ is the porosity of the powder bed defined by

$$\epsilon = \frac{\text{volume of voids}}{\text{volume of bed}}$$

The permeability coefficient α is defined by D'Arcy's law for viscous flow

$$Q_v = \alpha \frac{\Delta P}{L} \cdot \frac{1}{\eta}$$

in which Q_v is the volumetric rate of flow of fluid through the cross section of the bed in $cc/sec/cm^2$, $\Delta P/L$ is the pressure drop per unit length of bed in $dynes/cm^2/cm$ and η is the viscosity of the fluid in poises.

A similar equation applies for the volumetric specific surface S_v in cm^2/cm^3:

$$S_v = \sqrt{\frac{1}{5\alpha} \frac{\epsilon^3}{(1-\epsilon)^2}} = 0.45 \sqrt{\frac{\Delta P}{L} \cdot \frac{1}{Q_v \eta} \cdot \frac{\epsilon^3}{(1-\epsilon)^2}}$$

for a powder from a metal whose density is ρ. In order to determine speciic surface by permeametry using Carman's equation, two values should be measured:

    (1) The porosity of the bed of powder and

    (2) the permeability coefficient of the powder bed which is given by the rate of flow through the bed of air of known viscosity measured with a manometer under a given constant pressure head.

However, in the Fisher Subsieve Sizer these values are not measured

FIGURE 3-21 Front view of Fisher Subsieve Sizer (Courtesy: Fisher Scientific Co.).

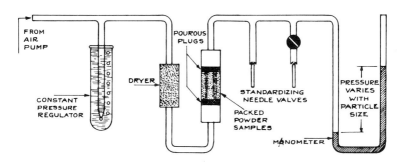

FIGURE 3-22 Schematic of Fisher Subsieve Sizer operation.

individually and the specific surface calculated from them. Instead, the instrument is equipped with a chart by means of which the instrument readings are converted into an equivalent particle size. The chart is shown in Figure 3-21, which is a front view, while Figure 3-22 is a schematic representation of the instrument. In order to make a measurment the tube containing the packed powder sample is prepared by placing a small filter disk over one end of the tube, forcing the filter into the tube with a porous metal plug. The tube is inverted and a sample of powder, equal in weight to the specific gravity of the material, is filled into the tube through a funnel. A second filter disk and plug are inserted in the open end. The tube is placed in a compression device, seen in the front view of the instrument, where the powder is compacted with a rack and pinion. The height of the packed powder bed, which is related to its porosity, is indicated by a porosity indicator. On the basis of the position of this indicator, the chart of the Fisher Subsieve Sizer is aligned horizontally to its proper location for the given porosity of the powder bed. The sample tube is then clamped into place in the air line. A switch starts a pump which passes a flow of dry air through the powder bed under a constant pressure head. The rate of flow determines the level of the manometer of the instrument. When this level has become constant the "Fisher number" is read from the chart by aligning a pointer with the manometer level.

As was indicated previously, for a powder consisting of spherical particles of uniform size the particle diameter x in μm is related to the specific surface S_w in cm^2/g of the powder:

$$x = \frac{6 \times 10^8}{S_w \cdot \rho}$$

where ρ is the density of the material. For such powders the specific surface measured by permeametry can therefore by converted directly to particle diameter. The chart of the Fisher Subsieve Sizer makes this conversion even though the particles in most powders are not uniform in size and not spherical. Under these conditions the "particle diameter" in μm which is read on the instrument is no longer clearly related to any dimensions which can be measured on the powder particles. The value should therefore properly be called a Fisher number, rather than a diameter. The Fisher number can, of course, be readily reconverted into specific surface. The relationships between Fisher number and specific surface and details regarding the calibration of the instrument are discussed by Haertlein and Sachse.[17]

(g) The gas adsorption (BET) method for measuring
    specific surface[6]

In order to determine specific surface by the BET method, the amount of
gas adsorbed in a monomulecular layer on the surface of the powder and the
area occupied by a molecule of the adsorbed gas must be known. The latter
value is generally taken as $16.2 \times 10^{-20} m^2$ for nitrogen, the most widely used
adsorbate. The amount of gas adsorbed in a monomolecular layer V_m in m^2
is calculated from an adsorption isotherm, i.e. a series of measurements of
the volume V of gas adsorbed as a function of the pressure p. From these
measurements the volume V_m may be calculated using a relationship
derived by Brunauer, Emmett and Teller:[6]

$$\frac{p}{v(p_o-p)} = \frac{1}{V_m C} + \frac{(c-1)p}{V_m C p_o}$$

In this equation p_o is the vapor pressure of the adsorbed gas at the adsorption temperature and C is a constant. A plot of $p/v(p_o-p)$ vs p/p_o generally
gives a straight line. From the slope and intercept of this line V_m the amount
of gas adsorbed in a monomolecular layer is determined. Quite often C is
large enough so that the intercept of the straight line can be taken as 0.

A sketch of the adsorption apparatus developed by Emmett and still
widely used is shown in Figure 3-23. Before measurements are made, any
gas adsorbed on the metal powder sample must be desorbed by heating the
sample in the vacuum of a diffusion pump. After desorption, the low
temperature bath, generally liquid nitrogen in a Dewar, is raised to sur-
round the sample. The use of Emmett's apparatus has been described by
Allen[3]: "Adsorbate gas is taken into the burette and its pressure measured
on the manometer. The stopcock between the sample and the burette is then
opened and the new pressure, after allowing time for the equilibrium to be
established, is read on the manometer. The volume of the gas admitted to
the sample bulb is proportional to the difference in the pressures before and
after opening the stopcock. This latter pressure is also the equilibrium
adsorption pressure. The volume adsorbed is equal to the volume admitted
less the volume of gas required to fill the dead-space in the sample bulb and
the burette connections. To obtain more adsorptions points, the mercury
level is raised to the next volume mark and a new pressure established.
Helium is used to calibrate the dead-space. In all adsorption calculations a
correction for the non-ideal behavior of nitrogen at liquid nitrogen temperature is included." When low specific surfaces of less than 1 m^2/g are to be
determined, krypton may be used as an adsorbate instead of nitrogen.

In recent years a modified BET method of measuring specific surface has
been developed using a continuous flow of gases. It is based on the gas

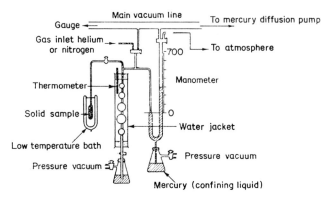

FIGURE 3-23 Schematic of BET apparatus for determining specific surface.

chromatographic technique. An example of such a continuous flow apparatus is the Perkin-Elmer Shell model 212D sorptometer.[19] It operates as follows:

A helium-nitrogen stream of known composition flows continuously over a sample cooled by liquid nitrogen; the sample adsorbs a quantity of nitrogen proportional to its surface area. When the coolant is removed and the sample warms, the nitrogen adsorbed on its surface is desorbed and transferred to the flowing gas stream. Its concentration there is measured with a thermal conductivity cell and displayed as a peak on a strip-chart recorder; peak size is a direct measure of the quantity of nitrogen liberated. Repeating this procedure at two additional helium-nitrogen ratios gives the data needed for a standard Brunauer-Emmett-Teller plot to determine the surface area of the sample. If ultimate accuracy is not required, one measurement often yields satisfactory results.

The BET method of determining specific surface is widely used for catalysts. Its use for metal powders is primarily for very fine powders, particularly those of the refractory metals and for characterizing the total surface area of porous powders as discussed in subsection D.

IV. Specific Tests to Determine the Processing Behavior of Metal Powders

A. Flow Rate and Apparent Density

When metal powders are compacted on automatic presses, the die cavity of the press is filled with a given volume of powder which flows from the filling

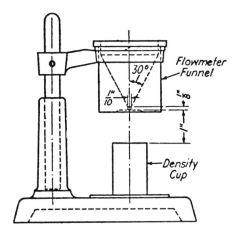

FIGURE 3-24   Hall flow meter.

shoe into the die cavity. A uniform and reproducible amount of powder should fill the die cavity from stroke to stroke of the press. For this purpose, the powder must flow freely into the die cavity and the volume of the loose powder filling the die cavity must have the same mass. This means that the so called "apparent density" of the powder must be controlled.

Flow and apparent density of metal powders are determined by standard methods developed by ASTM and MPIF. The apparatus most commonly used for measuring both flow rate and apparent density of metal powders is the Hall flowmeter shown in Figure 3-24, taken from ASTM standard B 212 and MPIF standard 4 for apparent density and from ASTM standard B 213 and MPIF standard 3 for flow rate. To determine flow the operator places a finger under the bottom of the funnel to close it and pours 50 g of powder into the funnel. He removes the finger to start the flow of powder and simultaneously starts a stopwatch. The watch is stopped when the last powder leaves the funnel and the elapsed time in seconds is recorded as the flow rate.

To determine apparent density powder is poured into a Hall funnel mounted on the stand above the density cup and allowed to flow into the cup which has a volume of 25 cm^3. When the cup is full, the funnel is swung aside, the powder is leveled off on top of the cup by means of a spatula. The mass of the powder in the cup multiplied by 0.04 is equal to the apparent density in g/cm^3.

Both the flow rate and the apparent density tests using the standard Hall flowmeter have certain drawbacks. With regard to the flow rate test, ASTM standard 213 and MPIF standard 3 can be used only for powders which are free flowing, although many powders and particularly powder mixtures containing a lubricant, which fill the die cavities in automatic presses quite satisfactorily, will not readily flow through the standard Hall flowmeter. The flow of powder from the filling shoe into the die is aided by the back and forth motion of the shoe, which is not available in the Hall flowmeter. Suggestions have been made to test flow by filling a small hopper full of powder and sliding it over a steel block with holes having a series of sizes. The smallest size opening which during filling maintains uniform apparent density is used as a measure of flow rate. No standardized tests based on this principle have yet been developed.

With regard to the test for apparent density the so-called Carney funnel which is similar to the Hall funnel, but has an orifice of .020″ diameter instead of .010″ has been standardized in ASTM standard B 417 and MPIF standard 28 for powders which do not flow through the Hall funnel. According to these standards the non-free flowing powder may be poked through the orifice of the flowmeter with a wire.

However, apparent densities determined with flowmeters using the Hall or the Carney funnel are lower than the densities obtained when die cavities are filled from an oscillating or reciprocating filling shoe, again because of the lack of motion of the flowmeter. An apparent density test, the so-called Arnold meter, in which the action of the shoe filling the die cavity is simulated, is being developed by ASTM. A steel block with a round hole having a volume of 20 cm^3 is placed on a sheet of glazed paper. A bronze bushing filled with powder is slid over the hole giving it a slight twisting motion. The powder in the hole is removed and weighed. Apparent densities obtained with this test are quite close to those measured on powder which has filled a die cavity from a filling shoe in an automatic compacting press.

For refractory metals and compounds two other methods for determining apparent density and tap density have been standardized. They are:

> ASTM standard B 329 Apparent density of powder of refractory metals and compounds by the Scott volumeter
>
> ASTM standard B 527 and MPIF standard 46 Tap density of powders of refractory metals and compounds by Tap-Pak volumeter.

In the Scott volumeter the powder is poured through a funnel over a series of inclined glass plates and then into a square cup with a capacity of 1 cu. in. In the tap density test, a weighed quantity of powder is filled into a graduated cylinder, shaken down with a standardized tapping device and its volume determined.

Both flow and apparent density of a powder are dependent on particle shape, particle size and particle size distribution. Spherical powders, normally produced by atomizing, have high flow rates and high apparent densities of the order of 50 percent of the specific gravity of the metal. Nevertheless, powders with spherical particle shapes are not widely used for compacting on automatic presses for reasons which will be discussed later in this section. Flake powders generally do now flow and have very low green densities. They are pressed into compacts only for exceptional applications, such as the so-called SAP (sintered aluminum powder). The particles of most powders used for compacting have irregular, more or less equiaxed shape with flow rates intermediate between spherical and flake powders. The apparent densities of these powders are commonly in the range from 25 to 35 percent of the specific gravity of the metal, although very fluffy, i.e. low apparent density powders, which are porous or have a highly irregular surface are used for special applications. Subsieve powders, i.e. those with all particles finer than 44 μm, generally do not flow. This is the reason why such powders are not used as such for compacts which are pressed on automatic presses. In order to make very fine powders flow they must be agglomerated. Very fine powders have lower apparent densities than coarser powders, but the apprent density also depends upon particle size distribution. Mixtures of coarse, i.e. +50 μm powder and fine, i.e. –50 μm powder may have higher apparent densities than the coarse or fine powders by themselves.

B. Compressibility (Compactibility), Compression Ratio and Green Strength of Powders

The terms compressibility and compactibility are used synonymously for metal powders. Standard methods of tests for compressibility have been issued by ASTM as standard B 331 and by MPIF as standard 45. Compressibility is generally expressed as the green density after ejection of a compact pressed from the powder in a confining die of specified dimensions with a specified pressure under specified conditions. It is a measure of a material characteristic inherent in the powder and widely used as a quality control test for metal powders. In the standard test either a rectangular specimen with a 1¼ by ½" cross section or a 1" diameter cylindrical specimen may be pressed applying pressure with both upper and lower punch. Lubrication is necessary to assist ejection of the compacted specimen from the die. Either a specified method of die wall lubrication or lubrication by mixing lubricant with the powder may be used, but since the results depend upon the method of lubrication, it must be specified. Green density at a single specified pressure, often 60,000 psi for powders used in structural parts, or at a series of specified pressures may be determined. In the latter case a compressibility curve such as the one shown in Figure 3-25 is obtained. The importance of

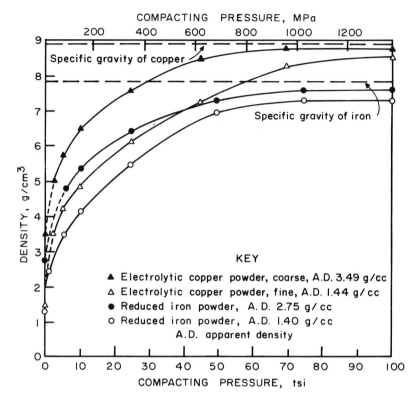

FIGURE 3-25 Compressibility (compactibility) curves for two iron and two copper powders.

high compressibility especially for iron powders used for structural parts has been discussed in chapter 2.

Compression ratio is defined as the ratio of the volume of the loose powder to the volume of the compact made from it. Obviously, this ratio must be known for a powder or powder mixture in order to correctly design tooling for compacting the powder. The ratio of the apparent density as determined in the Hall flowmeter and the green density at a given compacting pressure may be used as a measure of compression ratio, but the ratio so determined will generally be larger than the actual ratio in a tooling set because, as pointed out above, the Hall flowmeter apparent density will be lower than the density of the loose powder in the die cavity of an automatic press. Furthermore, the green density will depend upon the shape of the compact as will be discussed in chapter 4. For this reason, the exact compression ratio must generally be determined for each set of compacting conditions.

Standard test methods for the green strength of compacted metal powder specimens are specified in ASTM standard B 312 and MPIF standard 15. Metal powder compacts are almost never used in the green, i.e. unsintered condition. Nevertheless, they must be strong enough in this condition to resist the abrasion and breakage while they are transferred from the compacting press to the sintering furnace. Compacts from spherical powders, at least those of the higher melting metals, have such a low green strength that they cannot be handled without abrasion and breakage and spherical powders are, therefore, seldom used for cold compacted parts. The standard green strength test consists of a transverse bend test of a 1¼ by ½″ rectangular specimen ¼″ thick. The load at which the specimen breaks is determined and the corresponding strength is calculated using the beam formula. Not only the breaking strength, but also the resistance to abrasion of green compacts is important; they generally, but not always, go in parallel. At present no standard tests for abrasion resistance of green compacts are available. Compacts from mixtures of base metal powders and a lubricant generally have considerably lower green strength than those from the powder alone; in other words, the lubricant which is necessary to facilitate ejection does not act as a binder. Green strength tests should therefore be performed not only on the powder, but also on a powder-lubricant mixture.

C.   Dimensional Changes for Base Metal Powders Due to Sintering

Changes in dimensions of compacts from a given powder depend not only upon the characteristics of the powder, but also upon a large number of other factors, such as the green density of the compact, the type of furnace used, the heat-up rate, the sintering time, the sintering temperature, the cooling rate and the sintering atmosphere. For this reason the test for dimensional changes, issued by ASTM as standard B 610, is a comparative test. Compacts with a rectangular cross section 1¼ by ½″ and ¼″ thick are pressed from the powder to be tested to a uniform specified density. These specimens are sintered together with bars from an approved standard of the same powder under conditions representative of the actual production conditions. The dimensional changes in the bars from the powder to be tested are compared with those of the standard test bars in order to determine whether the tested powder is satisfactory.

References

1. C. Orr and J. M. Dallavalle, Fine Particle Measurement, New York, (1959).
2. R. R. Irani and C. F. Callis, Particle Size: Measurement, Interpretation and Application, New York, (1963).
3. T. Allen, Particle Size Measurement, London, (1968).
4. Ref. 2, p. 26.
5. E. L. Gooden and C. M. Smith, Ind. Eng. Chem., Anal. ed., vol. 12, p. 479, (1940).
6. S. Brunauer, P. H. Emmett and E. Teller, J. Amer. Chem. Soc., vol. 60, p. 309, (1938).
7. F. V. Lenel and T. Pecanha, Observations on the sintering of compacts from a mixture of iron and copper powders, Powder Metallurgy, vol. 16, No. 32, p. 351-365, (1973).
8. J. Haertlein and J. F. Sachse, Sieve Analysis, in "Handbook of Metal Powders," ed. by A. R. Poster, N.Y., p. 68-77, (1966).
9. Buckbee Mears Co., St. Paul, Minn., Specifications for Precision Micromesh Sieves, ASTM E 161.
10. P. S. Roller, Proc. ASTM vol. 32, Part II, p. 607, (1932). J. Amer. Cer. Soc., vol. 20, p. 167, (1937). ASTM Standard B 283 and MPIF Standard 10.
11. T. D. Sharples, Particle size distribution analysis by the micromerograph, in "Handbook of Metal Powders," ed. by A. R. Poster, N.Y., p. 44-55, (1966).
12. W. A. Buerkel, Turbidimetric particle size analysis as applied to tungsten metal powder and the carbide industry, in "Handbook of Metal Powders," ed. by A. R. Poster, N.Y., p. 20-37, (1966).
13. J. P. Olivier, G. K. Hickin and C. Orr, Jr., Rapid automatic particle size analysis in the subsieve range, Powder Technology, vol. 4, p. 257-263l, (1970/71).
14. G. C. West and S. Kinsman, Particle size distribution including Coulter Counter Analysis, in "Handbook of Metal Powders," ed. by A. R. Poster, N.Y., p. 38-43, (1966).
15. H. A. Kupfer, Automatic particle size distribution analysis of powdered material by the light blockage principle, Powder Metallurgy Int., vol. 10, No. 2, p. 96-97, (1978).
16. E. L. Weiss and H. N. Frock, Rapid analysis of particle size distribution by laser light scattering, Powder Technology, vol. 14, p. 287-293, (1976).
17. J. Haertlein and J. F. Sachse, The Fisher Subsieve Sizer, a rapid method for estimating particle size, in "Handbook of Metal Powders," ed. by A. R. Poster, N.Y., p. 63-67, (1966).
18. P. C. Carman, J. Soc. Chem. Ind. (London), vol. 58, p. 1 (1939). For a critique of the "Carman equation" and its range of applicability see T. Allen, ref. 3, p. 171-188.
19. H. W. Daeschner and F. H. Stross, An efficient dynamic method for surface area determinations, Anal. Chem. vol. 34, p. 1150-1155, (1962).

CHAPTER

4

Compacting

Chapter 4 is concerned with four topics:

(1) Mixing of powders, which precedes compacting and includes producing mixtures of different metal powders, e.g. the mixtures of copper and tin powder used in self-lubricating bearings. Also covered in this section will be mixing of metal powders with lubricants, a procedure which almost universally precedes compacting of parts in rigid dies. The reason why lubricants are mixed with metal powders and the advantages and disadvantages of the method will be presented.

(2) The behavior of metal powders under pressure. In this section the relationship between the pressure applied and the density of the compact produced is discussed. It has been studied more thoroughly in compacts pressed in rigid dies than in those isostatically pressed.

(3) The distribution of density and the distribution of stresses in compacts pressed in rigid dies.

(4) Tonnage and stroke capacity of presses.

I. Mixing of Metal Powders

Unless a single metal powder is to be compacted without the addition of lubricant, a mixing operation must precede compacting. The most common type of mixing equipment for base metal powder mixes are double cone mixtures of the type shown in Figure 4-1. The important consideration in mixing is that the powder must not fall freely through the air during any stage of mixing because this will always cause segregation. For this reason the cylindrical part of double cone mixers is kept short. This point must also be considered when other types of mixers, e.g. barrel type mixers with baffles, are used.

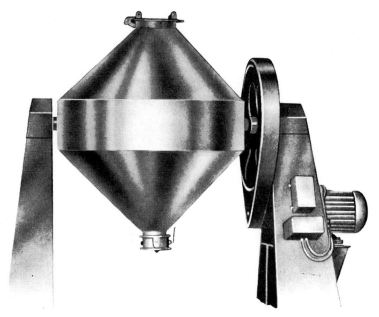

FIGURE 4-1 Double cone mixer.

When parts are pressed in rigid dies, lubrication must be provided to reduce friction between the compact being pressed and diewall and core rod. The lubricant which has a low shear strength keeps the metal surfaces apart. Complete separation is not possible even with well lubricated surfaces and there will be friction due to contacts between metal asperities which puncture the lubricant film. Lubricants are chosen which attach themselves strongly to the metal surfaces and are not easily penetrated. The most common lubricants for base metal powders are stearic acid, metal stearates such as zinc and lithium stearate and synthetic waxes, such as accrawax. Without lubrication the pressure necessary to eject compacts from the die would increase rapidly; after a few compacts had been pressed, they would seize in the die during automatic compacting.

The lubricant is most commonly introduced as a fine powder mixed with the metal powder or metal powders. The amount of lubricant added (generally ½ to 1% by weight) depends upon the shape of the compact. Complex shapes require larger amounts of lubricant to achieve a reasonably low ejection pressure. The mixing time and the intensity of mixing powder and lubricant will affect such properties of the powder mixture as flow and apparent density. For most base metal powder mixes mixing times of 20 to 40 minutes are common. In the United States mixing of metal powders and lubricant powders is quite often done in large quantities in the plants of the metal powder producers which furnish their customers with what is called "premixes." This means that the control of powder properties such as flow and apparent density of the finished mix and in some cases even the dimensional changes of the compacts during sintering become primarily the responsibility of the powder producer. Powder "premixes" furnished by the powder producer are less commonly used in cases where the premixes would be in transit for a considerable time.

The question why lubrication for pressing parts in automatic compacting should be provided by mixing the lubricant in powder form with the metal powder or powders rather than by lubricating the walls of diebarrel and core-rod has been frequently discussed. There is a small advantage in mixing the lubricant with the powder because it provides lubrication not only at the walls but also between the powder particles. Data on the relationship between green density of compacts and the amount of lubricant mixed with the powder substantiates this. However, the relative green densities of compacts pressed with diewall lubrication without lubricant addition to the powder are only slightly lower than those of compacts from metal powder-lubricant mixtures. In Figure 4-2 the green density and the ejection force are plotted as a function of the amount of zinc stearate added to the Swedish Sponge Iron grade MH 100.24 in pressing in a steel die compacts 1″ in diameter and 1″ long at three different pressures.[1] The density of zinc stearate is much lower than that of iron powder and the density of the compacts as percent of theoretical density should therefore decrease with increasing amounts of lubricant added. The curves show, however, that the green density measured in g/cm^3 actually first increases with increasing addition of lubricant and decreases only with larger additions of lubricants. The maximum green density achieved depends on the compacting pressure applied. Lubrication by mixing metal and lubricant powders has, however, serious disadvantages. The added lubricants must be decomposed during the sintering operation and the decomposition products thoroughly swept out of the preheating zone of the furnace, even though small residues of the lubricants which stay in the compacts are generally not considered harmful. This sweeping out of decomposed lubricants requires a relatively high flow of protective atmosphere through the furnace and makes housekeeping and maintenance of furnace equipment more complex. Compacts to be sintered in a vacuum furnace or by high frequency induction must often be "dewaxed" in a separate operation. Lubricants also drastically reduce the green strength of compacts, often to half the green strength of powder compacts without lubricants.

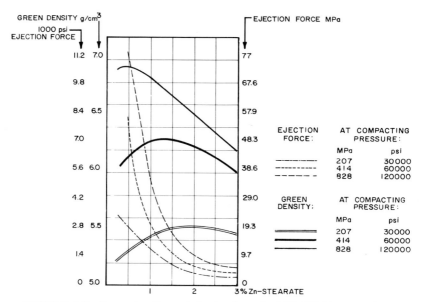

FIGURE 4-2 Green density and ejection force for MH 100.24 sponge iron powder as a function of the amount of zinc stearate added.

Automatic compacting with diewall lubrication is technically feasible and diewall lubrication systems have been patented. However, in view of the problems of applying exactly the right amount of lubricant as a solution or suspension in a liquid carrier and of eliminating the carrier rapidly and completely between lubricant application and filling the die with powder, the more conventional method of mixing the lubricant with the powder is still almost universally being used. Mixing equipment and the lubricants used for cemented carbides are different from those for base metal powder mixes and are described in chapter 16 on cemented carbides. Ball mills, employed in milling cemented carbide mixes, are also used in producing flake aluminum powder from atomized powder in producing special aluminum oxide bearing aluminum compacts.

II. Behavior of Metal Powders Under Pressure

When metal powders are pressed in a die, the resulting compacts generally are strong enough so that they can be handled without breaking. The green strength will, of course, depend upon the type of metal powders—those from soft metals having higher strength—and upon the pressure being applied. For soft metal powders quite low pressures, less than 35 MPa

(5000 psi), are sufficient to produce compacts which can be handled. For harder powders higher pressures are necessary. The question of what physical forces produce adhesion between metals and, for that matter, between other solids is at the bottom of understanding the green strength of metal powder compacts. It has been reviewed in some detail in Jones' "Fundamental principles of powder metallurgy," but only a few basic observations will be presented here. When clean metal surfaces are made to touch each other, the adhesion between them is small because the real area of contact between them is small. The area of contact between the surfaces can be increased when pressure is applied. The pressure produces elastic deformation, which for geometries such as a sphere on a flat surface can be calculated from Hertz' classical equations. For most practical cases of adhesion of surfaces under pressure, the amount of elastic deformation is negligibly small since the weight of the specimen alone will already cause plastic flow. Under these circumstances, the area of contact, regardless of the particle type or shape of surface asperities, will be roughly proportional to the force applied, but to produce complete contact extremely high loads are required. The analysis of adhesion on a fundamental basis is complicated by the fact that metal surfaces, and in particular, the surfaces of metal powder particles, are generally not clean, but are covered with an oxide film. In addition layers of gas molecules are adsorbed on these surfaces. The oxides themselves can be cold welded, but the strength of the bond is generally low compared with that of metals. On the other hand, when metals are rubbed together, which is what happens during compacting of metal powders, the oxide films are penetrated or rubbed off and metal to metal contact is established.

The processes occurring when a column of loose powder is compacted in a die have been described qualitatively by Seelig and Wulff[2] who postulate a series of stages. The first stage is restacking of the powder particles in the column, also called packing, in which the bridging which always occurs in a randomly arranged stack of particles is at least partially eliminated. The second stage involves elastic and plastic deformation of the particles. As discussed above, elastic deformation plays only a minor role. How much plastic deformation occurs depends on the ductility of the metal and may be of minor significance in very hard powders such as tungsten or tungsten carbide. In most metals plastic deformation causes work hardening, which diminishes the amount of further deformation under stress. It may eventually lead to the third stage in which the particles fracture under the applied load and form smaller fragments. This third stage is more important during compacting of nonmetallic powders. The three stages of Seelig and Wulff are not clearly separated, but overlap.

With increasing applied pressure the density of the powder column will increase, or its porosity will decrease. The relationship between applied pressure and the density or porosity of the powder compact has been studied by a large number of investigators, who have attempted establishing mathematical relationships between pressure and average relative density. Much of this work has been done on compacts pressed in rigid dies in spite of the fact that the density through the compact may vary considerably in such compacts, while the density distribution in isostatically pressed compacts is more uniform. The equations relating compacting pressure and relative density have been critically reviewed by Bockstiegel and Hewing.[3]

The first widely quoted relationship between pressure and average compact density was suggested by M. Yu Balshin[4] and reads

$$\ln P = AV + B$$

in which P is the applied pressure, V the relative powder volume, i.e. the ratio of the volume of the powder compact and the volume of solid metal of the same mass, and A and B are constants. Balshin considered the constant A as a "pressing modulus" analogous to a modulus of elasticity. The fact that with a finite pressure relative powder volumes smaller than 1 would be predicted clearly shows that the equation cannot be valid at high pressures. Plots of Balshin's data on the pressure-relative powder volume relationship show that in all cases the equation applies only over limited ranges of pressure.

The second widely used relationship between pressure and porosity was first suggested on the basis of a study of the porosity of a soil as a function of the depth of the soil below the surface.[5] Applying it to the pressure-porosity relationships, R. W. Heckel[6] derived an equation

$$P = \frac{1}{K} \left[\ln \frac{1}{(1-D)} + B \right]$$

in which D is the density of the compact, P the applied pressure and K and B constants.

In careful experiments Heckel showed that the equation applies to a large number of powders from different metals and produced by different methods. Figure 4-3 shows a plot of $\ln \frac{1}{(1-D)}$ vs. pressure for four powders. Except for the lowest pressures a straight line is obtained; the constant K is derived from its slope, the constant A from the intercept of the straight line

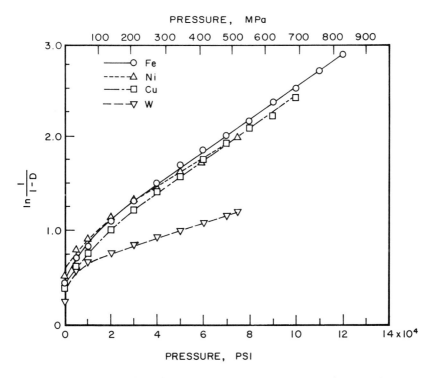

FIGURE 4-3 Relationship between pressure and green density for an iron, a nickel, a copper and a tungsten powder.

portion of the curve. The value of A is larger than $\ln \frac{1}{(1-D_o)}$ where D_o is the apparent density of the powder, as illustrated in the schematic representation of Figure 4-4. Heckel believed that the deviation of his curves from a straight line could be explained by attributing the low pressure part of his curves to the restacking stage of pressing postulated by Seelig and Wulff,[2] but Bockstiegel[3] showed that this is not correct. Studies by Morgan and Sands[7] and by Greenspan[8] indicate that the linear relationship between pressure and $\ln \frac{1}{(1-D)}$ also applies to compacts compacted isostatically, but that the slope of the lines is shallower than for compacts pressed in rigid dies, which means that at a given compacting pressure isostatically pressed compacts are denser than those pressed in rigid dies. This is shown in Figure 4-5 from the work of Morgan and Sands.[7] At very high pressures above 100,000 psi these authors found a deviation from the straight line relationship of Heckel's equation.

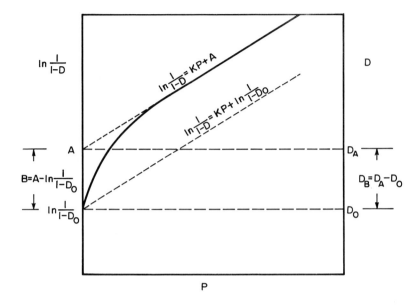

FIGURE 4-4 Schematic of relationship between ln 1/1-D and P at low pressures.

On the basis of a mathematical analysis of the plastic deformation of a hollow sphere under the influence of an applied pressure, equations somewhat similar to those of Heckel may be obtained, in which K is related to the reciprocal of the yield strength of the powder.[9] However, plotting Heckel's results of 1/K vs. yield strength for various powders does not give a straight line. There is also the question what value for yield strength should be used, since the metal powders work harden during compacting and their yield strength increases progressively.

Other attempts to derive equations for the pressure-density relationship are discussed by Bockstiegel and Hewing.[3] These authors came to the conclusion that all of the equations must be considered empirical rather than based on principles of mechanics. Nevertheless they are quite useful for the purpose of interpolating green densities as a function of pressure.

III. Density and Stress Distribution in Compacts Pressed in Rigid Dies

In any discussion of density and stress distribution in powder compacts it is useful to recall the difference in the transmission of pressure in a liquid and

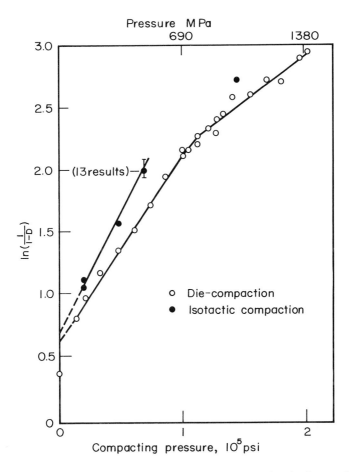

FIGURE 4-5 Relationship between pressure and green density for powders pressed isostatically and in rigid dies.

in a confined powder column. When a liquid is subjected to hydrostatic pressure inside a confined die the pressure is transmitted evenly upon any area of the die regardless of whether the liquid must flow only in the direction of the applied force or whether it must flow around corners. Powders do not behave this way. When they are compressed in a confined die they flow mainly in the direction of the applied pressure.

As illustrated in Figure 4-6, when pressure is applied from top, the powder in the sidearm of the die will not be compressed at all, but will stay as loose powder. It is important to note that at the present time all

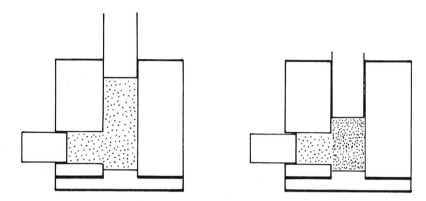

FIGURE 4-6 Density distribution of compacts pressed in dies with a sidearm.

compacting operations in rigid dies are arranged in such a way that pressure is applied either only from the top or from top and bottom but not from the sides. Presses which would apply pressure from the side could be built. The dies in such a press would have to be taken apart in order to remove the pressed piece from the die; also it would be difficult to avoid a plane of weakness in the compact where the powder is compressed from the top and from the side.

Because pressure is applied only from top and bottom in compacting metal powders in rigid dies, the shapes which can be compacted are limited. Pieces with reentrant angles, with holes at an angle to the direction of pressing or screw threads can, in general, not be produced by compacting.

Another consequence of the fact that pressure is transmitted primarily in the direction of pressing is the observation that when compacts with different levels of thickness are pressed in dies with a single lower punch, the density in the two levels will vary. Only when two lower punches are used and the compression ratio in both levels is kept equal will the pressed compact have uniform density. This is illustrated in Figure 4-7.

In the first section of this chapter it was emphasized that when compacts are pressed in rigid dies, lubrication must be provided to lower the friction between powder and die. Although the coefficient of friction is drastically lowered compared with unlubricated dies, there is still considerable friction, which has a strong influence upon the density distribution in compacts pressed in rigid dies. The most obvious effect is the difference in density at the top and at the bottom of compacts pressed only from the top.

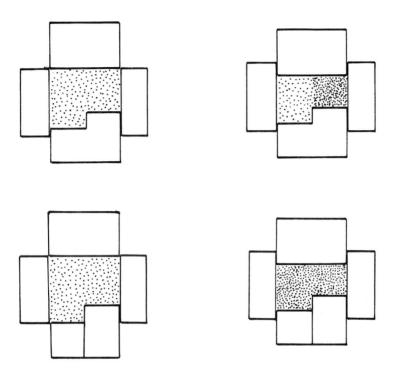

FIGURE 4-7 Density distribution in a two-level compact pressed with one and with two lower punches.

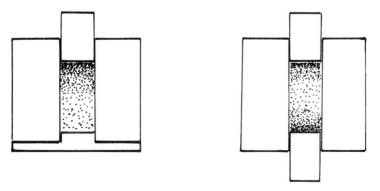

FIGURE 4-8 Density distribution in compacts pressed only from the top and from both top and bottom.

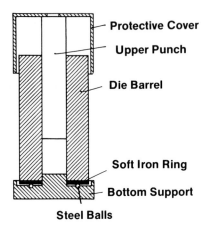

        Protective Cover

        Upper Punch

        Die Barrel

        Soft Iron Ring

        Bottom Support

Steel Balls

FIGURE 4-9 Schematic of apparatus to measure die-well friction during compacting.

Compacts pressed from both top and bottom also show density variations, but the lowest density will be in the center of the compact. This is illustrated schematically in Figure 4-8.

In order to measure how much of the force exerted by a punch upon a compact is transmitted through the compact, Unckel[10] developed the apparatus shown in Figure 4-9. It consists of a die with a fixed lower punch and a movable upper punch. The diebarrel rests upon three hardened steel balls which in turn rest on soft iron rings. The Brinell type impressions which the balls make on the ring measure how much of the force exerted to the upper punch is lost due to friction between compact and diewall. The size of the impressions can be translated into units of force by a suitable calibration.

The loss in force transmitted by a punch to a compact will depend upon the shape of the compact. The smaller the pressing area upon which the punch exerts pressure and the larger the wall area of the compact, the greater will be the loss and the lower will be the density of the compact. In other words, compacts which are long and slender in the direction of pressing will be less dense than those which are short and squat. For simple shapes with no variation in section thickness in the pressing direction Squire[11] found empirically the approximate relationship:

$$\text{Density} = -A \frac{\text{Wall area}}{\text{Pressing area}} + B$$

FIGURE 4-10   Radiograph of tin compact with inserted lead grid, pressed from both ends at 20,000 psi (138 MPa).

where A and B are constants for a given powder, a given pressure and a given method of lubricating.

When the density distribution in cylindrical compacts is determined, it is found that density gradients exist not only in the vertical, but also in the radial direction. Unckel[10] measured both density and hardness distribution on cylindrical compacts by slicing the compacts in small sections both radially and axially. A more rapid method to determine distribution was developed by Kamm, Steinberg and Wulff.[12] They fitted into the loose powder in a cylindrical die prior to pressing a lead grid. The grid is made of a thin lead sheet with square or round holes and inserted in the pressing direction. It must not touch either diewall or punch surfaces. When the compact is pressed, the inserted grid will stay parallel to the direction of pressing and the vertical grid columns will not be bent. The area of the grid openings, which are determined by x-ray radiography of the pressed compacts, will be a direct measure of the density of the compacts in the particular area. A radiograph of a grid is shown in Figure 4-10.[13]

A third method of obtaining a density distribution is to slice a green compact vertically and measure the hardness distribution over the sliced surface. The hardness values can be translated into density values by a calibration curve of hardness and density in very thin compacts pressed at a series of pressures. The density distribution determined by this method in a green nickel powder compact, 20 mm in diameter and 17.5 mm high pressed at 103,000 psi, is shown in Figure 4-11.[14] It is typical for compacts pressed from one end. The denest part of the compact is at the outer circumference at the top where wall friction has caused the maximum relative motion between particles. The least dense part is at the bottom at the outer circumference. Near the cylindrical surfaces of the compacts density decreases uniformly with height from top to bottom.

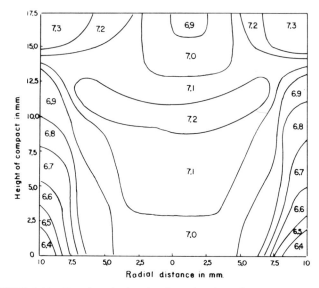

FIGURE 4-11   Density distribution in cylindrical nickel powder compact.

The results of the measurements of density distribution in metal powder compacts clearly show that the distribution of both magnitudes and directions of stresses during pressing of metal powder compacts in rigid dies must be quite complex. The early theoretical treatment of stress distribution in cylindrical powder compacts assumed that the vertical stresses are distributed uniformly over the cross section of the cylinders, that the radial stresses are a constant fraction of the vertical stresses and that the coefficient of friction is independent of the pressure. Since these assumptions are evidently incorrect, the equations derived for stress distribution are also incorrect. Bockstiegel and Hewing[3] describe attempts to improve the original equations, but these attempts have so far been unsuccessful.

From an experimental point of view, the deformation of the lead grids observed in the investigation of Kamm, Steinberg and Wulff[13] makes it possible to determine strain distributions and stress trajectories from lead grid radiographs, but the calculation of coefficients of friction from these data involves large errors because the grids cannot be extended to regions near the surface. Duwez and Zwell[15] were able to determine the radial distribution of axial stresses for copper powder compacts pressed from one end. For this purpose they build piston-type dynamometers into the fixed bottom punch of the die. In their arrangement, shown in Figure 4-12, they compensated as much as possible for the motion of the dynamometers

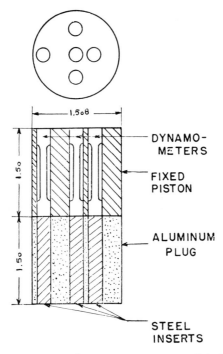

FIGURE 4-12 Arrangement of gauges to measure radial distribution of axial stresses.

relative to the rest of the bottom punch. They placed aluminum plugs under the part of the bottom punch without dynamometers, while the dynamometers rested on steel plugs. The results of the experiment are shown in Figure 4-13 for copper compacts pressed at 50,000 psi. The abscissa is the distance from the center of the fixed lower punch expressed as a fraction of the radius of the punch. The transmitted stress clearly decreases from the center to the periphery of the punch and also decreases with increasing thickness of the compact.

Of greater interest is the distribution of radial stresses at the diewall. Duwez and Zwell[15] tried to measure these stresses with their dynamometer arrangement, but were, in this case, unable to compensate for the movement of the dynamometer in the diewall which disturbs uniform flow and the ability of the powder to support shear. For this reason their results are invalid. Wulff and Shank[16] determined approximate values for the radial and shear stresses at the diewall by a method in which they measured strains in a relatively thin diewall using strain gauges. They found that in compacts pressed from both ends the radial stress is highest in the center of

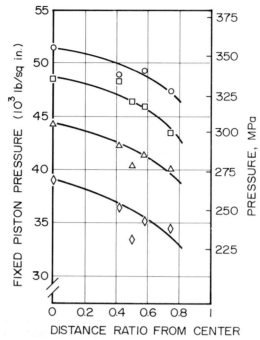

FIGURE 4-13 Radial distribution of axial stresses on fixed piston for copper compacts of various thickness to diameter ratios, (○) 0.11, (□) 0.35, (△) 0.61, (◇) 0.88, pressed at 50,000 psi (345 MPa).

the compact and decreases towards both ends of the compacts.

It may be well to emphasize once more that the goal in pressing compacts from metal powders in rigid dies is to obtain as uniform a density distribution as possible. As discussed before, the highest and most uniform density at a given compacting pressure is obtained in short and squat compacts, but in producing structural parts from metal powders the shape to be compacted is fixed. In this case, the principal method of obtaining uniform density distribution is to provide good lubrication.

IV. Tonnage and Stroke Capacity of Presses

The capacity in MN or in tons a press must have in order to produce compacts in rigid dies at a given pressure in MPa or in tons/in² depends upon the size of the part to be pressed and is equal to the pressure multiplied by the projected area of the part in m² or in² respectively. The

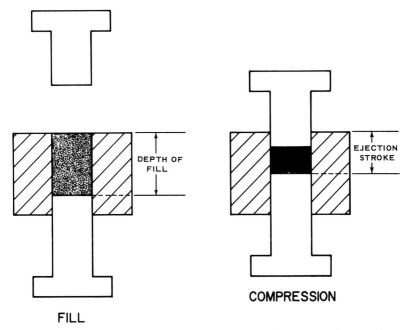

FIGURE 4-14 Depth of fill and length of ejection stroke for pressing compacts.

compacting pressure depends upon the desired green density of the part which in turn is determined by the requirements for the mechanical and physical properties of the sintered part. Compacting pressures may be as low as 70-140 MPa (5-10 tons/in^2) for tungsten powder compacts or as high as 550-830 MPa (40-60 tons/in^2) for high density steel parts. When a part is pressed from both top and bottom the press should apply the required load to both upper and lower ram of the press. In order to eject the pressed compact an ejection capacity must be available which is sometimes divided into the load for the break-away stroke, i.e. the first 1 to 12 mm (1/32 to ½ inch) of the ejection stroke and the load for a sustained stroke which is generally ¼ to ½ of the break-away stroke.

In addition to tonnage capacity of a press its stroke capacity, i.e. the maximum ram travel, is important because it determines the length of a part which can be pressed and ejected. In presses used for automatic compacting the stroke capacity is related to the length available for diefill and for the ejection stroke. As shown in Figure 4-14, the depth of fill is the distance from the top of the surface of the die to the top surface of the lower punch in the filling position. It is equal to the length of the part multiplied by the compression ratio of the powder.

The ejection stroke is the distance from the top of the lower punch to the top of the die in the compression position, as seen in the right hand sketch of Figure 4-14. The press must be able to eject the pressed part by moving the lower punch upwards, until it is flush with the die, or by moving the die downwards until its upper surface is flush with the lower punch.

References

1. Iron Powder Handbook, Hoeganaes Corp., Riverton, N.J., Section C, chapter 20, p. 3.
2. R. P. Seelig and J. Wulff, The pressing operation in the fabrication of articles by powder metallurgy, Trans. AIME, vol. 166, p. 492-504, (1946).
3. G. Bockstiegel and J. Hewing, Critical review of the literature on the densification of powders in rigid dies, Arch. Eisenhüttenwesen, vol. 36, p. 751-767, (1965).
4. M. Yu Balshin, "Theory of compacting," Vestnik Metalloprom., vol. 18, p. 127-137, (1938).
5. L. F. Athy, Density, porosity and compaction of sedimentary rocks, Bull. Am. Assoc. Petrol., Geologists, vol. 14 (1), p. 1-24, (1930).
6. R. W. Heckel, An analysis of powder compaction phenomena, Trans. AIME, vol. 221, p. 1001-1008, (1961).
7. V. T. Morgan and R. L. Sands, Isostatic compacting of metal powders, Metallurgical Reviews No. 134, Journal of the Institute of Metals, p. 87, (May 1969).
8. J. Greenspan, "Metals for the Space Age," ed. by F. Benesovsky, p. 163, Vienna, (1964).
9. C. Torre, Theory and Behavior of pressed powders, Berg u. hüttenmännische Monatshefte, Montan. Hochschule Leoben, vol. 93, p. 62-67 (1948).
10. H. Unckel, Mechanical properties of sintered iron for porous bearings, Arch. Eisenhüttenwesen, vol. 18, p. 125, (1944).
11. A. Squire, Density relationship of iron powder compacts, Trans. AIME, vol. 171, p. 485-503, (1947).
12. R. Kamm, M. Steinberg and J. Wulff, Plastic deformation in metal powder compacts, ibid., p. 439-453.
13. R. Kamm, M. Steinberg and J. Wulff, Lead-grid studies of metal powder compaction, Trans. AIME, vol. 180, p. 694-706, (1949).
14. G. C. Kuczynski and I. Zaplatynky, Density distribution in metal powder compacts, Trans. AIME, vol. 206, p. 215, (1956).
15. P. Duwez and L. Zwell, Pressure distribution in compacting metal powders, Trans. AIME, vol. 185, p. 137-144, (1949).
16. M. E. Shank and J. Wulff, Determination of boundary stresses during the compression of cylindrical powder compacts, Trans. AIME, vol. 185, p. 561-570, (1949).

5

Automatic Compacting

As explained in chapter 1, producing from metal powders socalled "structural parts" to closely controlled dimensions is one of the most important commercial applications of powder metallurgy. Its economic success depends upon being able to compact powders rapidly using presses and dies designed for automatic operation with a minimum of labor. The press and die designs developed for this purpose will be presented in this chapter.[1] Historically, the first products pressed on automatic compacting presses were pills for the pharmaceutical industry. On these pill presses powder is metered into a die cavity, pressed into a pill and the pill ejected and removed from the die cavity in an automatic cycle not requiring an operator. The first powder metallurgy products compacted on such pill presses were self-lubricating bearings. Eventually, the pill presses were extensively redesigned for operation at higher compacting pressures and for precise alignment of both press and die components necessary for pressing metal powder parts of complex design to closely controlled dimensions. Mechanical and hydraulic presses and presses using a combination of mechanical, hydraulic and pneumatic forces are used for compacting metal powders.

The first section of this chapter is concerned with the principles of design for compacting dies. The components of compacting dies and the sequence of motions of these components during automatic compacting are presented. How these motions of the die components are actuated by the rams of the presses is indicated in the second section. The third section is concerned with the construction of compacting tools, the materials used and the clearance requirements.

I. Design of Compacting Dies

The design of dies for four classes of parts of increasing complexity are discussed:

Class 1 parts: Single level components with the compacting force applied from one direction only.

Class 2 parts: Single level components with the force applied from two directions.

Class 3 parts: Two level components pressed with forces from two directions.

Class 4 parts: Multilevel parts pressed with forces from two directions.

In presenting this sequence of die designs, the two principal methods of ejecting parts, ejection by the lower punch or punches and ejection by withdrawal are explained. Compacting of class 1 and class 2 parts is presented in sufficient detail so that the principles of automatic compacting become clear. No attempt is made to discuss exhaustively the design for more complex parts. Instead a few approaches are discussed which demonstrate how the problems involved in designing dies may be solved.

A. Compacting of Class 1 Parts

Figure 5-1 shows an assembly of class 1 parts. They have a single level of thickness in the direction of pressing. On the other hand, they may have

FIGURE 5-1 Assembly of class 1 parts.

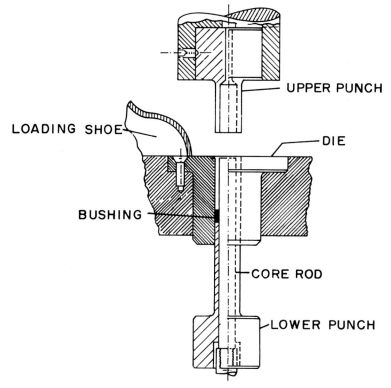

UPPER PUNCH

LOADING SHOE

DIE

BUSHING

CORE ROD

LOWER PUNCH

FIGURE 5-2 Die components for pressing short bushing on automatic press.

quite complex contours in the plane perpendicular to the direction of pressing. The variation in density from top to bottom during pressing from one direction only, which was explained in chapter 4, is within acceptable limits because the parts are thin. As an example of pressing class 1 parts, the compacting of a short bushing is chosen. The die components are shown in Figure 5-2, which is a drawing of a tool assembly. The die components consist of:

> a stationary die-barrel fitted into the table of the press,
> a stationary core rod concentric with the inner diameter of the die-barrel,
> a hollow upper punch fitted into the upper ram of the press,
> a hollow lower punch moving between die-barrel and core rod and fitted to the lower ram of the press.

119

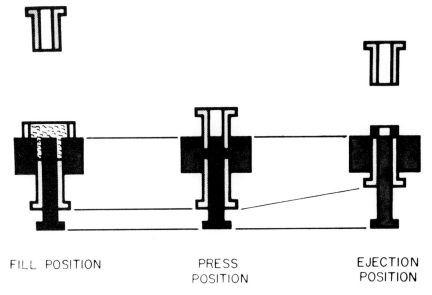

FILL POSITION   PRESS   EJECTION
                     POSITION   POSITION

FIGURE 5-3 Sequence of operations for pressing class 1 part on automatic press with ejection by lower punch.

The inner diameter of the die-barrel, the diameter of the core rod and the upper surface of the lower punch define the die cavity in which the bearing is pressed.

The automatic sequence of operations, shown schematically in Figure 5-3, involves the following steps:

A loading shoe filled with powder moves over the die cavity. An oscillating shoe is indicated in Figure 5-2. A schematic outline of a reciprocating shoe is shown in Figure 5-3. Powder drops from a gate in the lower surface of the shoe and fills the die cavity. The shoe oscillates or reciprocates over the cavity several times to insure reproducible filling and then moves out of the way. This is the "fill position" in Figure 5-3.

The upper punch descends and enters the die cavity, while the lower punch remains stationary. When the upper punch has reached the "press position" in Figure 5-3, a compact has been pressed.

The upper punch moves upward and out of the die, the lower punch rises until it is flush with the die-table, thereby ejecting the bushing. This is the "ejection position" in Figure 5-3. As the shoe moves over the die cavity starting the next cycle, it also moves the pressed part aside. The loading shoe in Figure 5-2 is filled with powder from a hopper not shown in the

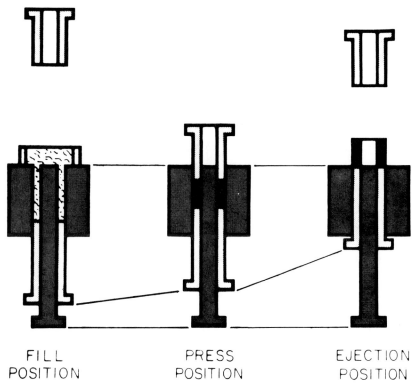

FILL   PRESS   EJECTION
POSITION   POSITION   POSITION

FIGURE 5-4   Sequence of operations for pressing class 2 part on automatic press with ejection by lower punch.

figure. The oscillating or reciprocating motion of the loading shoe is actuated by a mechanism built into the press.

An alternative way to compact a class 1 part by single action pressing is to close the top of the die by a sliding anvil instead of the upper punch, after the die cavity has been filled with powder. Pressure is then applied by upward motion of the lower punch.

B.   Compacting of Class 2 Parts

A longer bushing than the one shown in Figure 5-2 would be a class 2 part. In order to get reasonably uniform density distribution, it is pressed from two directions, as shown in Figure 5-4. After the die cavity has been filled with powder, both upper and lower punches move, the upper punch entering the die cavity and moving downward, the lower punch moving

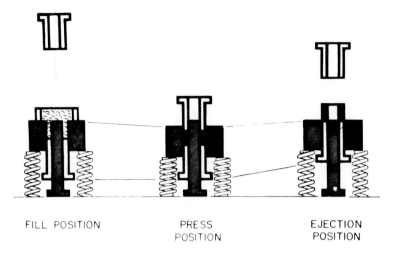

FILL POSITION   PRESS POSITION   EJECTION POSITION

FIGURE 5-5 Sequence of operations for pressing class 2 part with floating die and ejection by lower punch.

upward, until they have reached the "press position" in Figure 5-4. If upper and lower punch move simultaneously at the same rate during pressing, the plane of minimum density in the bushing, called the neutral axis, will be at its center.

Double action may also be obtained by having the lower punch stationary during pressing, but have the die-table mounted on springs so that it floats. This is illustrated in Figure 5-5. When the upper punch enters the die cavity, friction developing between powder and inside surface of the die causes the die-table to move downwards against the resistance of the springs on which it is mounted. Instead of springs, the die-table may also be mounted on a hydraulic cylinder. The movement of the die-table has the same effect as moving the lower punch during pressing, causing pressure application from both top and bottom. If the die-barrel is constrained to move simultaneously but at only one half the rate of the upper punch, the neutral axis is at the center of the compact.

In the tool designs shown so far, the compacts are ejected by the lower punch moving upward flush with the die-barrel. This design approach was and is used primarily in the United States. A different approach, illustrated schematically in Figure 5-6, is called the withdrawal system. It was developed in Germany during the second World War and is now used all over the world. It appears to gradually supersede the system with ejection upward by the lower punch. An integral feature of compacting by the

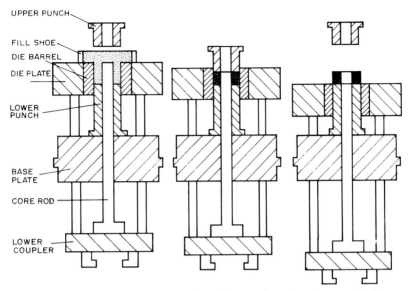

UPPER PUNCH

FILL SHOE

DIE BARREL

DIE PLATE

LOWER PUNCH

BASE PLATE

CORE ROD

LOWER COUPLER

FIGURE 5-6 Sequence of operations for pressing class 2 part with the withdrawal system.

withdrawal system is the use of a removable or built-in tool holder or dieset. A photograph of a tool holder is shown in Figure 5-7. Tool holders are sometimes also used in compacting with ejection by the lower punch, but for compacting with the withdrawal system they are obligatory.

The tools consist of a base-plate mounted on the bed of the press and two movable plates, the die-plate and the lower coupler, which are rigidly connected with each other by columns passing through bearings in the base-plate. What moves during compacting with the withdrawal system are, on the one hand, the upper punch and, on the other hand, the assembly of die-plate and lower coupler.

As shown in Figure 5-6, the lower punch is mounted on the base-plate and therefore does not move during the compacting cycle. The die-barrel is mounted in the die-plate, the core rod on the lower coupler. The die cavity is filled with powder by the shoe. After the upper punch has entered the die cavity, both upper punch and die-plate move downwards so that double action compacting is obtained. After the compact has been pressed, the upper punch moves up, but die-plate and lower coupler move further down until the top of the die-plate is flush with the lower punch. The compact is ejected and can be moved out of the way by the loading shoe. Die-plate and lower coupler then move back into the filling position and the cycle repeats. The movement of the upper punch and of the die is shown schematically in

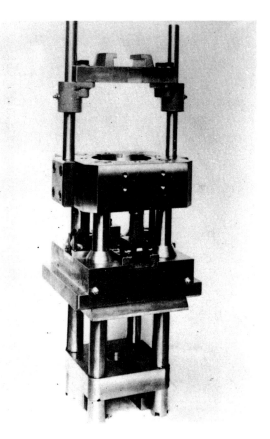

FIGURE 5-7   Photograph of tool holder used in withdrawal system of compacting.

the cycle diagram, Figure 5-8. Two alternate motions of the die are shown to indicate that the die travel should be adjusted with reference to that of the upper punch in order to control the density-distribution along the length of the part.

C.   Compacting of Class 3 Parts

As an example of a class 3 part a component with a flange, as shown in Figure 5-9, is chosen. Using the system of compacting with ejection by the lower punches, such a flanged component will generally be pressed on a press which has, in addition to an upper punch, two independently actuated lower punches. As shown in the figure, the motion of the lower punches is adjusted so that the compression ratio (i.e. the ratio between the height of the columns of loose powder and the height of the sections in the green compact) is the same for the body and the flange. Both body and flange are pressed from two directions. In ejecting the component the outer and the inner lower punch move together until the outer lower punch is flush with the die-barrel, then the inner lower punch completes the ejection of the part.

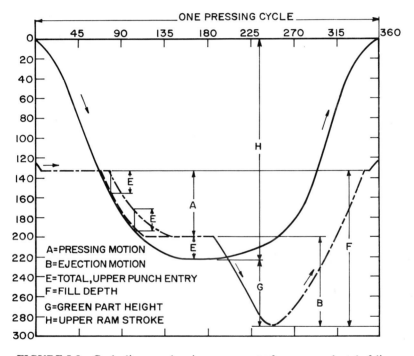

FIGURE 5-8 Cycle diagram showing movement of upper punch and of die during compacting with withdrawal system.

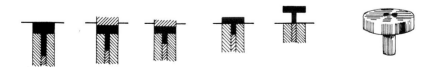

FIGURE 5-9 Sequence of operations for pressing flanged part with ejection by lower punches.

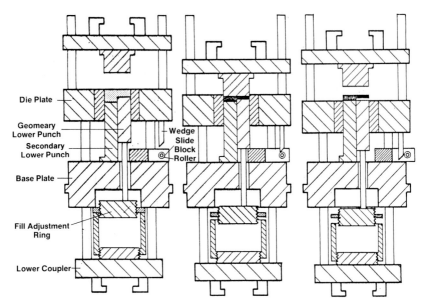

Die Plate

Geomeary
Lower Punch

Secondary
Lower Punch

Wedge
Slide
Block
Roller

Base Plate

Fill Adjustment
Ring

Lower Coupler

FIGURE 5-10 Sequence of operations for pressing flanged part with withdrawal system.

Compacting of a flanged component using the withdrawal system of tooling is illustrated in Figure 5-10. The press actuates the motion of the upper punch and that of the assembly of die-plate and lower coupler. The left hand or stationary lower punch which produces the lower part of the flanged component is mounted on the base plate and does not move during compacting. The right hand or secondary lower punch which produces the flange is supported, through a link, by the fill adjustment ring for the secondary lower punch. In this way equal compression ratios for the two levels of the component, i.e. ratios of the height of the loose powder columns in the fill position to the thickness of the levels in the pressed position, is assured.

When the loading shoe (not shown in the figure) has filled the die cavity and moved out of the way, the upper punch enters the die cavity and descends to the pressing position, while die-plate and lower coupler also descend. At the end of the pressing stroke the right hand (auxiliary) lower punch is supported by a slide block which regulates the compacted length of the flange. As seen in the figure, a wedge is mounted underneath the die-plate which fits into a slot in the slide block. During ejection the die-plate and lower coupler move down until the die-barrel has cleared the flanged compact. The movement of the die-plate causes the wedge to engage the

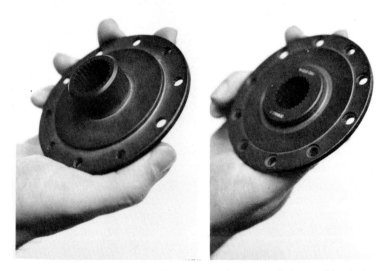

FIGURE 5-11 Photograph of an automotive transmission turbine hub as an example of a four-level part.

roller which moves the slide sideways. This allows the right hand (auxiliary) lower punch to move downward so that it clears the compacted flanged component. The component is moved out of the way by the loading shoe. Die-plate and lower coupler move back into the filling position and the cycle is repeated. While in the system of compacting with ejection by the lower punches, the upper punch and the two lower punches are independently actuated, only two motions, that of the upper punch and that of the combination die-plate and coupler, are available in the withdrawal type of system shown for pressing the flanged component. The motion of the auxiliary slide block compensates for the missing third independent motion in this case.

The principal advantage of the withdrawal system of tooling is the fact that the lower punches are relatively short and are well supported during compaction and particularly during ejection. Withdrawing the die-barrel downward exposes the green part without moving it. When there are multiple lower punches, as many of them as possible rest directly on the base-plate. Withdrawal tooling can be built for very complex parts. On the other hand, in the systems of tooling with ejection by the lower punches, the motions of the punches are built into the multiple action presses. In many cases no tool holders are required.

D. Class 4 Components

The practice of compacting multilevel parts with ejection by the lower punch or punches has declined in recent years. It has been replaced more and more by withdrawal type compacting, and is, therefore, not discussed in this section. Instead a description of compacting the automotive transmission turbine hub, shown in Figure 5-11, as an example of compacting a class 4 multilevel part using ejection by withdrawal is presented. Figure 5-12a shows a sequence of 10 positions of the tool components which follow each other during compacting. The tooling system is more complex than those shown for the withdrawal tooling in Figures 5-6 and 5-10. A press is required which has two auxiliary plates upon which some of the tool components are supported. The five tool components supported from below are marked, A, B, C, D and E in Figure 5-12b, which shows the details of the tool components in position 2. Going from the outside to the center these five components are

- A. The die barrel which forms the outside of the part. It is mounted in the die plate.
- B. The punch to which the pins forming holes in the outer flange are attached. It is supported by the upper auxiliary plate.
- C. The punch which forms the inner flange of the part. It is mounted on the lower auxiliary plate.
- D. The punch which produces the body of the part. It is mounted on the base plate and is stationary during compacting.
- E. The core rod forming the inside diameter of the part.

The die set has a dual upper punch. The outer part, marked F, moves with respect to the inner part, marked G.

In position 1 the shuttle feeder seen on the left approaches the compact. A pick-up tool is attached to the feeder which holds the compacted part during the filling operation. In position 2 the part is removed by the pick-up tool, while the die members assume the fill position. In position 3 the shuttle feeder has moved over the die cavity and is filling it with powder. Position 4 is an underfill position in which the powder columns are retracted to avoid spilling of powder during compacting. In position 5 pressing begins with powder being transferred into the dual upper punch. In position 6 the part is fully pressed. Ejection begins in position 7 with the die-barrel moving downward exposing the outside diameter of the part. Ejection continues in position 8 in which the punch forming the outer flange is withdrawn by movement of the upper auxiliary plate and in position 9 in which die-barrel, both outer punches and core rod are withdrawn. The outer part of the upper dual punch is retracted, leaving the part held between the innermost

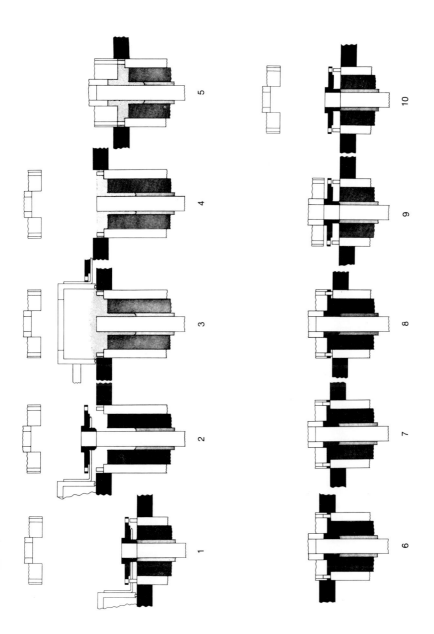

FIGURE 5-12a Sequence of operations for pressing four-level part using withdrawal system and press with two auxiliary platens. See text.

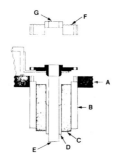

FIGURE 5-12b   Details of position 2 in Figure 5-12.

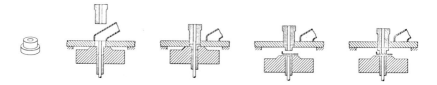

FIGURE 5-13   Tooling for compacting with die-barrel divided into two parts, one of them movable with respect to the other in the direction of punch movement (Olivetti system).

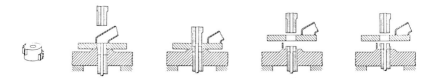

FIGURE 5-14   Schematic of compacting part with undercut using the Olivetti system.

stationary punch and the inner part of the upper dual punch. In position 10 the entire upper punch is raised leaving the part ready to be picked up as in position 1.

In compacting parts as complex as the transmission turbine hub the elastic deflections of the various punches must be carefully balanced. The calculations, necessary in designing adapters and tooling taking these deflections into account, are facilitated by suitable computer programs. A microprocessor is used for the sequence control in pressing complex parts, such as the turbine hub.

A variation of the design for automatic compacting of multilevel parts has been patented by the Olivetti Company.[3] Presses are being built under license by the company. In this design, the die-barrel, which in other compacting tool designs is a single unit, is divided into two parts, one of which is movable with respect to the other in the direction of pressing. During pressing the two parts of the die-barrel are held together. During ejection they are separated and the compacted part is ejected between the two parts. In tool designs, which conventionally require two lower punches, the outer lower punch is replaced by the lower part of the die-barrel, so that only a single lower punch is necessary. This is illustrated in Figure 5-13. This tool design eliminates several limitations in parts geometry which exist when parts are pressed with conventional tool design:

> Parts which have undercuts such as the one shown in Figure 5-14.
> Parts which would require impractically thin punches as the one in Figure 5-15.

In the examples used as illustrations for compacting, the top and bottom surfaces for single level parts were essentially flat and, in multi-level parts,

FIGURE 5-15 Part with nearly tangent surfaces which can be compacted using Olivetti system.

the surfaces of each level were flat. The question arises whether compacting on automatic presses is also possible for parts with curved top or bottom surfaces. It will be difficult or impossible to obtain completely uniform density underneath a punch with a curved surface, since the ratio of loose powder column height to thickness of the compact underneath such a punch will not be uniform. This difficulty is accentuated by the fact that automatic presses provide for level fill of powder across the die cavity, although non-level fill is possible in special cases. How non-uniform the density of a part with a curved surface will be is often difficult to predict. The assumption that powder particles move only in the direction of pressure application is not strictly true; there is a certain amount of sideways motion during compacting. The density variation will also depend upon the radius of curvature of the curved surface - the larger the radius the less variation in density - and upon the compression ratio of the powder -the larger the ratio the less density variation. In many cases the variation in density may be quite acceptable. In other cases the uneven density distribution in a part as pressed may be evened out by a repressing operation after the part has been sintered. Compression of curved-faced compacts has been discussed in detail by Goetzel.[4]

II. Presses

In the first section of this chapter the motions of the components of compacting dies are presented. These motions are actuated by the rams of the compacting presses. Details of the construction of these presses go beyond the scope of this book. Instead, the types of presses used in automatic compacting are listed and, for the mechanical presses, the press action is illustrated in schematic diagrams. Presses used in compacting are:

> Mechanical presses,
> > Eccentric or crank type,
> > Toggle or knuckle type,
> > Cam type which includes rotary presses,
> Hydraulic presses,
> Presses which use a combination of mechanical, hydraulic and
> pneumatic forces.

The action in crank type presses is illustrated schematically in Figure 5-16, that in toggle presses in Figure 5-17 and that in cam type presses in Figure 5-18. In many mechanical presses the rams are actuated by a combination of the various mechanisms. Photographs of mechanical presses are shown in Figures 5-19, 5-20 and 5-21. The press in Figure 5-19 is built for tooling with ejection by the lower punches, the one in Figure 5-20

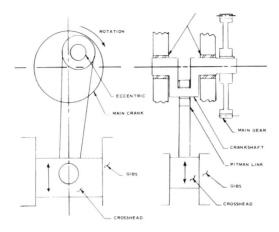

FIGURE 5-16 Schematic of action of crank type (eccentric type) mechanical press.

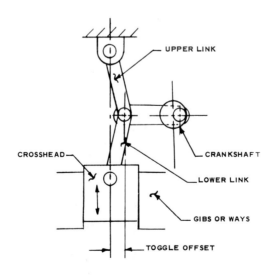

FIGURE 5-17 Schematic of action of press with toggle mechanism.

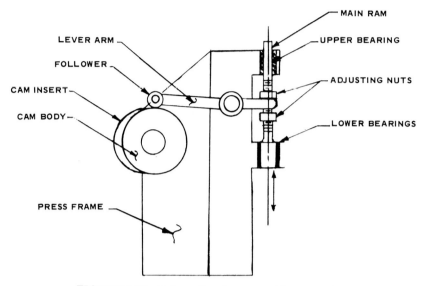

FIGURE 5-18   Schematic of a press with cam action.

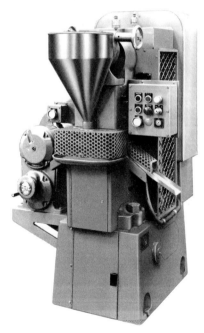

FIGURE 5-19   Crank type mechanical press, Stokes Model 530 (Courtesy: Stokes-Merrill Div. of Pennwalt Corp.).

FIGURE 5-20 Mechanical press using withdrawal system, Dorst Model TPA-50 (Courtesy: A. C. Compacting Presses, Inc.).

for withdrawal type tooling. Figure 5-21 shows a press with two auxiliary platens used for compacting parts like the transmission turbine hub shown in Figure 5-11.

Rotary presses are a special type of cam operated press for high speed compacting of relatively small compacts. Figure 5-22 shows such a press. Rotary presses have a rotating table with a series of openings, each one fitted with top and bottom punches. Both bottom and top punches ride on tracks. As the table rotates, the upper punch is raised to clear the feed funnel; further rotation causes the upper punch to descend. The lower punch runs on part of a track which is adjustable to control the powder fill. At the point where the actual compacting takes place, top and bottom punches are engaged by appropriate pressure rollers which squeeze the powder into the shape of the compact. Both upper and lower pressure rolls are adjustable by eccentrics. After compacting, the tracks let the upper

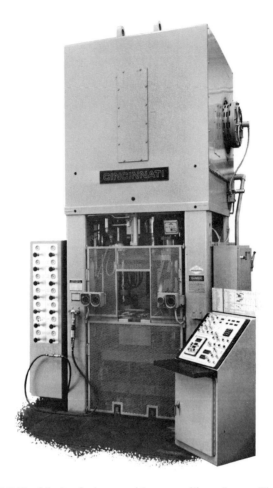

FIGURE 5-21 Mechanical press with two auxiliary platens, Cincinnati 220 ton Rigid Reflex (Courtesy: Cincinnati, Inc.).

punch recede and the lower punch rise to eject the finished compact. The operation of a rotary press is shown schematically in Figure 5-23.

Hydraulic presses are relatively compact even for very high pressures. By supplying a suitable pumping system providing high volumes of pumping fluid at low pressures and low volumes at high pressures, fast advance of the rams during part of the stroke is possible, so that hydraulic presses have production rates on the same order as mechanical presses. Hydraulic

FIGURE 5-22 Rotary press, Courtoy type R 53 (Courtesy: A. C. Compacting Presses, Inc.).

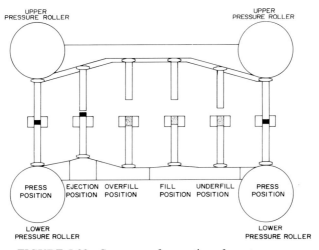

FIGURE 5-23 Sequence of operations for rotary press.

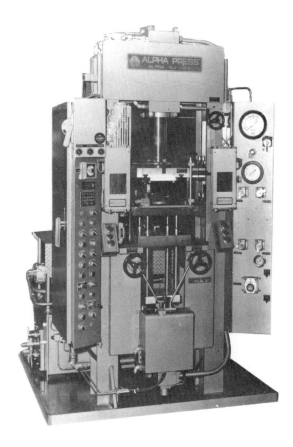

FIGURE 5-24 Hydraulic compacting press, Alpha 60 ton (Courtesy: Alpha Press Company).

presses for automatic compacting are being built with capacities from less than 100 tons to as high as 2500 tons. A hydraulic press is shown in Figure 5-24.

Mechanical presses may make use of auxiliary pneumatic or hydraulic devices. Examples are the pneumatic cylinders used on the press shown in Figure 5-21, on which the transmission turbine hub of Figure 5-11 is compacted. The pneumatic cylinders are used to control the motion of the auxiliary platens. Other examples are the hydraulic overload devices which regulate the maximum pressure applied to the punches of mechanical presses and prevent overloading due to overfilling of the die. On the other hand, hydraulic presses are usually equipped with limit switches or stop collars so that parts with closely controlled length dimensions can be pressed.

The production rate which can be achieved on automatic compacting presses will depend upon the size of the press and also on the configuration of the part. The maximum rates at which small mechanical presses with up to 20 ton capacity can be operated may be 100 or more compacts per minute; for larger presses with 100 ton capacity the maximum rates may be up to 25 compacts per minute. However, most presses have variable speed drives and are not operated at their maximum rates because at these rates the die cavity cannot be filled with powder uniformly from stroke to stroke. The actual rates will therefore be generally less than the maximum rate at which the press can operate. Rotary presses have much higher production rates. For small bearings rates of up to 400 compacts per minute are not unusual.

III. Construction Materials and Clearance of Tool Components

The die-barrel components of compacting tools usually consist of a wear-resistant insert or inserts which are held in a retaining ring made of steel with high toughness. Most commonly the insert or inserts are shrunk into the retaining ring which has the advantage that the compressive stresses induced in the die-barrel during shrinking counteract the radial stress during compacting. Kuhn[5] has published a paper in which equations are provided to determine the optimum dimensions of a die-barrel assembly consisting of a wear resistant insert shrunk into a single stress ring, into two stress rings and inserted into a ribbon wound die. The optimum dimensions are somewhat different for cemented carbide inserts than for tool steel inserts since cemented carbides should not be subjected to tensile stresses. Also estimates are given of the fraction of axial stress transmitted in the radial direction: "a radial stress approximately 0.5 of the axial stress would be a reasonable design rule for most powders." Generally, lateral pressure during compaction acts only over part of the total die length, but Kuhn[5] points out that "for most compacting processes the effects of added constraints by the unpressurized die lengths can be neglected." The inserts are less frequently clamped into the retaining ring. The material of the retaining ring is generally a low-alloy non-deforming steel heat treated to a hardness between $R_c 25$ and $R_c 40$. The inserts are produced from high-carbon, high chromium oil or air hardening tool steels for medium

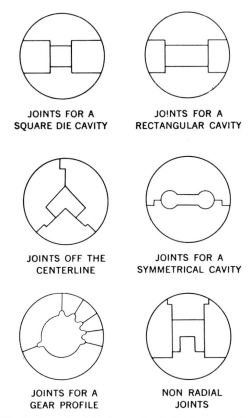

JOINTS FOR A
SQUARE DIE CAVITY

JOINTS FOR A
RECTANGULAR CAVITY

JOINTS OFF THE
CENTERLINE

JOINTS FOR A
SYMMETRICAL CAVITY

JOINTS FOR A
GEAR PROFILE

NON RADIAL
JOINTS

FIGURE 5-25   Design of carbide inserts for fitting into retaining ring of die-barrel.

production requirements and are heat treated to a hardness of R_c60-64 to give them adequate abrasion resistance. For high production volumes cemented carbide inserts are universally used. The grade of carbide used depends on the shape of the part:

94%WC - 6%Co for round dies,
88%WC - 12%Co for simple contours,
80-85%WC and 15-20%Co for gear dies and difficult contours.

Although quite complex internal contours can be produced in single piece inserts by electric discharge machining, it is more common to fit several carbide inserts together as illustrated in Figure 5-25.

The entering edges to the die cavity in the die-barrel are bevelled or radiused. In order to minimize expansion (spring-back) strains during ejection, the die-barrels are sometimes made taper, but the taper should be less than the spring-back of the part during ejection.

The core rod or core rods are a tool component which has to have high wear resistance in those portions which are in rubbing contact with the powder during pressing. On the other hand, core rods, particularly when they are thin, should be tough enough to avoid breakage. The wear portions of core rods are occasionally made of cemented carbide which is brazed or mechanically attached to the lower part of the core rod. For larger core rods a cemented carbide sleeve is fitted to the core rod. The part of the core rod below the wear portion is made of steel heat treated to have lower hardness and higher toughness. Punches which form the end faces of the parts to be compacted are less subject to wear, but must have sufficient compressive yield strength so that they do not deform. Long and slender punches must also have considerable toughness to avoid breakage due to non-axial loading. For this reason silico-manganese, nickel-chromium and nickel-chromium-molybdenum non-deforming tool steels heat treated to a hardness of R_c56 or less are used for punches.

Punch and core rod dimensions are generally relieved behind the forming face making the close fitting portion as short as possible to permit the escape of powder. In the production of compacting tools in the tool room, punches should be finished to a given clearance with respect to die-barrel and core rod. The diametral clearance may be as low as 0.0002 inches for bearings, but may be more liberal from 0.0005 to 0.001 inches for other applications.

References

1. Powder Metallurgy Equipment Manual, 2nd edition, Metal Powder Industries Federation, Princeton, N.J., p. 5-32, 53-60, (1977).
2. Cincinnati, Inc., Cincinnati, O., 45211.
3. The Olivetti Compacting System.
4. C. G. Goetzel, Treatise on Powder Metallurgy, New York, p. 325-330, (1949).
5. H. A. Kuhn, Optimum die design for powder compaction, Int. J. of Powder Metallurgy, vol. 14, No. 4, p. 259-275, (1978).

6

Compacting Other Than in Rigid Dies

In this chapter several methods of consolidating metal powders are presented. They include:

> Cold isostatic compacting,
> High energy rate and triaxial compacting,
> Powder rolling or roll compacting,
> Endless strip by the Emley and Deibel method,
> Extrusion of metal powder with a plasticizer,
> Injection molding of metal powders.

I. Cold Isostatic Compacting[1]

In isostatic compaction a uniform pressure is applied simultaneously to all the external surfaces of a powder body. For this purpose the powder is sealed in a flexible envelope and the assembly is immersed in a fluid which is pressurized. There are variations in isostatic pressing and the pressure may not be always completely uniform, but the friction between powder and die, characteristic for other methods of pressing, is always absent in isostatic pressing. The process is shown schematically in Figure 6-1. The advantages

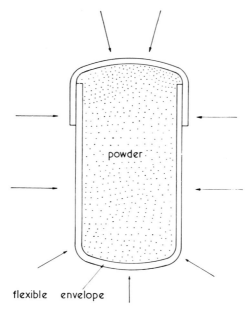

FIGURE 6-1 Schematic of isostatic pressing with powder sealed into flexible envelope and the assembly subjected to uniform pressure.

of isostatic over other forms of pressing are:

1. Much greater uniformity in density is achieved because of the uniform pressure
2. Shapes with high ratios of length to diameter which cannot be readily pressed in rigid dies can be easily isostatically pressed
3. Parts with reentrant angles and undercuts can be pressed
4. Parts with thinner wall section can be pressed in isostatic pressing
5. The equipment for cold isostatic pressing, in particular the dies, are less costly than those for rigid die pressing
6. Lubricants do not have to be mixed with metal powders before pressing

On the other hand, isostatic pressing has definite disadvantages

1. Dimensional control of the green compacts is less precise than in rigid die pressing
2. The surfaces of isostatically pressed compacts are less smooth
3. In general the rate of production in isostatic pressing is considerably lower; attempts to develop variations of

isostatic pressing which are faster will be discussed at the end of this section

4. The flexible molds used in isostatic pressing have shorter lives than rigid steel or carbide dies.

The type of isostatic pressing in which the powder sealed in its flexible envelope is completely surrounded by a liquid is called wet-bag isostatic pressing. It may be performed in pumped equipment illustrated in Figure 6-2, in which a liquid is contained in a pressure vessel which is pressurized by an external pump. The pressure vessel is generally fabricated from medium carbon forged and heat treated alloy steel; its ratio of bore to outside diameter must be sufficient to withstand the internal pressure. It is closed by a breechblock or continuous thread closure (as in Figure 6-2) and sealed by an O-ring. Figure 6-3 illustrates wet-bag isostatic pressing in which the pressure vessel is built into a hydraulic pump. Again an O-ring provides the seal between the piston of the press and the liquid, such as water-soluble oil mixtures, hydraulic oils and glycerin.

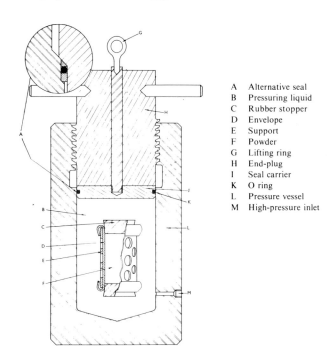

A   Alternative seal
B   Pressuring liquid
C   Rubber stopper
D   Envelope
E   Support
F   Powder
G   Lifting ring
H   End-plug
I   Seal carrier
K   O ring
L   Pressure vessel
M   High-pressure inlet

FIGURE 6-2 Schematic of pumped equipment for isostatic pressing showing alternative types of seal.

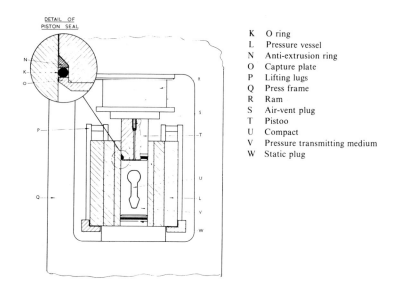

K	O ring
L	Pressure vessel
N	Anti-extrusion ring
O	Capture plate
P	Lifting lugs
Q	Press frame
R	Ram
S	Air-vent plug
T	Pistoo
U	Compact
V	Pressure transmitting medium
W	Static plug

FIGURE 6-3 Schematic of direct compression equipment for isostatic pressing.

A number of elastomers are being used for the flexible envelope. Depending on the application, disposable or reusable molds are employed. Disposable molds of polystyrene or polyethylene provide a rigid, thin wall envelope which prevents breakage of the green compact, but makes the process more expensive because of the cost of the mold. Reusable molds are made of urethane, silicone rubber, neoprene or natural rubber. These molds have a greater wall thickness and provide good shape control, but in complex parts may cause the part to break upon pressure release. The mold material must be sufficiently flexible, i.e. have high elasticity with low permanent set to accommodate the volume change during compacting; it must have high resistance to abrasive wear, which generally means high hardness, in the range of 65-95 Shore A. High hardness generally also means that the elastomer is rigid enough not to penetrate between powder particles. The envelopes are sealed either mechanically or with rubber adhesives. They may or may not be evacuated before isostatic compaction.

For special applications a solid pressure transmitting medium may be chosen instead of a liquid.[2] In this case the pressure transmitting material also acts as the powder container. Highly plasticized polyvinyl chloride, a reversible gel which can be cast into the desired shape, has been used for this purpose. Since the material serving to transmit pressure and as powder envelope is more rigid than other envelope materials, this method of isostatic pressing gives better control of the shape of the component. It is used in isostatic pressing in hydraulic presses.

Flexible, thin-sectioned envelopes may deform when packed with powder and support must be provided to control the compact shape. A simple support which is a container with holes is illustrated in Figure 6-4. Another method of controlling the shape of isostatically pressed compacts is by means of formers. The use of formers is illustrated in Figure 6-5.

Wet-bag isostatic pressing is relatively slow since it involves filling the envelope with powder and sealing it outside the pressure vessel, immersing it in the liquid, applying and releasing pressure, removing envelope and compact from the pressure vessel and stripping the green compact from the envelope. In order to accelerate the manufacturing cycle, dry bag isostatic pressing has been developed, which is illustrated in Figure 6-6. The envelope is permanently sealed into the pressure vessel. The mold cavity is filled with powder sealed with the cover plate, pressure is applied and released, the cover plate is raised, the green compact removed and the cycle repeated. Pressure vessels which have cover plates both on top and on the bottom of the mold cavity may be used so that the powder can be introduced at the top and the compact ejected at the bottom. Only one compact is pressed at a time in dry bag isostatic pressing, while with wet-bag tooling in sufficiently large pressure vessels several compacts having

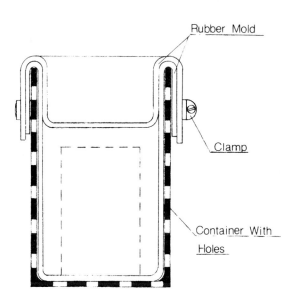

FIGURE 6-4 Support for isostatic pressing consisting of container with holes.

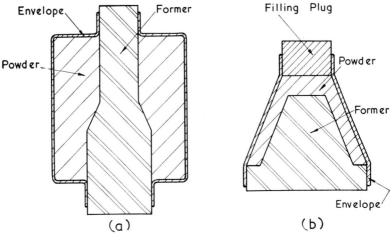

FIGURE 6-5 Examples of formers for producing shaped components by isostatic pressing.

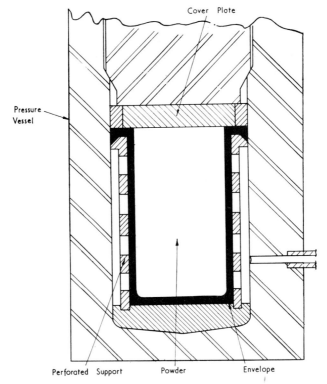

FIGURE 6-6 Schematic of equipment for dry-bag isostatic pressing.

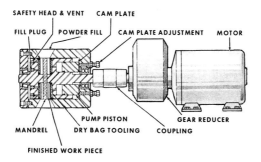

FIGURE 6-7 Rotopress for high production isostatic pressing.

FIGURE 6-8. Hydraulic fitting; isostatic preform from titanium powder.

different shapes may be compacted simultaneously. Higher quality or thicker envelopes are required in dry bag tooling because the same enveleope is used for many compacts. The selection of shapes is more limited in dry bag tooling, in which compacts with undercuts cannot be pressed. Even faster rates of compacting than with the usual dry bag tooling are possible in specialized equipment such as the rotopress,[3] illustrated in Figure 6-7. The powder is filled from the top and as the unit rotates 180°, the part is pressurized and compacted and then ejected at the bottom of its cycle. For small parts and moderate pressures up to 30,000 psi, 10 to 15 pieces per minute may be pressed on this machine.

Isostatic pressing is less widely used for metal than for ceramic powders, where the automatic fabrication of such components as spark plugs has been highly mechanized. Nevertheless, increasing quantities of metal powder compacts are isostatically pressed making use of the advantages listed above. These applications include:

1. Complex shapes which cannot be pressed in rigid dies. Good examples of such shapes can be found particularly in powder metallurgy products from relatively expensive metals, e.g. titanium[4] where material savings are of prime importance. Figure 6-8 shows a hydraulic aircraft fitting from a mixture of titanium powder and an aluminum-vanadium masteralloy blend which is isostatically pressed before being sintered and hot forged. The material saving advantage of isostatic pressing is also made use of for the funnel shaped components of Figure 6-9 again having a Ti - 6% Al - 4% V composition. The sintered compact and the finished machined component are shown. Figure 6-10 shows isostatically pressed titanium balls used as ball valves for sea water. Isostatic pressing makes it possible to compact them in the cored condition. Finally the relatively thin-walled canister including its bottom, shown in Figure 6-11, can be isostatically pressed in one piece from titanium powder.

Another example of complex shapes, which are produced by cold isostatic pressing, are cutting tool shapes compacted from a very fine high speed steel powder [5] and sintered to full density by the Fuldens process (chapter 24). Figure 6-12 shows examples of such compacts.

A process called "semi-isostatic compacting" has been developed for compacts with interior threads.[6] As shown in Figure 6-13 a mandrel with an exterior thread corresponding to the interior thread to be molded is held

FIGURE 6-9 Lens Housing (Gould), isostatic preform from titanium powder and finsihed product.

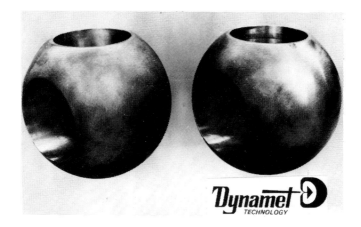

FIGURE 6-10 Canister isostatically pressed from titanium powder (Dynamet).

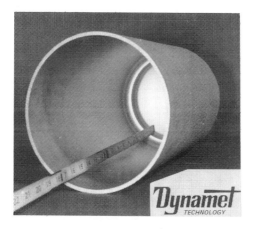

FIGURE 6-11 Ball valves, isostatically pressed from titanium powder (Dynamet).

between two covers. The powder is filled around the threaded cylinder and surrounded by the elastomeric envelope which is sealed to the covers. When isostatic pressure is applied to the assembly, the powder is compacted in the direction perpendicular to the threaded mandrel. After the pressure is released, the compact is unscrewed from the mandrel and sintered. To obtain a higher density the sintered compact may be screwed again on a sizing core-rod and isostatically pressed once more. The density distribution in "semi-isostatic compacting" is not as uniform as in the usual

FIGURE 6-12 High speed steel tools isostatically pressed from water atomized and annealed high speed steel powder and sintered (Courtesy: Consolidated Metallurgical Industries, Inc.).

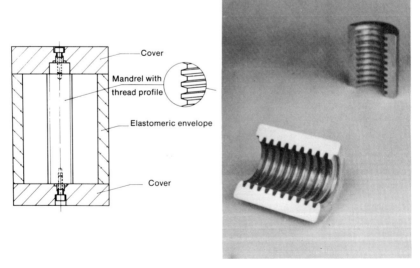

FIGURE 6-13 Tool design for isostatic pressing of threaded bushing and bushing cut in half.

isostatic compacting, since the density of the sections of the threaded part near the covers is somewhat lower than the center. The photograph on the right of Figure 6-13 shows the part cut in half.

2. Shapes, such as long, slender hollow cylinders, are often used for porous filters because of their large surface area to volume ratio. Such cylinders are quite generally isostatically pressed from powder. Both stainless steel and titanium powder are common raw materials. An isostatically pressed stainless steel filter is shown in Figure 6-14, a titanium filter in Figure 6-15.

3. Large shapes from tungsten powder, such as rocket nozzles, are isostatically pressed because the quantities involved do not warrant building rigid dies which would be very expensive. Also the nozzle shape can be readily isostatically pressed.

4. The shapes of compacts whose further fabrication is by hot consolidation quite often lend themselves to cold isostatic pressing. Examples are compacts from tungsten and molybdenum and their alloys which are to be hot rolled into sheet or hot forged into the shape of dies. In order to obtain compacts of high speed steel powder having sufficient density so that they can be hot isostatically pressed to theoretical density, cold isostatic pressing is used as the first step. Compacts in the form of rods

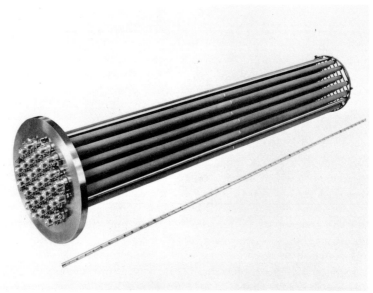

FIGURE 6-14 Assembly of isostatically pressed porous tubing from stainless steel powder (Courtesy: Mott Metallurgical Co.).

153

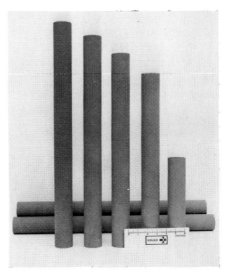

FIGURE 6-15   Isostatically pressed porous tubes from titanium powder (Gould).

from stainless steel powder which are to be extruded into tubes are first cold isostatically pressed.

     5.   A potential application is in pressing isostatically preforms[7] to be sintered by high-frequency induction heating and then forged into dense components. The main advantage of isostatic pressing in this case would be that the preforms can be pressed from powder to which no lubricant has been admixed. In such preforms the requirements for dimensional control are less stringent than in ordinary structural parts.

II.   High Energy Rate and Triaxial Compacting

Development work in high energy rate metal forming has led to experiments in which metal powders were compacted at high velocities, rather than the low rates employed in the usual commercial compaction. The advantages claimed for this process are higher green densities, greater green and sintered strength of the compacts and more uniform density. They are attributed to the generation of high pressure waves at the compact interface which propagate through the compact. One type of high energy rate forming uses dies similar to those in conventional compacting, but the upper punch is in the form of an impactor which moves at high velocity through a barrel to compact the powder.[8,9] The impactor or projectile may be actuated by an explosive charge[10] or by compressed gas. The barrel is generally evacuated before the powder is compressed. The relative density of compacts produced by this method can be related to the energy imparted to the compactor.

In another type of high energy rate compacting which may be called explosive isostatic compacting,[11,12,13] the powders are loaded into a mild steel tube, the ends of the tube welded shut, sheet explosives taped to the tube and tapered at one end to attach an explosive cap. The compact densities obtained in such experiments are uniform and considerably higher than in conventional compacts. In tungsten powder compacts densities as high as 97.6% of theoretical have been achieved. The possibility of applying explosive isostatic compacting to such shapes as cones or hollow cylinders has been explored. Experimental work on explosive compacting at high temperatures[14] has also beeen reported. None of these types of high energy rate compacting have been developed for commercial use.

Another experimental technique to press powders to higher densities than those obtained in conventional compacting is triaxial compression.[15,16] Isostatic pressure is applied on the circumference of a cylindrical powder specimen confined in a flexible envelope and an axial load is superimposed upon the isostatic pressure through a vertical piston, as shown in Figure 6-16. In this way shear stresses are introduced and denser compacts are obtained than by isostatic compaction alone. This method of triaxial stress application is adapted from a test in soil mechanics to determine the ultimate strength of soils. Still another approach to obtain

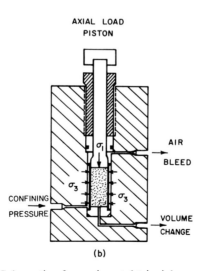

FIGURE 6-16 Schematic of experimental triaxial compaction chamber.

higher and more uniform density in metal powder compacts is also based on inducing shearing motion during the compression of the powder. For this purpose the die-barrel is rotated while the powder is pressed.[17] This rotation almost completely eliminates the friction between powder and diewall. Experiments in which the die was rotated while a compact was pressed were first made with coal powder briquettes.

III.   Powder Rolling or Roll Compacting

A.   Introduction to Powder Rolling

Powder rolling, also called roll compacting, is the most important of a series of P/M processes used to manufacture shapes in which one dimension is a great deal longer than the other ones. W. D. Jones has coined the term "Continuous Powder Metallurgy" for these processes. In powder rolling metal powder is fed from a hopper into the gap of a rolling mill and emerges from the gap as a continuous compacted strip or sheet. The rolls of the mill may be arranged one above the other as in conventional rolling mills with the powder sliding from the hopper through an adjustable gate down a chute, as shown in Figure 6-17. More commonly the rolls are arranged horizontally. The powder may fall directly from the hopper into the roll gap. This is called "saturated feed," which can be adjusted by the head of powder above the roll gap. In "starved" feed the amount of powder entering the roll gap is controlled by adjustable gates. These types of powder rolling are shown schematically in Figures 6-18 and 6-19.

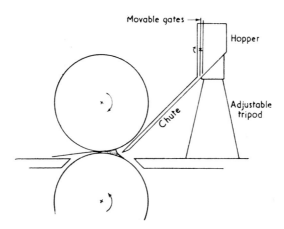

FIGURE 6-17   Schematic of powder rolling with rolling mill rolls arranged vertically.

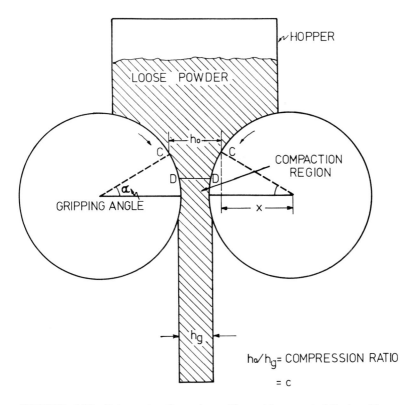

FIGURE 6-18 Schematic of powder rolling with saturated feed; rolling mill rolls arranged horizontally. For explanation of symbols, see text.

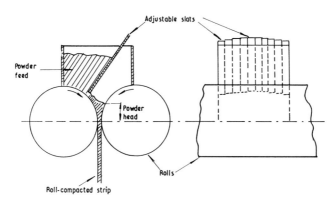

FIGURE 6-19 Powder rolling with feed of powder controlled by adjustable gates.

The consistency of the strip produced by powder rolling corresponds to that of a green powder compact. In order to produce a useful product the green strip must be sintered. The sintered strip is, however, still porous and if strip of full density is to be obtained, it must be rolled again, either hot or cold. Hot rerolling will produce a strip which can be used as such. Cold rerolling must generally be followed by annealing, a second rerolling step, etc. Conventional rolling starts with a fairly thick slab. In order to produce thin sheet or strip a large number of rolling passes are necessary. It is often necessary to anneal the strip several times between passes to make it soft enough for further rolling. One of the principal attractions of powder rolling is the possibility of producing strip or sheet of a thickness close to that needed in the final product and thereby reducing the number of passes necessary.

Powder rolling was described in a number of early patents, [18] but the first detailed investigation of the process was undertaken by Naeser and Zirm[19] in Germany who published an account of their work on rolling RZ iron powder into strip in 1950. Powder rolling of copper powder was investigated and described by Franssen[20] in Germany and by Evans and Smith[21] in England in 1954. During the years from 1958 to 1968 powder rolling received considerable attention in Great Britain, the United States, Canada, the Soviet Union and Japan. The principal interest was in combining low-cost processes for producing such metal powders as copper, nickel, cobalt, iron and aluminum with the powder rolling process so as to develop methods of fabricating thin sheet or strip more economically than by the conventional approach. One result of these efforts was the Sherritt Gordon commerical process in Canada for rolling nickel and cobalt powders into strip and sheet.[22] Pilot plants were also built for rolling copper,[23] aluminum[24] and iron powders[25,26] into sheet, but these efforts have not led to commercial production. An installation for rolling water atomized copper-nickel alloy powder into strip in Canada is near the production stage.

In addition to using powder rolling for the large scale production of strip and sheet of base metals and alloys, the technique has also been employed for a number of specialty type strip materials, several of which are produced commercially. They include special nickel and cobalt alloys for electronic and magnetic applications, porous powder rolled strip, in particular stainless steel strip for filters and nickel strip for electrodes, strip consisting of a fissionable material in a metallic matrix, and an aluminum alloy strip for bearing applications.

Following this introductory part of the section on powder rolling, the principles of powder rolling in comparison with conventional rolling and the methods suggested for sintering roll compacted sheet and strip are discussed in the second part. The method of commercial production of nickel sheet by Sherritt Gordon is included in this part. In the third part some modified methods of powder rolling, in particular hot aluminum powder rolling and rolling of powder applied as a slurry to a substrate, are presented. The fourth part is concerned with the economics of large scale production of sheet and strip by powder rolling. In the last part, methods of producing specialty types of rolled strip from powder are described.

B. Principles of Powder Rolling[27, 28]

It is instructive to compare powder rolling with rolling with rolling of a slab of solid metal in a rolling mill, as illustrated in Figure 6-20. A slab of metal with the thickness h_o is drawn by the rolls with diameter D into the roll gap. The emerging sheet has a thickness h_g. The velocity of the slab as it enters the rolls in V_o, it accelerates and reaches the velocity of the roll surface at position B, the neutral point, and leaves the roll at the greater speed V_f. The roll pressure p increases from p_A, where the slab enters the gap to the maximum value p_B at the neutral point B and decreases to p_C as it leaves the rolls. In order to draw the slab into the gap, there must be sufficient friction between the slab and the rolls. The coefficient of friction μ which depends on the material of rolls and slab, the surface finish of the rolls and the presence of lubricant on the roll surface determines the maximum angle of entry α which is equal to $\tan^{-1} \mu$. In conventional rolling a range of strip thickness can be rolled in a rolling mill with a given diameter D depending upon the setting of the roll gap.

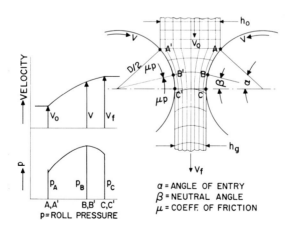

FIGURE 6-20 Schematic of deformation, pressure distribution and velocity during rolling of solid metal sheet.

Powder rolling differs from conventional rolling in several respects:

1.   The raw material is powder which does not withstand any tensile stresses. For this reason the powder cannot be pulled into the rolls but must flow under gravity or, less commonly, be forced under pressure.

2.   During passage through the rolls the material will change not only its mechanical properties, but also its volume and density.

A schematic presentation of powder rolling with a saturated feed is shown in Figure 6-18. The nip angle or gripping angle α determines the plane identified by the section CC = h_o in the figure. Above this plane the density of the powder mass is supposed to be equal to the apparent density of the powder, below this plane, in the compacting zone, the rolls apply pressures to the powder particles decreasing the void volume between particles first by rearrangement and then by plastic deformation of the particles. This is, however, an oversimplified picture. Powder particles adjacent to the roll surface are drawn into the roll gap through friction between rolls and particles in a feed zone above the plane CC. As the particles are pulled down, adjacent layers of particles are being dragged along the first layer. The force with which the particles are dragged depends upon the internal friction between particles which strongly depends upon their shape and size and may vary widely between powders. An attempt to treat the process in the feed zone quantitatively was made by Lee and Schwartz.[29] As in conventional rolling the velocity of the compacted powder increases, reaches the velocity of the roll surface at D and exceeds it, as the compacted powder strip emerges from the roll gap. The pressure distribution along the roll surface during roll compacting of powder was measured by Russian investigators[30,31] as a function of the particle size and strip thickness using dynamometers. The distribution is similar to that in conventional rolling. In their theoretical treatment of pressure distribution Lee and Schwartz found qualitative agreement with the results of the Russian investigators.

In contrast to conventional rolling the thickness of the strip which can be rolled in powder rolling is closely limited by the diameter of the rolling mill rolls. In order to form a coherent strip the powder must be compacted to a sufficiently high green density and green strength. The degree of compaction, i.e. the compression ratio, is given by the ratio of the density of the strip, as it emerges from the roll gap, to the apparent density of the powder. This change in density is accomplished as the powder is transported through the feed zone and the compacting zone. The length of these zones is determined by the diameter of the rolls (D), the internal friction between the powder particles and the friction between powder and rolls. With the simplified geometry of Figure 6-18, the nip angle may be defined as $\cos\alpha = \frac{x}{D/2}$. The dimension h_o would be equal to $D(1-\cos\alpha) + h_g$. The strip thickness h_g is equal to h_o/c, where c is the compression ratio. The strip thickness would then be

$$h_g = \frac{D(1-\cos\alpha)}{c-1}$$

Naeser and Zirm[19] in their pioneering work on powder rolling presented a diagram, shown in Figure 6-21, in which they plotted the thickness and density of strip rolled from RZ powder at a roll speed of 6.3m/min. in a rolling mill with a 70 mm roll diameter as a function of the width of the roll gap. They also plotted the power applied to the rolls. As the width of the roll gap is decreased from 0.35 to 0.02 mm, the thickness of the strip decreases from 0.75 to 0.35 mm and the density increases from 4.5 to 7.5g/cm³. For practical powder rolling, strip thickness in the middle of this range near 0.55 mm, with a green density of 5.8g/cm³ equal to 74% of theoretical density, can be most readily handled. Thicker and thinner strips cannot be readily coiled, thicker ones because they are too fragile, thinner ones because they are too stiff. Also for the thin strips inordinately high power is required. Rolling strip 0.55 mm thick in a mill with a diameter of 70 mm

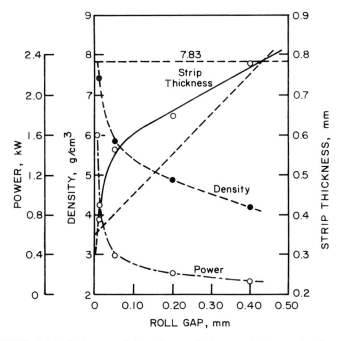

FIGURE 6-21 Influence of the roll gap width upon thickness, density and power requirement in rolling RZ powder into strip.

gives a ratio of roll diameter to strip thickness of 127:1 and corresponds to a nip angle of 8°. In order to roll thicker strip, mills with a larger roll diameter would be required. Figure 6-22 shows the relationship between roll diameter and optimum green strip thickness. It is based on the experiments of Naeser and Zirm[19] who rolled RZ powder not only in the mill with a 70 mm roll diameter, but also in several mills with larger rolls. Also included are the results of the production work of Sherritt Gordon[22] on rolling into sheet nickel powder produced by hydrogen reduction of a nickel ammine sulphate solution. In this work it was found that to produce thicker strip, it was not only necessary to use a mill with a larger roll diameter, but also to use a coarser powder. The ratios of roll diameter to optimum green strip thickness vary widely for different powders. They depend upon the flow and apparent density of the powder and on the coefficient of friction between rolls and powder. For the very fluffy carbonyl nickel powder ratios near 600:1 were optimal[32], electrolytic copper powder[33,34] ratios near 200:1 and for the RZ iron powder and the Sherritt Gordon powders ratios somewhat above 100:1. Increasing the coefficient of friction by roughening the surface of the rolls is often quite effective, but in large scale production it is difficult to maintain the character of roughened surfaces. On the other hand, variations in particle shape and an increase in the temperature at which powders are rolled are very effective ways of increasing the

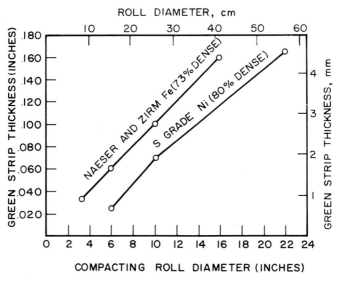

FIGURE 6-22 Effect of compacting roll diameter on green strip thickness for RZ iron powder and S grade nickel powder.

coefficient of friction, as will be discussed in connection with the hot powder rolling of aluminum powder.

The green strip rolled from powder should have uniform thickness and density across the width of the strip and its edges should be well formed and as dense as the center of the strip. Numerous devices have been developed, e.g. a patented guide plate assembly to distribute powder feed across the width of the strip[35] and edge devices to prevent powder from flowing laterally out of the roll gap.[36]

The output of strip in powder rolling depends upon the speed of the rolling mill. Up to a given speed, which Tunderman and Singer[37] called the "flow transition speed," the flow rate of powder increases linearly with roll speed. Above this "flow transition speed" relatively less and less powder flows into the roll gap, until at speeds considerably above the transition speed continuous strip can no longer be rolled. The "flow transition speed" varies with the character of the powder. Tunderman and Singer[37] determined flow transition speeds for a series of iron powders. For optimum setting of the roll gap of the mill, i.e. for optimum ratio of roll diameter to strip thickness, they found flow transition speeds in the range from 25 to 100 ft/min with the highest transition speed for a coarse reduced sponge iron powder. Rolling speeds in the neighborhood of 100 ft/min have been found the maximum practicable in any work on rolling powders at room temperature. These speeds are much lower than those used in conventional rolling of sheet and strip. For fine powder of a fluffy character, such as carbonyl nickel and reduced copper powders, much lower maximum rolling speeds have been found possible. For these powders the flow of powder into the roll gap and therefore the rate of rolling is limited by the fluidization of the powder by air being released in the compacting zone which pushes up into the feed zone and suspends the powder in the air stream. Fluidization can be diminished by rolling in a gas with a lower viscosity than air. When Worn and Perks[38] rolled carbonyl nickel powder into strip in hydrogen, which has only half the viscosity of air, they obtained 50% thicker strip at a given roll speed. Vinogradov and Fedorchenko[39] made experiments on powder rolling in vacuum, which would completely eliminate fluidization of the powder.

The green rolled strip must be sintered and for all applications of strip from powder except porous strip must be rerolled. A number of different schemes have been proposed for these operations. In their original work on rolling RZ iron powder Naeser and Zirm[19] suggested treatment of the strip as indicated schematically in Figure 6-23. The strip proceeds from the rolling mill directly into the sintering furnace, is sintered, cooled to room temperature, enters a cold rolling mill, proceeds from there into an annealing furnace, where it is annealed and cooled to room temperature. Two additional cold rolling and annealing steps are suggested to obtain completely dense strip. The principal problem with this scheme is to synchronize the rate of rolling with that of sintering and annealing. Naeser and Zirm suggested that their RZ iron powder strip may be sintered at temperatures as high as 1400°C where a sintering time of one minute would be sufficient. They also showed that strip may be sintered by passing an electric current through it.

Two schemes for treating the sintered strip, quite different from that of Naeser and Zirm, were suggested by Franssen[40] which are illustrated in Figures 6-24 and 6-25. In the first of these schemes (Figure 6-24) the rolled strip is fed into a continuous sintering furnace in which it is supported by an endless woven wire mesh belt. As it emerges from the cooling zone of the

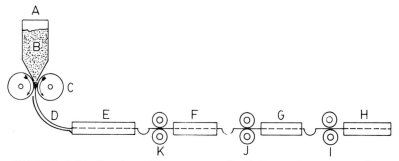

FIGURE 6-23 Powder rolling process using fully continuous operation with an endless strip. A feed hopper, B metal powder, C rolling mill, D strip, E, F, G and H furnaces, K, J and I cold reduction of strip after first, second and third sintering treatment.

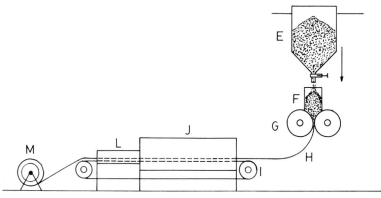

FIGURE 6-24 Powder rolling process with strip reeled into individual coils after first sintering treatment. E feed bunker, F powder feed, G rolls, H strip, I conveyor band, J furnace, L cooling zone, M coiler.

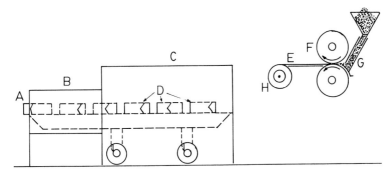

FIGURE 6-25 Powder rolling process with vertical arrangement of rolls, coiling of green strip and sintering coils in rocker bar furnace. B cooling zone, C rocker bar furnace, D coils, E strip, F rolls, G powder feed, H coiler.

furnace it is wound on a coiler. The sintered strip is rerolled and annealed in separate operations. In the second scheme (Figure 6-25) the green strip emerging from the rolling mill is coiled. The strip is then sintered in the form of coils. Franssen suggests a walking beam furnace for this operation. Since the heat transfer in sintering and then cooling strip through a furnace, in coil form is considerably slower than in feeding a simple strip through a furnace, sufficient furnace equipment must be provided. In the schemes suggested by Franssen difficulties with synchronization are either partially or completely avoided.

The only large scale commercial operation in which strip is rolled from powder is the production of nickel and cobalt strip by Sherritt Gordon Mines Ltd. in Canada.[22] Sherritt Gordon S grade nickel powder is rolled in a 22″ diameter rolling mill into strip approximately 0.140″ (3.5 mm) thick and 13 3/4″ (349 mm) wide. The green strip exits vertically downward from the mill, is turned 90 degrees and wound into coils. The coil is unwound and the strip introduced into a continuous sintering furnace in which it is heated to a temperature above 1600°F (880°C) and then fed into a hot rolling mill in which its thickness is reduced by twice the amount necessary to theoretically obtain complete densification. This amount of reduction insures that the strip is completely dense on emerging from the hot rolling mill. From the mill it enters into an atmosphere controlled chamber in which superficial oxide produced during rolling is reduced. The advantage of hot rolling over cold rolling the sintered strip is that complete densification can be obtained with one rerolling step rather than repeated cooling, rerolling, annealing etc. and that one can start with a thinner green strip. The sintering time for the green strip is 20 minutes, even though sintering for five minutes would be sufficient to make the strip strong enough for hot rolling. The additional sintering time is needed to reduce the sulfur content of the nickel. The nickel strip produced from powder has a higher chemical purity than commercial nickel strip produced by fusion metallurgy. The small differences in mechanical properties of nickel strip from powder and conventional nickel strip are due to this higher purity of strip from powder. One of the principal uses for the nickel strip is in producing coins from it.

C. Modified Methods of Powder Rolling

Before discussing the economics of producing dense strip from powder two processes, which differ from the powder rolling process discussed so far, should be briefly reviewed; both of them were developed to the pilot plant stage. In one of these processes aluminum or aluminum alloy powder is rolled into sheet at a temperature high enough so that the sheet emerging from the mill is completely dense. It therefore requires neither sintering nor rerolling. The process called the "Compacted Sheet Process" was developed by Reynolds Metals Co.[24] A characteristic of the process is the special type of aluminum or aluminum alloy powder having coarse acicular particles which is produced by pouring molten metal into a spinning perforated cylinder. Both the characteristic acicular shape of the powder particles and the elevated temperature of rolling increase the coefficient of friction between powder and rolls and between powder particles, and thereby make it possible to roll aluminum sheet at considerably higher rates than for other powder rolling processes. Rates greater than 200 ft/min have been achieved in pilot plant production. A schematic drawing of the process is shown in Figure 6-26.

Careful control of the conditions of producing the powder, generally in the size range -10 + 70 mesh, and of the temperature to which the powder is preheated, generally near 600°F (316°C), is necessary to produce sheet requiring acceptable separating force between the rolls, acceptable power consumption and producing minimum edge cracking of the sheet. Because of the character of the powder particles and the elevated temperature of rolling, an 0.125" thick sheet can be rolled on a mill with 6" diameter rolls, i.e. a low ratio of approximately 50:1. The mechanical properties of alloy sheet produced by the compacted sheet process are quite similar to those of conventional alloy sheet, except for a slight increase in strength of the compacted sheet compared with conventional sheet because of the presence of a small amount of very finely dispersed oxide particles in the structure. A modification of the process for hot rolling aluminum powder into strip, in which an insulated electric filament is incorporated into the strip, has been described by H. H. Hirsch.[41]

A process for rolling thin steel strip from powder, the so-called BISRA process, has been investigated in the laboratories of the British Steel

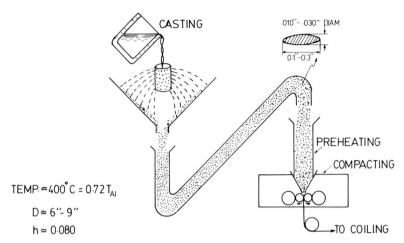

STRIP SPEED ≈ 400 FT/MIN

PRODUCTION RATE (FOR 3 FT WIDTH) ≈ 40 TONS/HR.

FIGURE 6-26 Schematic of the Reynolds Metal process for hot rolling aluminum powder.

Corporation and a pilot plant built at their Sketty Hall laboratories.[26] In contrast to the processes described before, the BISRA process uses an unsaturated feed system. The iron powder, mixed with an organic binder, is formed into a slurry. The slurry, containing 70 weight percent iron powder, 0.6 percent of binder and balance water, is applied to a substrate generally a stainless steel belt to the desired thickness. The slurry is dried forming a non-adherent selfsupporting film. The film is removed from the substrate and roll compacted into green strip. The strip is sintered, simultaneously eliminating the binder. The sintered strip is cold rolled, resintered and given a final temper roll. With this process much higher speeds of powder rolling of several hundred feet per minute are possible, if suitable methods of high speed roller coating the substrate with a slurry of suitable viscosity are available. The aim in developing the process was to produce strip in the range of gauges from 0.012 to 0.004 in. (0.3 to 0.1 mm). A somewhat similar slurry process for making thin gauge sheet from metal powder has been described by G. E. Wieland and E. M. Rudzki of Bethlehem Steel Corporation's Research Department.[42] Producing iron sheet from a slurry of a super-concentrated iron oxide ore with intermediate reduction of the oxide is mentioned as well as producing ferritic and austenitic stainless steel sheet either from powder mixtures (iron powder and ferrochrome powder or iron powder, ferrochrome powder, nickel powder and molybdenum powder) or from prealloyed powders.

D. Economics of Powder Rolling

In considering the economics of producing from powder dense sheet and strip of base metals and alloys, such as iron and steel, copper and copper alloys, aluminum and aluminum alloys, stainless steel and nickel, it must be remembered that, in order to be competitive with conventional manufacturing facilities for sheet and strip, plants built for this purpose would have to process much larger quantities of material than plants in which structural parts are produced from metal powder. In the cost estimates for rolling steel strip by the BISRA process[26] a plant with an annual output of about 100,000 tons is considered. In assessing the economics of stainless steel strip production by roll compacting Shakespeare[43] considered a plant for the production of 340,000 tons of strip per year, while Smucker[23] made estimates for a plant to roll copper powder into strip at a rate of 50 tons/day.

The cost estimates for producing sheet and strip from powder must include

1. the cost of producing the powder
2. the cost of converting the powder into finished strip or sheet.

The interest in rolling sheet and strip from powder evolved in the first place from the expectation that metal powders suitable for rolling could be produced at a cost equal to or lower than the cost of the metal in ingot form. This expectation was actually realized in the Sherritt Gordon process for producing nickel powder. Furthermore the nickel powder producing process could be readily modified for optimum use of the powder for powder rolling. In the case of copper, a number of processes are available for producing powder from acqueous solutions of copper salts, which are intermediate products in the hydro-metallurgy of copper. However none of these processes has been developed to the point where a powder is available which is low in cost, of sufficient purity and with a particle size and shape suitable for powder rolling. In the case of iron and steel large numbers of direct reduction processes exist, but up to the present no suitable low cost iron powder for rolling into strip or sheet has come on the market. An economic risk analysis of a miniplant for production of steel strip from powder has been given by Zecca et al.[44] They assumed that the powder would be produced by atomizing molten steel. They came to the conclusion that "the profitability of a P/M steel plant appears to be barely adequate, but with a substantial risk of a poor rate of return... Lower cost powder produced from ore or low grade scrap by hydrometallurgical methods appear to hold the key for the future of P/M steel (strip) plants."

The success of the Sherritt Gordon process for rolling nickel powder into sheet appears to indicate that most of the problems involved in the technology of converting powder into strip or sheet can be solved. It seems likely that the cost of conversion can be made sufficiently low to make the powder rolling process economical, once low cost methods of producing suitable powder have become available. Processes in which the sintered strip is completely densified by one hot rolling step will probably be lower in cost than processes involving a series of cold rolling and annealing steps. However, as Shakespeare[43] has pointed out, hot rolling without oxidation a sintered strip from stainless steel powder, which is still porous, will be considerably more difficult than hot rolling sintered strip from iron, nickel or copper powder. Rolling strip and sheet from powder will be particularly advantageous for those metals and alloys which readily work harden and which therefore require intermediate anneals during conventional rolling and those in which considerable quantities of scrap are produced in converting an ingot into finished sheet or strip.

E. Specialty Applications of Powder Rolling

Powder rolling is of interest not only for the large scale production of dense base metal sheet and strip, but also for a number of specialty applications, where the economic considerations for mass produced sheet and strip outlined above do not apply. A number of special nickel and cobalt alloys for electrical and magnetic applications are rolled into strip from powder.[45] This applies particularly for nickel strip used in cathodes for electron tubes. The chemical purity of the strip rolled from carbonyl powder and sintered can be controlled much better than of strip made by fusion metallurgy. Controlled addition of magnesium, titanium and silicon in quantities of a few hundredth of a percent are possible for some grades of electronic nickel strip. Other grades must be extremely pure and have very low contents on the order of .001% max. of such impurities as copper and sulfur. After being sintered the strip is repeatedly cold rerolled and annealed until it has become completely dense. Hot rerolling is not feasible for this strip because it may introduce impurities. The alloys may be strengthened by small additions (0.05-0.2%) of thorium oxide to the powder mixture.[46] Alloys of cobalt with 7% iron, which are more ductile than pure cobalt and have magnetic applications, have been rolled into strip.[47] Another use of this alloy strip if for tubular cobalt-base hard facing welding rod, in which the strip is roll formed and filled with a blend of metal powders to yield a composite rod with the chemical composition of Stellite R-6.

Roll compacting powder generally yields a porous strip which has to be rerolled after sintering to make it dense. Roll compacting is therefore a very good method for producing porous strip. Producing porous stainless steel filters in sheet and strip form has been suggested. Such filter materials generally have porosity of less than 50 percent. Porous sheet and strip of pure nickel with higher porosities are required for alkaline battery and fuel cell electrodes.[48,49] In order to produce these very porous materials special methods have been developed. While for electrodes with porosities of 40-50 percent, equiaxed fine carbonyl powders are used which give electrodes with a narrow pore spectrum, for electrodes with higher porosities filamentary carbonyl nickel powders are employed, which after roll compacting have better green strength. For the highest porosities, i.e. more than 70%, a "spacing agent" such as methyl cellulose may be mixed with the carbonyl nickel powder in quantities up to 40 volume %. The spacing agent is evaporated before the material is sintered. Finally an electroformed nickel mesh may be incorporated into the rolled strip to produce highly porous electrodes with adequate green strength. The arrangement for rolling strip with a mesh is shown in Figure 6-27. Roll compacted electrode material from filamentary carbonyl nickel powder is generally sintered at low temperatures of 700°C in dissociated ammonia producing an acceptable sintered material of the desired porosity. The less porous electrodes from equiaxed fine carbonyl powder may be sintered at higher temperatures up to 1150°C.

Roll compacting has been suggested for material in strip form consisting of a fissionable material, generally UO_{2}, in a metal matrix, e.g. stainless steel or tungsten.[50]

Roll compacting may be used not only for producing sheet or strip material which is uniform in composition across its thickness, but also for sandwich material. An example of such a roll compacted sandwich material is bimetallic strip used in producing main and connecting rod bearings.[51] The strip has a layer consisting of an aluminum-lead prealloyed powder while the other layer is pure aluminum. The arrangement necessary to roll compact such a sandwich material is shown in Figure 6-28. Two powder hoppers are necessary with a powder flow blade controlling the flow of the two powders into the roll gap. The coil of composite strip is sintered and its pure aluminum layer roll bonded to steel strip reducing the sandwich strip 50% in thickness.

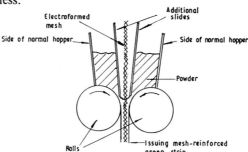

FIGURE 6-27 Rolling of strip from carbonyl nickel powder designed to incorporate mesh into strip.

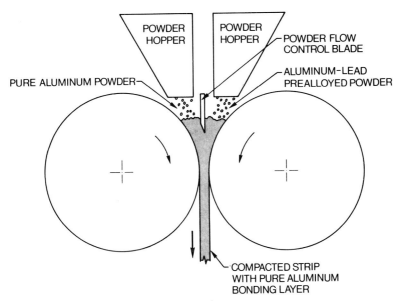

FIGURE 6-28 Rolling of strip consisting of layers of aluminum lead prealloyed powder and of pure aluminum powder for bearings.

IV. Endless Strip by the Emley and Deibel Method.[52,53]

Emley and Deibel suggested a "continuous powder metallurgy" method of producing endless strip in 1959. The method is shown schematically in Figures 6-29 and 6-30. The powder is pressed in a channel, the bottom and sides of which are B and C in Figure 6-29. The compacting punch has a flat upper surface and a lower surface angled to the powder surface. The sequence of the operation is indicated in Figure 6-30. The punch is raised clear of the pressing area, loose powder is charged into the channel, a starting block contains the powder in the channel during the first compacting step. In this step the "finishing area," i.e. segments A and B, are compacted to the desired final green density, while segments C, D, E and F are pressed to increasingly lower densities. The punch is then raised and the powder is moved a distance less than the finishing area and again pressure is applied. The process may be adapted to automatic production of bars of unlimited length, if reciprocating type presses are built for it. The endless bar will then have to be sintered. To obtain full density it will have to be rolled. As in powder rolling, a hot rolling step or a sequence of cold rolling and annealing steps may be used.

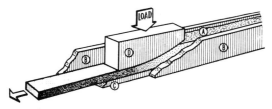

FIGURE 6-29 Cutaway schematic view of continuous compaction apparatus.

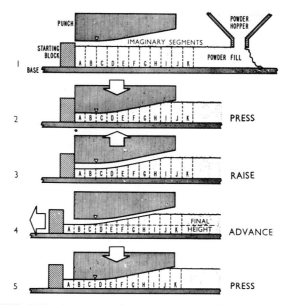

FIGURE 6-30 Sequence of operations in continuous compaction.

V. Extrusion of Metal Powder with a Plasticizer

Development work on extruding metal powders mixed with a plasticizer goes back to the beginning of the 20th century when Just and Hanamann used it to produce tungsten wire.[54] Very fine tungsten powder was mixed with organic additions, such as solutions of dextrin, caramel, sugar and tragant and a kneadable plastic mass produced, which could be extruded through diamond dies. The extruded filaments were wrapped on cartons, dried and cut into "hairpins." The hairpins were heated by passing current through them in a hydrogen atmosphere. Moisture and the organic additions evaporated, the remaining carbon was removed by decarburization and the resulting porous tungsten skeleton sintered by grain growth to a relatively dense metallic body. This so-called "paste method" of producing tungsten filaments has been completely superseded by the Coolidge process.

The present commercial applications of extrusion of metal powders with a plasticizer are quite limited. The process is used for producing thin walled tubing and other long thin shapes of cemented tungsten carbide. The principal problem is the control of dimensions because of the large amount of shrinkage when the extruded material is sintered. The relative density of the extruded material is low, the plasticizer has to be eliminated and the density of the remaining carbide material raised to near theoretical density.

Another application is in producing tubing of porous stainless steel. Mixtures of stainless steel powder and plasticizer can be extruded into tubing having smaller wall thicknesses than tubing produced by isostatic pressing. Again the control of dimensions during elimination of the plasticizer and subsequent sintering in dry hydrogen is the principal problem in this technique.

A third commercial example of extrusion of metal powder with a plasticizer is in the production of fountain pen tips.[55] The powder is an alloy of 97½ percent ruthenium and 2½ percent platinum, in which the ruthenium particles are coated with platinum. The plasticizer is polystyrene to which a mixture of diphenyl and diphenyl-ether has been added. The rubbery mass of metal powder particles coated with the plasticizer is extruded at 320-350°F (160-175°C) into a rod. The rod is molded into spherical particles between reciprocal steel punches with a hemispherical cavity. The spheres are first freed of plasticizer at 750°F (400°C) and then finally sintered at 2900°F (1600°C) to nearly theoretical density.

VI. Injection Molding of Metal Powders[56]

Quite complex shapes including shapes with wall thicknesses from 5 mm (0.2″) down to 0.6 mm (0.023″) and less and shapes with cross-cored holes, which are impossible to compact by conventional powder metallurgy processes, can be produced by applying to metal powders (and ceramic powders) the technology of injection molding of plastics. The metal powder is compounded with an organic binder. The mixture is injected into the mold of a plastic injection molding machine at a temperature of 135 to 205°C (275 to 400°F) and an applied pressure of 140 MPa (20,000 psi). The viscosity of the plastic binder is low enough so that the mixture of metal powder and binder fills the mold completely. Molding cycles range from less than 10 seconds to 90 seconds depending on the size of the part. During cooling the binder solidifies and the part can be stripped from the mold. The binder is removed and the part sintered in an operation similar to

FIGURE 6-31 Screw seal with discontinuous internal thread for aircraft wing flap ball screw assembly, injection molded from nickel powder.

conventional sintering of metal powder parts. Because of the relatively large volume percentage of binder in the original mixture, the parts shrink 20 to 25 percent to a final density in the range from 94 to 98 percent of the solid density of the metal or alloy.

As will be discussed in chapters 8 and 11 on sintering, very fine powders and long sintering times are necessary to achieve the required shrinkage to near solid density. In spite of the large shrinkage the dimensions of the sintered parts can be reasonably accurately controlled in the range from $\pm$ 0.08 to 0.12 mm ($\pm$ .003 to .005″).

The injection molding process developed by R. E. Wiech is commerically used by the Parmatech Corporation. No details of the nature of the organic binder used and the methods of removing the binder have been disclosed. An example of a part produced by injection molding from nickel powder is shown in Figure 6-31. It is a 2″ diameter screw seal with a discontinuous internal thread. The seal is used in an aircraft wing flap ball screw assembly.

References

1. V. T. Morgan and R. L. Sands, Isostatic compaction of metal powders, Metallurgical Reviews #134, Journal of the Inst. of Metals, May, (1969).
2. R. A. Powell, Isostatic compaction of metal powders in conventional molding tools, Int. J. of Powder Metallurgy, vol. 1, No. 3, p. 13-18, (1965).
3. H. C. Jackson, Isostatic pressing for high-volume production, Mod. Dev. in Powder Metallurgy, vol. 1, ed. by H. H. Hausner, p. 188-193, (1966).
4. R. E. Garriott and E. L. Thellmann, Titanium powder metallurgy, a commerical reality, Mod. Dev. in Powder Metallurgy, vol. 11, ed. by H. H. Hausner and P. W. Taubenblat, p. 63-78, (1977).
5. T. Levin and R. P. Hervey, P/M alternative to conventional processing of high speed steel, Metal Progress, vol. 115, No. 6, p. 31-34, (1979).
6. L. Albano-Müller and R. Röhlig, Threads by P/M isostatic compaction, Int. J. Powder Metallurgy, vol. 15, No. 3, p. 191-197, (1979).
7. S. Mocarski and P. C. Eloff, Equipment considerations for forging powder preforms, Int. J. of Powder Metallurgy, vol. 7, No. 2, p. 15-25, (1971).
8. J. W. Hagemeyer and J. A. Regalbuto, Dynamic compaction of metal powders with a high velocity impact device, Int. J. of Powder Metallurgy, vol 4, No. 3, p. 19-25, (1968).
9. R. M. Rusnak, Energy relationships in high velocity compaction of copper powder, Int. J. of Powder Metallurgy, vol. 12, No. 2, p. 91-99, (1971).
10. O. V. Roman, Some theoretical and practical aspects of high energy rate forming, Mod. Dev. in Powder Metallurgy, vol. 4, ed. by H. H. Hausner, p. 513-523, (1971).
11. Reference 1, p. 94.
12. S. W. Porembka, Explosive compacting, Ceramic Age, (Dec. 1963).
13. R. A. Cooley, Explosive forming of metal and its potential application to compacting metal powders, in "Powder Metallurgy" ed. by W. Leszynski, N.Y., p.525-542, (1961).
14. V. G. Gorobtsov and O. V. Roman, Hot explosive pressing of metal powders, Int. J. of Powder Metallurgy, vol. 11, No. 1, p. 55-60, (1975).
15. R. M. Koerner and F. J. Quirus, High density P/M compacts utilizing shear stresses, Int. J. of Powder Metallurgy, vol. 7, No. 3, p. 3-9, (1971).
16. R. M. Koerner, Triaxial stress state compaction of powder, in "Powder Metallurgy Processing" ed. by H. A. Kuhn and A. Lawley, N.Y., p. 33-50. (1978).
17. L. F. Hammond and E. G. Schwartz, The effect of die rotation on the compaction of metal powders, Int. J. of Powder Metallurgy, vol. 6, No. 1, p. 25-36, (1970).
18. W. D. Jones, Fundamental Principles of Powder Metallurgy, London, p. 924, (1960).
19. G. Naeser and F. Zirm, Stahl und Eisen, vol. 70, p. 995-1003, (1950).
20. H. Franssen, Z. f. Metallkunde, vol. 45, p. 238, (1954).
21. P. E. Evans and G. C. Smith, The continuous compacting of metal powders, in "Symposium on Powder Metallurgy," The Iron and Steel Institute, London, Special Report N. 58, p. 131-136, (1956).

22. M. H. D. Blore, V. Silins, S. Romanchek, T. W. Benz and V. N. Mackiw, Pure nickel strip by powder rolling, Metals Engin. Quart., vol. 6, No. 2, p. 54-60, (1966).

23. R. A. Smucker, Perfected and practical methods of processing powder into commercial strip, Iron and Steel Engineer, vol. 36, No. 7, p. 118-123 (1959).

24. T. S. Dougherty, Aluminum sheet from finely divided particles, Journal of Metals, vol. 16, No. 10, p. 827-830, (1964).

25. S. R. Crooks, The rolling of strip from iron powder, Iron and Steel Engineer, vol. 39, No. 2, p. 72-76, (1962).

26. I. Davies, W. M. Gibbon, A. G. Harris, Thin steel strip from powder, Powder Metallurgy, vol. 11, No. 22, p. 295-313, (1968).

27. H. S. Nayar, Powder Metallurgy Review 4, Strip products by powder metallurgy, Powder Metallurgy Int., vol. 4, No. 1, p. 30-36, (1972).

28. V. A. Tracey, The roll compacting of metal powders, Powder Metallurgy, vol. 12, No. 24, p. 598-612, (1969).

29. R. S. Lee and E. G. Schwartz, An analysis of roll pressure distribution in powder rolling, Int. J. of Powder Metallurgy, vol. 3, No. 4, p. 83-92, (1967).

30. A. P. Chekmarek, P. A. Klimenko and G. A. Vinogradov, Investigation of specific pressure, specific friction and the coefficient of friction during metal powder rolling, Poroshkovaya Metallurgiya, No. 2 (14), p. 26-30, (1963).

31. G. A. Vinogradov and P. Katashinskii, Study of specific pressure during the rolling of metal powders, Poroshkovaya Metallurgiya, No. 3 (15), p. 30-36, (1963).

32. D. K. Worn, The continuous production of strip by the direct rolling process, Powder Metallurgy, No. 1/2, p. 85-93, (1958).

33. P. E. Evans and G. C. Smith, The compaction of metal powders by rolling I- The properties of strip rolled from copper powder, Powder Metallurgy, No. 3, p. 1-25, (1959).
P. E. Evans and G. C. Smith, The compaction of metal powders by rolling II- An examination of the compaction process, ibid., p. 26-44.

34. D. G. Hunt and R. Eborall, The rolling of copper strip from hydrogen-reduced and other powders, Powder Metallurgy, No. 5, p. 1-23, (1960).

35. A. R. Boughton, D. K. Worn, R. P. Perks, U.K. Patent 842,443, (1960).

36. See reference 28.

37. J. H. Tunderman and A. R. E. Singer, The flow of iron powder during roll compacting, Powder Metallurgy, vol. 11, No. 22, p. 261-294, (1968).

38. D. K. Worn and R. P. Perks, Production of pure nickel strip by the direct-rolling process, Powder Metallurgy, No. 3, p. 45-71, (1959).

39. G. A. Vinogradov and I. M. Fedorchenko, Poroshkovaya Metallurgiya, vol. 1, No. 1, p. 61, (1961).

40. H. Franssen, Production of copper strip, Metal Industry, p.277, (March 25, 1955).

41. H. H. Hirsch, The production of aluminum strip heater using a novel hot powder rolling technique, Mod. Dev. in Powder Metallurgy, vol. 4, ed. by H. H. Hausner, p. 537-547, (1971).

42. G. E. Wieland and E. M. Rudzki, A slurry process for making thin-gauge sheet

from metal powder, Int. J. of Powder Metallurgy, vol. 12, No. 2, p. 103-113, (1976).

43. C. R. Shakespeare, The economics of stainless-steel strip production by roll compacting, Powder Metallurgy, vol. 11, No. 22, p. 379-399, (1968).

44. A. R. Zecca, G. E. Dieter and R. D. Thibodeau, Economic risk analysis of a miniplant for production of steel strip from powder, in "Powder Metallurgy Processing," ed. by H. A. Kuhn and A. Lawley, N.Y., p. 173, (1978).

45. W. E. Buesher, R. Silverman and L. S. Castleman, Rolling of metal strip with tailored properties, Progress in Powder Metallurgy, Proc. 18th annual P/M Technical Conference, N.Y., p. 42-47, (1962).

46. W. D. Lafferty, W. E. Buesher and L. P. Clare, Powder rolled, oxide-strengthened cathode nickel, Mod. Dev. in Powder Metallurgy,, vol. 4, ed. by H. H. Hausner, p. 583-590, (1971).

47. R. W. Fraser, D. J. I. Evans, The properties of cobalt-iron alloys prepared by powder rolling, Powder Metallurgy, vol. 11, No. 22, p. 358-378, (1968).

48. D. M. Stephens and G. Greethan, The production of porous nickel sheet by direct rolling of powder, Powder Metallurgy, Vol. 11, No. 22, p. 330-341, (1968).

49. N. J. Williams and V. A. Tracey, Porous nickel for alkaline battery and fuel cell electrodes, Production by roll compacting, Int. J. of Powder Metallurgy, vol. 4, No. 2, p. 47-62, (1968).

50. T. J. Ready and H. D. Lewis, A technique for powder-rolling tungsten and tungsten - 45% UO_2 dispersions, Int. J. of Powder Metallurgy, Vol. 1, No. 2, p. 56-63, (1965).

51. M. L. Mackay, Innovation in P/M: an engine bearing material, Metal Progress, vol. 111, No. 6, p. 32-35, (1977).

52. F. Emley and C. Deibel, A new method for compacting metal or ceramic powders into continuous sections, Proc. 15th annual meeting, Metal Powder Industries Federation, p. 5-13, (1959).

53. C. Deibel, D. R. Thornburg, F. Emley, Continuous compaction by cyclic pressing, Powder Metallurgy, No. 5, p. 32-44, (1960).

54. R. Kieffer and W. Hotop, Pulvermetallurgie und Sinterwerkstoffe, Berlin, p. 222, (1943).

55. M. F. Pickus, Tips for fountain pens - a case history on the adaptibility of the powder metallurgy process, Int. J. of Powder Metallurgy, vol. 5, No. 3, p. 7-15, (1969).

56. Anonymous, Product Engineering, p. 27-33, (August 1969).

CHAPTER

7

Sintering Practice

In this chapter sintering furnaces and sintering atmospheres are discussed. Sintering furnaces are presented in three sections:

Continuous furnaces used in large scale sintering in protective atmospheres of structural parts, self-lubricating bearings and occasionally cemented carbides,

Batch type furnaces for sintering in protective atmosphere compacts of refractory metal powders and their alloys, cemented carbide compacts and compacts of other metal powders for which sintering in continuous furnaces would not be practical or economical,

Vacuum sintering furnaces, including batch type and continuous, for sintering structural parts, cemented carbides, refractory metals and Alnico magnets.

Furnaces used in hot consolidation processes, such as hot pressing, hot isostatic pressing, hot extrusion and hot forging are discussed in chapter 13. Specialized sintering processes used in sintering tungsten bars by direct current passage, friction materials and metal diamond tool materials are described in the chapters which deal with these applications.

The three sections on sintering atmospheres are:

Purpose of sintering atmospheres
Description of sintering atmospheres
Thermodynamic background of sintering atmospheres

I. Continuous Furnaces with Protective Atmospheres[1]

A. Design and Methods of Heating

Continuous furnaces for sintering in protective atmospheres developed from those used in copper brazing and in bright annealing, in which protective atmospheres are also required. The furnaces have three zones:

> a preheat zone, often called burn-off zone,
> a high heat zone,
> and a cooling zone.

The preheat or burn-off zone must be designed so as to expel the lubricant during compacting. The zone must be long enough so that the decomposition products of the volatilizing lubricant are completely removed before the compacts enter the high heat zone, since products such as carbon or zinc (from zinc stearate) affect the heating elements and the ceramic lining of the high heat zone. The rate of lubricant removal must be low enough so that the compacts do not spall or fracture.

The preheat zone is often part of the atmosphere controlled region of the furnace heated with electrical elements. With this design the volatilizing lubricant is exhausted with the gas of the protective atmosphere and does not burn until it emerges from the throat of the preheating zone.

In other cases it consists of a separate semimuffle chamber or a chamber heated with radiant tubes in which the compacts are heated to a maximum temperature of 800° F (430° C) at which they do not excessively oxidize even without a protective atmosphere. A newer development is a rapid burn-off (RBO) chamber,[2] shorter than the usual preheating zone, which uses a burner whose hot combustion gases impinge upon the compacts to volatilize and burn the lubricant.

The temperature in the high heat zone of continuous sintering furnaces must be closely controlled, since the temperature has a strong influence upon the dimensional changes of the compacts during sintering and upon their mechanical properties. In larger furnaces, with heating zones ten or more feet long, individual controls for each of several subzones may be provided.

One type of continuous furnace has a gas-tight muffle in the high heat zone. Such a gas-tight muffle is needed in furnaces which are heated by fuel (gas or oil) combustion because the products of combustion of heating fuel do not form a suitable atmosphere for sintering furnaces. Muffles may also be used in electrically heated furnaces when extremely close control of the atmosphere, particularly low dewpoints, is required. Only metallic muffles are really gas-tight. The muffle material is generally wrought inconel alloy 600 (75% Ni, 15% Cr, bal Fe) or 601 (60% Ni, 23% Cr, 1.5% Al, bal Fe).

They are useful for temperatures up to 2200° F (1200° C). Most electrically heated furnaces do not have a muffle, but a gas-tight outer sheet steel shell which is lined with refractories. The heating elements mounted on the inner surface of the refractory lining radiate directly upon the compacts to be sintered. The presence of carbon monoxide in many protective atmospheres must be taken into account when the furnace refractories are chosen. At intermediate temperatures between 800 and 1200° F (427-649° C) the carbon monoxide decomposes forming carbon and carbon dioxide. Iron oxide and certain other oxides are catalysts for this reaction. This makes it necessary to use high-temperature refractories which do not contain these oxides even in those zones of the furnace where the temperatures do not reach the values (2800°-3000° F, 1538°-1649° C) for which high temperature refractories are generally recommended.

For high temperature furnaces (2300°-2400° F, 1260°-1320° C) using hydrogen or dissociated ammonia atmospheres, in which a very low dewpoint is required, refractories which contain silica as a binder cannot be used because the silica is reduced by very dry hydrogen. In these furnaces pure alumina refractories are needed. There are five widely used types of electrical heating elements for continuous sintering furnaces with protective atmospheres:

> 80% Ni - 20% Cr alloy for temperatures up to 1150° C (2100° F)
> Fe-Cr-Al alloys, (e.g. 72% Fe, 23% Cr, 5% Al) for temperatures up to 1290° C (2350° F)
> Silicon Carbide for temperatures up to 1370° C (2500° F)
> Metallic Molybdenum for temperatures up to 1700° C (3100° F)
> Molybdenum Disilicide also for temperatures up to 1700° C (3100° F)

The metallic types of heating elements (Ni-Cr, Fe-Cr-Al, Mo) are most commonly arranged in loops of ribbon or rod along the sides of the rectangular opening of the furnace. Silicon carbide elements ("Globar" or "Hotrod") are in the form of rods which generally run perpendicular to the direction of travel of the compacts through the furnace, often with a row of rods above and another below the level of the trays on which the compacts pass through the furnace. While the Ni-Cr, the Fe-Cr-Al and the silicon carbide heating elements can be heated both in air and in the protective atmosphere, molybdenum heating elements must be heated practically from room temperature in hydrogen or dissociated ammonia. The molybdenum disilicide type of element has the principal advantage that it can be heated in air up to the temperatures of furnaces heated with metallic molybdenum heating elements. There are, however, still problems particularly in designing suitable lead-ins for molybdenum disilicide heating elements.

Instead of molybdenum heating elements in the form of loops of rod, elements consisting of bunches of molybdenum wire held together with molybdenum wire ties are used in Europe.

The cooling zone of continuous sintering furnaces containing a protective atmosphere must be long enough so that the compacts do not oxidize when they emerge from it. The usual design of a cooling zone consists of a two foot long insulated zone followed by a water jacketed zone. The temperature of the cooling water in the water jacket is controlled to avoid water condensation in the cooling zone. A newer development is the "convecool" type of cooling zone in which the rate of cooling is controlled by the circulation of the protective atmosphere through the cooling zone.[2]

B. Methods of Conveying

An important feature of continuous sintering furnaces is the method by which the compacts are conveyed through the furnace. The four methods most commonly used are:

> Mechanical pusher
> Mesh belt
> Roller hearth
> Walking beam

In the mechanical pusher type furnace, series of boats or trays one abutting on the next are pushed through the furnace by the pusher mechanism located at the entrance of the preheat zone. A schematic drawing of such a furnace heated by silicon carbide heating elements is shown in Figure 7-1. A pusher type furnace heated by loop-type molybdenum heating elements is shown in the photograph, Figure 7-2. In the foreground of the photograph is a belt mechanism returning the trays from the discharge to the entrance end of the furnace.

The trays or boats are generally of a ceramic material so that they can withstand, at the temperature of the hot zone, the stresses of being slid on the hearth of the furnace and of being pushed by a long string of trays behind them. Trays and boats which can be used at temperatures as high as 1370°C (2500°F) are available. They are relatively heavy so that the ratio of the net weight of compacts to the total weight of trays and compacts being conveyed is low.

Mesh-belt furnaces, shown schematically in Figure 7-3, have an endless woven belt generally made of a nickel-base high-temperature alloy wire which runs over large motor-driven drums at the ends of the furnace. Inside the furnace it slides over the refractory hearth in the hot zone. The

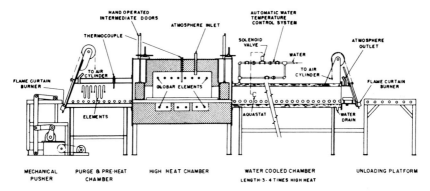

FIGURE 7-1 Longitudinal section of mechanical pusher continuous furnace with silicon-carbide heating elements.

FIGURE 7-2 Photograph of continuous pusher type furnace (Courtesy: C. I. Hayes, Inc.).

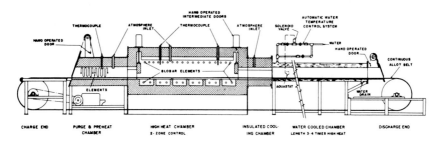

FIGURE 7-3 Longitudinal section of mesh-belt furnace with silicon carbide heating elements.

183

compacts to be sintered can be put directly on the belt or they can be put on trays often also fabricated from heat-resisting mesh. The doors in mesh-belt furnaces are continually open and the furnaces therefore require large amounts of protective atmosphere. A photograph of a mesh-belt furnace with a 12" wide belt, a 60" long presintering and 60" long high-heat zone is shown in Figure 7-4. A modification of the mesh-belt furnace, called a hump-back furnace, is shown schematically in Figure 7-5. These furnaces have inclined entrance and exit zones with the heating and cooling chamber at an elevated level. They are used for sintering materials, such as aluminum alloys and stainless steel, which need particularly dry atmospheres. The hump-back construction is intended to prevent contamination of sintering atmospheres, particularly those high in hydrogen, which,

FIGURE 7-4 Photograph of continuous 12" wide, mesh-belt sintering furnace, 60" long presintering, 60" long high heat zone (Courtesy: C. I. Hayes, Inc.).

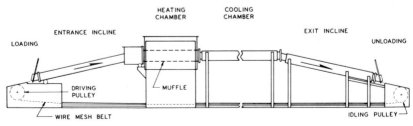

FIGURE 7-5 Section through hump-back mesh belt furnace; belt drive and take-up mechanisms omitted.

because of their tendency to rise, stay in the elevated sections of the furnace and do not escape readily. A photograph of a hump-back furnace for sintering aluminum structural parts is shown in Figure 7-6.

Roller hearth furnaces, shown schematically in Figure 7-7, have a hearth in the hot zone of the furnace which consists of a series of parallel rollers on 6 to 9″ centers made of a heat-resisting alloy. On the sides of the furnace, the rollers pass through the brickwork so that the bearings in which the rollers turn are outside the hot zone of the furnace and can be water cooled. The cold ends of the rollers are fitted with sprockets over which a chain runs which drives all the rollers at a uniform speed. The roller hearth is continued both into the entrance and the cooling zone. Ordinary steel rollers will be satisfactory in these zones. The rollers of the roller hearth must turn continually as long as the furnace is up to temperature because they would otherwise sag. The compacts are again conveyed over the roller-hearth in trays or boats. Provisions are made for opening the doors at the end of roller-hearth furnaces only when a tray is charged or discharged so that less protective atmosphere is needed than in mesh-belt furnaces.

Both mesh-belt and roller-hearth furnaces are limited in temperature because of the temperature limitations of the belt and roller material. They are generally used only up to 1150°C (2100°F).

Walking-beam furnaces, shown schematically in Figure 7-8, were first developed in Germany for use with molybdenum heating elements. In these furnaces boats or trays are conveyed by a mechanism which periodically lifts them, advances them a short distance and than lets them settle back on a ceramic ledge. The part of the lifting mechanism heated to the furnace temperature is made of refractory. This refractory is backed by alloy, but the alloy is at a lower temperature. Walking beam furnaces can therefore convey trays continually through a furnace heated to temperatures as high as 1315°C (2400°F). Low carbon steel boats are used in these furnaces which, however, are relatively long-lasting since they are not subjected to much stress.

C. Temperature Control

Temperatures in sintering furnaces are measured and controlled by thermocouples or for higher temperatures by radiation type pyrometers, in which a lens system is focused upon a characteristic location in the hot zone and a group of thermocouples (a thermopile) measures the temperature in the focus of the optical system which is directly related to the temperature of the characteristic location.

FIGURE 7-6 Photograph of hump-back furnace for sintering aluminum parts (Courtesy: C. I. Hayes, Inc.).

The simplest type of controller used with sintering furnaces is an on-off controller. The controller closes a valve (in gas fired furnaces) or opens a contact (in electrically heated furnaces) when the temperature is above the set point and opens a valve or closes a contact when the temperature falls below the set point.

A second more sophisticated type of control is the proportioning control. Proportioning controls for electric furnaces regulate the successive duration of on-time to off-time in accordance with either the size, the speed or the duration of temperature changes.

A third means for holding temperature very closely constant, which is particularly useful for furnaces with molybdenum heating elements, is the saturable core reactor. It has a special transformer which, in addition to the regular AC windings connected to the load, has also a DC winding which is connected with the temperature measuring instrument through an amplifier. Changes in the DC winding of the transformer, caused by changes in temperature, change the impedance of the transformer and thereby the input to the heating elements of the furnace. The input to a furnace provided with a saturable core reactor will never be greater than the amount needed to hold the furnace at a constant temperature and therefore avoids the temperature fluctuations in the heating elements and in the furnace inherent in other control systems.

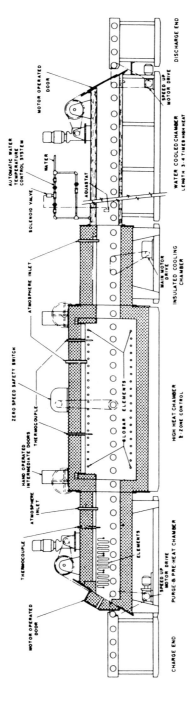

FIGURE 7-7 Longitudinal section of roller hearth furnace with silicon carbide heating elements.

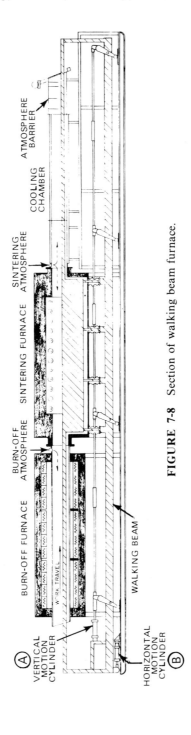

FIGURE 7-8 Section of walking beam furnace.

II. Batch Type Furnaces With Protective Atmospheres

Batch type furnaces are used for sintering in protective atmospheres when the quantities produced do not warrant installing continuous furnaces. Furnaces used may be similar to pusher type continuous furnaces except that the boats are stoked through the furnace by hand rather than with a mechanical stoker. A schematic of a box-type sintering furnace is shown in Figure 7-9. Furnaces of this type are used to sinter compacts of the refractory metals molybdenum and tungsten. The heating elements in this case are molybdenum or tungsten wire. In the clasical method of sintering cemented carbides the carbide bits are packed in alundum in graphite boats and stoked through a batch type furnace in an atmosphere of hydrogen. The furnaces have a D-shaped Alundum muffle, upon which molybdenum wire is wound on the outside. Only cemented carbides containing no other carbides than WC can be sintered satisfactorily in hydrogen. All grades containing TiC of TaC are sintered in vacuum furnaces. In recent years vacuum furnaces have also been widely used for cemented carbides containing only WC as the carbide phase.

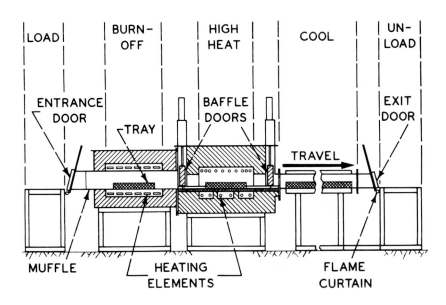

FIGURE 7-9 Box type, manually operated sintering furnace.

In addition to box type furnaces, bell type furnaces are also occasionally used for batch type sintering of metal powder compacts, particularly in cases where very good atmosphere control is required. Such a furnace is shown schematically in Figure 7-10. The compacts are arranged on a load supporting base with a removable sealed retort over the load to contain the protective atmosphere. A portable heating bell is lowered over load and retort for the heating cycle. In an alternative design the load, with the retort over it, would be lifted by an elevator into a stationary bell furnace.

III.   Vacuum Sintering Furnaces

Batch type vacuum sintering furnaces were first developed for sintering compacts from refractory and reactive metals such as tantalum, niobium, titanium, zirconium and their alloys, which react with any but the noble gases, argon, helium, etc., for sintering cemented carbides containing TiC and TaC and for sintering Alnico magnets. A schematic drawing of a high frequency batch-type furnace for sintering cemented carbides is shown in Figure 7-11.[3] The material to be sintered is arranged on graphite trays and is heated by a graphite receptor heated by the high frequency current passing through a water-cooled spool or by resistance heating of graphite heating elements. A photograph of a resistance heated vacuum sintering

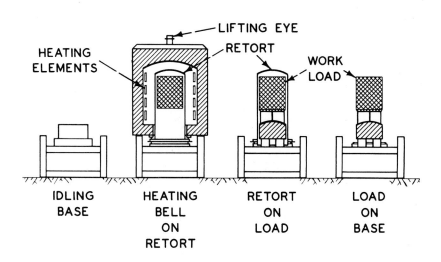

FIGURE 7-10 Bell type sintering furnace showing stationary bases, retorts and heating bell.

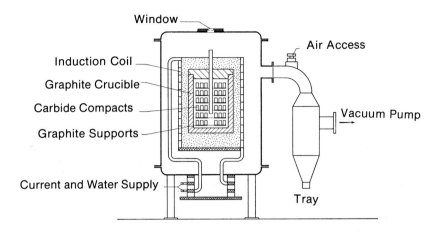

Window

Air Access

Induction Coil

Graphite Crucible

Carbide Compacts

Graphite Supports

Vacuum Pump

Current and Water Supply

Tray

FIGURE 7-11 Schematic of vacuum furnace for sintering cemented carbides.

furnace for cemented carbides is shown in Figure 7-12. It provides for die lubricant removal, presinter and final sinter. An inert gas recirculating fan accelerates cooling. A digital microprocessor controls temperature and vacuum through the cycle. Continuous vacuum sintering furnaces for cemented carbides have been developed, although batch type furnaces appear to provide better control of the sintering cycle than continuous furnaces. In continuous furnaces using a rotating table the material is dewaxed, sintered and cooled in vacuum,[4] in pusher type continuous furnaces it is dewaxed in hydrogen and sintered in vacuum with vacuum locks between dewaxing and high heat chamber and between high heat and cooling chamber.[3]

Batch type vacuum sintering furnaces are used for sintering tantalum capacitors. The heating elements in this case are often made from tantalum sheet with a series of radiation shields between the heating element and the cold wall of the furnace.

Vacuum sintering furnaces have also been introduced for sintering structural parts, first for those from stainless steel powder[5] and more recently those from plain carbon and low alloy steels. Batch type furnaces may be single chamber furnaces in which the work is heated in vacuum

FIGURE 7-12 Photograph of vacuum sintering furnace for tungsten carbide providing for lubricant removal, presinter and final sinter; cooling by inert gas recirculated by fan. Temperature and gas pressure controlled by digital microprocessor programmer (Courtesy: GCA Corp., Vacuum Industries Div.).

using graphite cloth or graphite rod heating elements and cooled in the same chamber with circulating nitrogen cooled by internal finned heat exchangers. They may also have dual chambers in which heating in vacuum and cooling in circulating nitrogen is done in separate chambers. The removal of the lubricant (preferably a synthetic wax and not zinc stearate) must be done in a separate operation.

The newest development in vacuum furnaces is a continuous furnace for sintering iron and steel parts in a vacuum chamber using graphite heating elements. These furnaces incorporate burn-off chambers, vacuum locks and cooling zones through which the compacts arranged on trays are conveyed continuously. A photograph of a furnace of this type is shown in Figure 7-13.

One of the principal reasons why vacuum sintering has aroused so much interest is the fact that the energy requirements for providing protective atmospheres are quite high and may be as much as one half of the total

FIGURE 7-13   Continuous vacuum furnace.

energy requirements for sintering. Vacuum sintering would therefore sub-stantially reduce the total energy consumption in the sintering process.

IV.   Purpose of Sintering Atmospheres

The primary purpose of a sintering atmosphere is to control the chemical reactions between the compacts and their surroundings. A secondary purpose is to flush out the decomposition products of the lubricants which are given off in the preheat ("Burn-off") zone of the furnace. The impor-tance of the control of chemical reactions becomes evident when it is recalled that most compacts in the as pressed ("green") condition are porous. The gases in the sintering atmosphere can therefore react not only with the outside surface of the compacts, but can penetrate the porous structure and react with the interior compact surfaces. If reaction with the interior surfaces is prevented, protective atmospheres may not be neces-sary. For instance, aluminum powder compacts pressed at high pressure without the addition of a lubricant may not have any interconnecting porosity and may be sintered in air. The oxygen of the air will form aluminum oxide on the outer surface of the compact, but the thin oxide layer prevents further reaction with the interior of the compact.

The most important reactions which must be controlled are:

1. Reduction of oxides on the surface of the powder particles in the compact, thereby making extensive metal to metal contact possible and the prevention of any further oxidation either in the high heat or the cooling zone of the furnace.
2. Carburization and decarburization in compacts of iron and steel.

Other gas-metal reactions, such as that of nitrogen with austenitic stainless steel alloys, play a minor role.

Because of their importance, the thermodynamics of oxidation-reduction and of carburization-decarburization reactions will be briefly discussed after the composition of those commercial sintering atmospheres most widely used has been presented.

V. Description of Sintering Atmospheres

Commercial sintering atmospheres are:

A. Hydrogen
B. Dissociated ammonia
C. Partially combusted hydrocarbon gases, such as methane or propane
D. Nitrogen without or with the addition of small amounts of hydrogen and/or methane

A. Hydrogen

Important methods for producing hydrogen are by electrolysis of aqueous solutions and by processes in which hydrocarbon gases are reacted with oxygen or steam to form mixtures of hydrogen and carbon monoxide according to the reactions:

$$CH_4 + \frac{1}{2}O_2 = 2H_2 + CO$$
$$CH_4 + H_2O = 3H_2 + CO$$

The resultant gas mixtures are catalytically reacted with steam to convert the CO to CO_2 which is then chemically absorbed to produce more or less pure hydrogen. Hydrogen as a sintering atmosphere is relatively expensive at the present time, particularly in small quantities, although more cost effective methods of producing the gas are under intensive study.

In using hydrogen, its tendency to form explosive mixtures with air, its low density, only 7 percent of that of air, and its very high thermal conductivity, almost seven times that of air, should be kept in mind. Depending upon the method of manufacture, hydrogen may contain

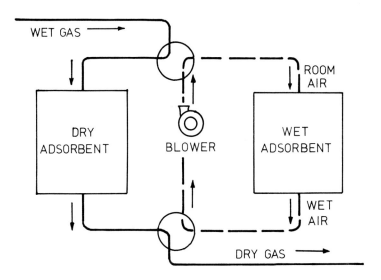

FIGURE 7-14 Apparatus for gas drying.

impurities, of which oxygen and water vapor are the most important. Oxygen can be removed by passing hydrogen over heated copper turnings or more completely by passing it through a palladium catalyst purifier. The water vapor may be absorbed by activated alumina or by zeolites ("molecular sieves") in an apparatus shown in Figure 7-14. The adsorbing towers are periodically reactivated by passing hot air through them. The water vapor content of hydrogen or any other gas is customarily expressed as a dewpoint, i.e. the temperature at which the water vapor in a gas with a given H_2O content starts to condense when the gas is cooled. The relationship between dewpoint and volume percent water vapor is shown in Table 7-1.

Table 7-1 Relationships between dewpoint and concentration of water vapor in volume percent

Dewpoint	°C	37	27	15	5	0	−7	−18	−29	−40
	°F	100	80	60	40	32	20	0	−20	−40
Volume % H_2O			7	3.5	1.75	0.82	0.60	0.37	0.15	0.06	0.02

Sintering atmospheres can be readily dried to dewpoints of –40 to –50°C in activated alumina driers. Even lower dewpoints on the order of –80°C can be obtained in zeolite ("molecular sieve") adsorbers.

B. Dissociated Ammonia

Dissociated ammonia containing 75 vol % H_2, 25 vol % N_2 is produced from gasified liquid ammonia by the reaction

$$2NH_3 = N_2 + 3H_2$$

in a dissociator. A schematic sketch of a dissociator in which the ammonia is dissociated at 750°C over a catalyst is shown in Figure 7-15. Dissociated ammonia as it comes from the dissociator is generally quite dry with dewpoints in the range from –40 to –50°C.

Until recently mixtures of hydrogen and nitrogen lower in hydrogen and higher in nitrogen than dissociated ammonia were produced either by partial combustion with air of dissociated ammonia or by catalytic conversion of mixtures of ammonia and air. The water vapor formed was subsequently removed from the gas by adsorption. These methods of producing hydrogen-nitrogen mixtures high in nitrogen have been largely superseded by diluting dissociated ammonia with nitrogen gas obtained by fractional distillation of liquid air.

AMMONIA   DISSOCIATOR

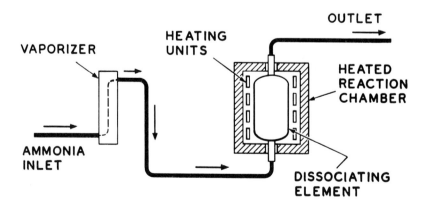

FIGURE 7-15   Flow diagram for ammonia dissociator.

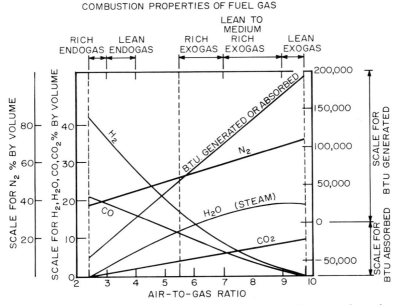

FIGURE 7-16 Percentages of the various products of combustion of methane as a function of the air-to-gas ratio.

C. Partially Combusted Hydrocarbon Gas

Methane, propane and other hydrocarbon gases may be partially combusted with air. In the case of methane the volume ratios of air to gas may range from 2.4:1 to 9.7:1. The products of combustion are water vapor, carbon dioxide, hydrogen, carbon monoxide and nitrogen with small amounts of uncombusted methane for low air-gas ratios. Figure 7-16[6] shows the amounts of each of these products of combustion as a function of the volume ratio of air and methane. The figure also indicates the range of air-gas ratios used in what is called "rich" and "lean" "exogas" and "rich" and "lean" "endogas." A schematic sketch of an "exothermic" gas generator is shown in Figure 7-17. In this type of generator the heat produced by the reaction is sufficient to keep the catalyst of the reaction chamber at the reaction temperature. Much of the water vapor produced in the reaction is removed by cooling the gas and condensing the water vapor. When the gas is cooled with cooling water, the dewpoint of the "exogas" is generally 6°C (10°F) above the temperature of the cooling water. If refrigeration is used, the dewpoint of the "exogas" may be lowered to 5°C (40°F). From the exothermic gas a gas quite low in CO_2 and H_2O, called "purified exothermic gas," may be produced by absorbing

197

these constituents. The purification may be in two steps, absorbing the CO_2 with ethanolamine and the water vapor in activated alumina or in one step using a zeolite ("molecular sieve") absorber.

For air-methane ratios lower than 6:1 the heat produced by the reaction is insufficient to keep the catalyst in the reaction chamber at the reaction temperature. An external source of heat for heating the reaction chamber is needed. A schematic sketch of a generator of this type called an "endothermic" gas generator is shown in Figure 7-18. Very rich "endothermic" gas with air to methane ratios lower than 2.7:1 is somewhat unstable and has a tendency to precipitate carbon in the form of soot. Therefore, "endothermic" gas at somewhat leaner ratios is generally produced. If it is necessary for carbon control, natural gas may be added to the endothermic gas before it enters the furnace chamber.

D. Nitrogen[7]

In view of the rising prices of natural gas and its unstable supply situation sintering atmospheres based on nitrogen produced by fractional distillation of liquid air have recently come to the fore. For most applications

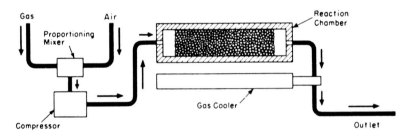

FIGURE 7-17 Flow diagram for exothermic gas generator.

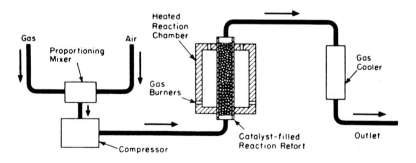

FIGURE 7-18 Flow diagram for endothermic gas generator.

these atmospheres consist of a minimum of 90% of nitrogen with the balance being made up of hydrogen and methane. The nitrogen gas and the hydrogen and methane are piped in separately before they are mixed and introduced into the furnace. Atmospheres of this type are used increasingly in particular to replace "endogas."

VI. Thermodynamic Background of Sintering Atmospheres

The reaction for the oxidation of a metal may be written:

$$M + O_2 = MO_2 \quad I$$

If it is assumed that the oxide has a constant chemical composition (the fact that this is not strictly correct for certain oxides does not seriously affect the thermodynamic argument), an equation for the standard free energy for reaction I may be written:

$$\Delta G_I = -RT \ln K_{pI} = RT \ln p_{o_2}$$

in which R is the gas constant, T the absolute temperature and K_p the equilibrium constant for the raction. p_{O_2} is the pressure of oxygen in the atmosphere, at which the metal and the oxide are in equilibrium with each other, i.e. the pressure at which the rate of oxidation of the metal is equal to the rate of decomposition of the oxide into metal and oxygen. A plot of the standard free energy for the oxidation of a series of metals vs. temperature, the so-called Richardson diagram[8] is shown in Figure 7-19. From the standard free energy at any temperature the oxygen decomposition pressure of the oxide can be readily calculated using the above equation. For the oxides in the diagram the standard free energy becomes less negative with increasing temperature, but even at a relatively high sintering temperature of 1000° C, the values are −172 kJ (−41 kcal) for Cu_2O corresponding to an oxygen pressure of 9.4×10^{-8} atmospheres of oxygen, for NiO −251 kJ (−60 kcal), corresponding to 5.2×10^{-11} atmospheres of oxygen and for FeO −352 kJ (−84 kcal), corresponding to 4.0×10^{-15} atmospheres of oxygen. This means that copper, nickel and iron will readily oxidize at 1000° C in air of atmospheric pressure ($p_O = 0.2$). To make the oxides decompose, the oxygen pressure would have to be reduced to extremely low pressures. Only the oxides of the noble metals, Ag, Au and Pt have decomposition pressures above 1 atmosphere, which means that these metals do not oxidize in air.

To reduce oxides like Cu_2O, NiO, FeO or even ZnO or Cr_2O_3 they must be reacted with reducing gases, such as hydrogen or carbon monoxide. The reducing potentials of these gases may be obtained by combining the standard free energy of the oxidation reaction of the metals at a given

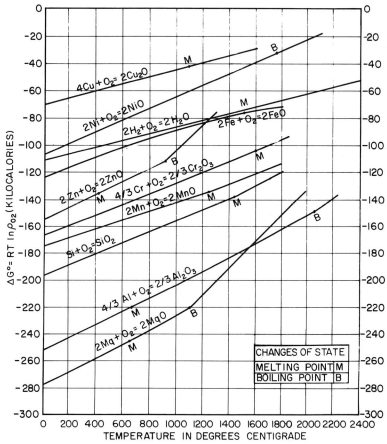

FIGURE 7-19 Standard free energy of formation of metal oxides.

temperature with that of hydrogen or carbon monoxide at the same temperature. In other words, the standard free energies of

$$2H_2 + O_2 = 2H_2O \quad \text{II} \quad \text{and of} \quad 2CO + O_2 = 2CO_2 \quad \text{III}$$

respectively must be subtracted from the standard free energy for $M + O_2 = MO_2$ giving the standard free energies of the reactions

$$M + 2H_2O = MO_2 + 2H_2 \quad \text{IV}$$

and

$$M + 2CO_2 = MO_2 + 2CO \quad \text{V}$$

The free energies of reactions IV and V are written as:

$$-\frac{\Delta G^{\circ}_{IV}}{RT} = K_{pIV} = 2\ln\frac{p_{H_2}}{p_{H_2O}}$$

$$\text{and } -\frac{\Delta G^{\circ}_{V}}{RT} = K_{pV} = 2\ln\frac{p_{CO}}{p_{CO_2}}$$

From the values of the standard free energies of these reactions at any given temperature, the ratios of the partial pressures of H_2 and H_2O and of CO and CO_2 respectively can be calculated at which the reactions are in equilibrium, i.e. at which the rate of reduction of the oxide with hydrogen or with carbon monoxide is equal to the rate of oxidation of the metal with steam or with carbon dioxide. Plots for these equilibrium ratios as a function of temperature for the reactions

$$Fe + H_2O = FeO + H_2$$

and

$$Fe + CO_2 = FeO + CO$$

are shown in Figure 7-20.[9] These ratios are the limits, above which oxidation of the metal takes place and below which any oxide is reduced to the metal. It is seen that FeO will be reduced at 1000°C by hydrogen if the ratio p_{H_2O}/p_{H_2} is not higher then 0.66, but at 500°C the maximum p_{H_2O}/p_{H_2} ratio is 0.26. This may be applied directly to the sintering of compacts from pure iron in exothermic gas. At an air-gas ratio of 6.7:1 the exothermic gas contains 10% H_2. If the dewpoint of the cooled gas is 80°F it will contain 3.5 vol % H_2O. The p_{H_2O}/p_{H_2} ratio is therefore 0.35 which is on the reducing side at 1000°C, but on the oxidizing side at 500°C. Compacts of iron will therefore readily sinter in this gas, but as they are cooled to room temperature in the cooling zone of the furnace, the exothermic gas atmosphere will become oxidizing and an oxide layer will be formed on the surface of the compacts. To keep the compacts from oxidizing in the cooling zone, the exothermic gas may be cooled to a dewpoint of 40°F, corresponding to 0.82 vol % H_2O and a p_{H_2O}/p_{H_2} ratio of 0.082 which is reducing down to temperatures where the rate of the oxidation reaction is negligibly slow. An alternative would be to use an endothermic gas with an air-gas ratio of 4:1 which contains 28% of H_2 and therefore a p_{H_2O}/p_{H_2} ratio of $0.035/0.28 = 0.125$ which is also reducing down to sufficiently low temperatures.

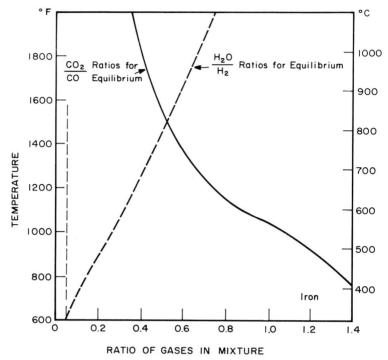

Figure 7-20 Equilibrium ratios for the reactions $Fe + H_2O = FeO + H_2$ and $Fe + CO_2 = FeO + CO$ as a function of temperature.

Compared with iron oxide the equilibrium p_{H_2O}/p_{H_2} ratio for the reduction of cuprous oxide is much higher, so that copper does not oxidize even if it is heated in an atmosphere of 99% H_2O and 1% H_2. On the other hand, to reduce zinc oxide, chromium oxide, manganese oxide or silicon oxide the water vapor content of a hydrogen atmosphere must be much lower than for the reduction of iron oxide. How low the water vapor content of hydrogen must be for such metals as Fe, Mo, Zn, Cr, Mn, V, Si, Ti, Al, Zr and Be to reduce the oxides at different temperatures is shown in Figure 7-21.[10] The data in this diagram were again derived from the standard free energy values in the Richardson diagram. The dewpoint of water vapor containing hydrogen is plotted vs temperature for the oxides of the 11 metals. Such a plot may be used to obtain an idea how dry a hydrogen atmosphere must be available to sinter compacts from austenitic steels containing 18% of readily oxidizing chromium. The diagram refers to pure chromium and shows that at 1000°C the maximum dewpoint to avoid oxidation is –32°C while at 1250°C the maximum allowable dewpoint is –14°C. In austenitic stainless steel the chromium does not exist as pure

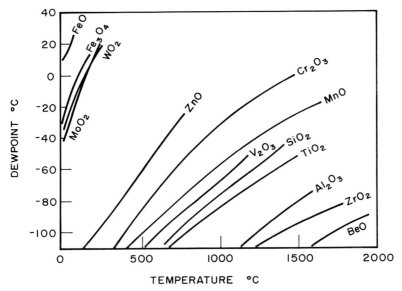

FIGURE 7-21 Relation of hydrogen dewpoint to equilibrium temperature for reduction of various metal oxides in hydrogen.

metallic chromium but as a solid solution in the alloy. For this reason, the thermodynamic activity of an alloy with 18% chromium in solid solution with respect to that of pure chromium enters the calculation and the requirements for dryness of the atmosphere for the alloy are less severe than they would be for pure chromium. The diagram clearly indicates, however, that with increasing sintering temperatures the atmospheres for sintering stainless steel successfully may have higher dewpoints. The same considerations apply to the sintering of compacts containing ferrosilicon powder.

Stainless steel powder compacts are often sintered in dissociated ammonia. The nitrogen in this atmosphere dissolves in austenitic stainless steels, making the compacts somewhat stronger and less ductile than when sintered in hydrogen or vacuum. This is in contrast to iron and low-alloy steels, for which nitrogen in the sintering atmosphere is completely inert.

A somewhat similar situation as for stainless steel exists in the case of atmospheres for sintering brass powder compacts. As Figure 7-21 shows, quite dry atmospheres are necessary to reduce zinc oxide to zinc. Because the chemical activity of zinc in brass is only a small fraction of that of pure zinc ($a_{Zn} = 0.05$ for brass with 20 atomic % Zn) the requirements as to dryness for atmospheres in which brass powder compacts are sintered are not as stringent as for pure zinc. However, sintering brass powder compacts presents another problem. When brass compacts are sintered on open trays (without a cover over the tray) there is a tendency for "dezincification," i.e. the zinc sublimes from the surface of the compacts and changes their chemical composition. Covering the sintering trays will inhibit dezincification, but the thin oxide skin on the compacts will react with the atmosphere and increase its water vapor content. In covered trays the water vapor is not as readily swept out as in open trays. Therefore in sintering compacts from brass powder the atmosphere must be dry enough so that even with some build-up of water vapor content due to limited circulation of the atmosphere, the water vapor content stays below that which causes oxidation. Brass powder compacts have been successfully sintered in atmospheres of dissociated ammonia and of "endothermic gas."

B. Carburizing-Decarburizing Reactions

Steels, i.e. alloys of iron and carbon with small amounts of other alloying ingredients, are an important powder metallurgy material. They are generally produced by compacting mixtures of iron powder and graphite powder and sintering the compacts at a temperature where the graphite dissolves in the iron, forming austenite, a solid solution of carbon in γ-iron. Sintering atmospheres for steel must be not only reducing, but must also be in equilibrium with the composition, i.e. the carbon content of the austenite being produced during sintering. They should be neither carburizing, a reaction in which the carbon content of the austenite is increased by interaction with the sintering atmosphere, nor decarburizing, a reaction in which the carbon content is decreased. A simple example of an atmosphere which reacts with sintering steel compacts is one of carbon monoxide and carbon dioxide with no other constituents in the mixture. The reaction may be written

$$Fe + 2\ CO = (Fe,C) + CO_2 \qquad VI$$

in which (Fe,C) stands for a solid solution of carbon in austenite. The equilibrium constant for the reaction may be written

$$K_{pVI} = (p_{CO_2} \times a_c)/(p_{CO}^2 \times a_{Fe})$$

in which a_{Fe} is the activity of iron in the iron-rich solid solution of carbon in austenite. This activity is near 1 and may be neglected, resulting in the equation

$$K_{pVI} = \frac{p_{CO_2}}{p_{CO}^2} \times a_C$$

a_C in this equation is the activity of carbon in austenite. At any given temperature each value of a_C corresponds to a given value of weight percent carbon in the steel up to a saturated solution of carbon in austenite. Therefore at any temperature, a value of the ratio p_{CO_2}/p_{CO}^2 which defines the "carburizing potential" of the sintering atmosphere corresponds to a given value of weight percent carbon in austenite. There are two important differences between the oxidation-reduction reaction $Fe + CO_2 = FeO + CO$ and the carburizing-decarburizing reaction:

1. Whether Fe is oxidized or FeO reduced depends upon the ratio of the partial pressure of CO to the partial pressure of CO_2, while carburization and decarburization depends upon the ratio of the partial pressure of CO_2 to the square of the partial pressure of CO.

2. For the oxidation-reduction reaction there is, at any temperature, one critical ratio of the partial pressures. At any ratio of P_{CO}/p_{CO_2} higher than critical, the atmosphere is reducing, while at any ratio lower than the critical it is oxidizing. In contrast, any given ratio of p_{CO_2}/p_{CO}^2 at a given temperature corresponds to a certain activity of carbon in the austenite, i.e. to a certain carbon content in weight % in the alloy. Increasing the p_{CO_2}/p_{CO}^2 ratio will lower the carburizing potential, shift the equilibrium to a lower activity, i.e. a lower carbon content and will cause decarburization. Lower ratios of p_{CO_2}/p_{CO}^2 will raise the carburizing potential, shift the equilibrium to a higher activity and carbon content and will cause carburization.

The fact that the activity of the carbon in the austenite can be controlled by the ratio of partial pressures p_{CO_2}/p_{CO}^2 follows directly from the phase rule:[11]

$$F = 2 + C - P$$

in which F are the degrees of freedom or the number of independent variables, P is the number of phases and C is the "number of components" which is equal to the number of chemical species minus the number of chemical equations connecting these species. For reaction VI:

P = 2, i.e. solid solution phase and a gas phase
C = 3, i.e. 4 chemical species Fe, C, CO and CO_2 minus the one
connecting chemical equation VI.

Therefore F = 3. Two of these independent variables are temperature and pressure, which are fixed. The pressure in flowing sintering atmospheres is atmospheric pressure. This leaves only one independent variable which means that when a ratio of p_{CO_2}/p_{CO}^2 is chosen, the carbon activity and therefore the weight percent of carbon in austenite is fixed and vice versa.

Steel compacts are quite often sintered in "endogas" atmospheres with a fairly rich air-methane ratio near 2.7 parts of air to one part of methane. As Figure 7-16 indicates, such an atmosphere contains near 40% H_2, near 20%

CO, minor quantities on the order of 1% or less of CO_2, H_2O and also of unreacted CH_4 with the balance—near 40%—of the mixture being N_2. At first glance, any control of the carburizing potential in such a mixture appears hopeless. However, using the phase rule, as in the case of CO-CO_2 mixtures, the problem can be simplified.

In the case of the CO-CO_2 mixtures there was one carburizing-decarburizing chemical reaction and one corresponding equilibrium constant, i.e.

$$Fe + 2CO = (Fe,C) + CO_2 \qquad VI$$

$$K_{p\ vi} = \frac{p_{CO_2}}{p_{CO}^2} \times a_c$$

For the endogas mixture there are two additional, independent carburizing-decarburizing chemical reactions and their corresponding equilibrium constants:

$$Fe + CH_4 = (Fe,C) + 2H_2 \qquad VII$$

$$K_{pVII} = \frac{p_{H_2}^2}{p_{CH_4}} \times a_c$$

$$Fe + CO + H_2 = (Fe,C) + H_2O \qquad VIII$$

$$K_{pVIII} = \frac{p_{H_2O}}{p_{CO} \cdot p_{H_2}} \times a_c$$

Applying now the phase rule to the endogas atmosphere containing H_2, CO, N_2, CO_2, H_2O and CH_4, there are again two phases (solid solution and gas phase). There are 8 chemical species (Fe, C and the 6 gases), from which 3 connecting equations, VI, VII and VIII, must be subtracted, making the number of components and also the number of independent variables equal to 5. Again, temperature and pressure are chosen as two of these variables. The partial pressure of hydrogen, p_{H_2}, and of carbon monoxide, p_{CO}, are chosen as two additional independent variables, because these two gases are present in large concentrations. Small changes in the partial pressures of the carburizing gas CH_4 and of the decarburizing gases CO_2 and H_2O will cause large changes in carburizing potential, but will leave p_{CO} and p_{H_2} practically unchanged. Having chosen four independent variables, P, T, p_{CO} and p_{H_2}, only one independent variable is left. This means that for a given carburizing potential, corresponding to a given activity or weight percentage of carbon in the alloy, there is only one possible combination of p_{CO_2}, p_{H_2O} and p_{CH_4} at a given temperature. In other words, the partial pressures of CH_4, CO_2 and H_2O cannot be changed independently of each other, but any change in the partial pressure of one of these gases will automatically produce a chemical reaction changing the partial pressures of the other two. It also means that the carburizing potential of the gas mixture can be measured by measuring just one of the partial pressures or concentrations of CO_2, H_2O and CH_4 in the endogas. Of these three concentrations, that of H_2O can be measured with a dewpoint meter. It is even more common to measure the carburizing potential by determining the concentration of CO_2 with an infrared gas analyzer. The relationship between the concentration of carbon in weight percent in the compact and the dewpoint of endogas at various temperatures is shown in Figure 7-22,[12] that between the carbon concentration in the compact and the percent volume concentration of CO_2 (equal to 100 times the partial pressure of CO_2 in atmospheres), in the endogas again at a series of temperatures is shown in Figure 7-23. The very low concentrations of H_2O and CO_2 in endothermic gas in equilibrium with steels with 0.2 to 0.7% C at sintering temperatures of 2000 and 2100° F should be noted.

The carburizing potential of endogas can not only be measured by measuring H_2O or CO_2 concentrations, but can also be automatically controlled by adjusting the carburizing potential to higher carbon values by small additions of CH_4 or to lower carbon values by small additions of air (which in the furnace immediately reacts to CO_2 and H_2O).

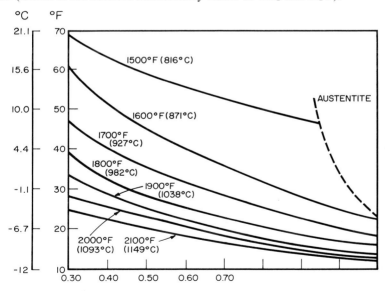

FIGURE 7-22 Equilibrium between dewpoint and weight % carbon in plain carbon steels in endothermic gas atmospheres for various sintering temperatures.

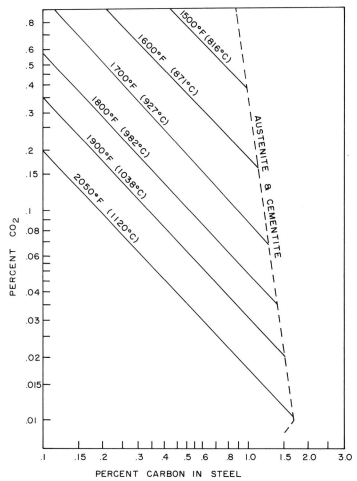

Figure 7-23   Equilibrium between volume % CO_2 and weight % carbon in plain carbon steel in endothermic gas atomspheres for various sintering temperatures.

An inspection of reaction VII between a solid solution of carbon in austenite and hydrogen forming low carbon austenite and methane would lead to the conclusion that sintering an iron-carbon alloy in pure hydrogen or dissociated ammonia, both of which are practically free of methane, should lead to decarburization of the iron-carbon alloy. On the basis of thermodynamics this reaction would certainly be expected. However it has been found that, if the hydrogen or dissociated ammonia is sufficiently dry, the decarburizing reaction is very slow. Iron-carbon compacts may be sintered in dissociated ammonia with only minimal decarburization during the usual sintering times, if care is taken that the dewpoint in the sintering furnace stays sufficiently low. The use of such an atmosphere involves a risk, since its carburizing potential is not in thermodynamic equilibrium with the activity (concentration in weight percent) of the carbon in the steel, while during sintering in endothermic gas thermodynamic equilibrium between atmosphere and the iron-carbon compacts may be maintained.

References

1. Powder Metallurgy Equipment Manual, Metal Powder Industries Federation, Princeton, N.J., 2nd edition, (1977).
2. Drever Company, Huntington Valley, Pa.
3. R. Kieffer and F. Krall, Furnaces for sintering cemented carbides, Planseeberichte für Pulvermetallurgie, vol. 18, No. 2, p. 111-123, (1970).
4. M. J. Blasko and S. W. Kennedy, Developments in vacuum sintering of P/M products, Mod. Dev. in Powder Metallurgy, vol. 9, ed. by H. H. Hausner and P. W. Taubenblat, p. 253-263, (1977).
5. G. Otto, Vacuum sintering of stainless steel, Int. J. of Powder Metallurgy, vol. 11, No. 1, p. 19-24, (1975).
6. W. D. Jones, Fundamental Principles of Powder Metallurgy, London, p. 587, (1960).
7. H. S. Nayar, Nitrogen-base sintering atmospheres (Part I), Mod. Dev. in Powder Metallurgy, vol. 9, ed. by H. H. Hausner and P. W. Taubenblat, p. 213-222, (1977).
8. F. D. Richardson and J. H. E. Jeffes, The thermodynamics of substances of interest in iron and steel making from 0°C to 2400°C, J. of Iron and Steel Inst., vol. 160, p. 261-270, (1948).
9. H. M. Webber, Sintering Furnaces in "Powder Metallurgy," ed. by J. Wulff, Cleveland, Ohio, p. 292-303, (1942).
10. W. D. Jones, Fundamental principles of powder metallurgy, London, p. 546, (1960).
11. E. G. DeCoriolis, O. E. Cullen and J. Huebler, Carbon concentration control, Trans. ASM, vol. 38, p. 659-687, (1947).
12. Powder Metallurgy Equipment Manual, Metal Powder Industries Federation, Princeton, N.J., 2nd ed., p. 119, (1977).
13. F. V. Lenel and L. W. Baum, Control of sintering atmospheres by infrared absorption analysis, Int. J. of Powder Metallurgy, vol. 5, No. 2, p. 13-18, (1969).

8

Sintering of a Single Phase Metal Powder — Experimental Observations

When Rhines[1] presented a seminar on the theory of sintering in 1946, he began his paper with a series of experimental observations on:

> The green density and green strength of compacts
> The effects of variations in powder properties, compacting and sintering conditions on density and dimensional changes in sintering
> The effects of sintering upon microstructure
> The effects of sintering upon mechanical properties

Based on these observations, Rhines developed his theory of sintering. In the last thirty years many model experiments on sintering have been performed and much analytical work has been done which have greatly increased our knowledge of the mechanism of sintering and put it on a more quantitative basis. Nevertheless, it seems appropriate to review once more experimental observations including many not yet available to Rhines. They are of value, because in many commercial applications of powder metallurgy, such as refractory metals, porous products and structural parts, compacts from single metal powders or from homogeneous solid solution alloy powders are pressed and then sintered requiring an overview of what happens when they are sintered. The observations also make clear what data any theory of sintering should be able to explain.

I. The Green Density and the Green Strength of Compacts

Loose powder particles adhere to each other to a limited extent. This observation parallels the well known observations described in detail by Jones[2] and Goetzel[3] on "cold welding" between metal surfaces freshly cleaved and brought back into contact. When a powder is compressed, the green density and green strength of the compact rise with increasing compacting pressure indicating that the area of contact between particles also increases. Nevertheless, the strength of green compacts is lower than that of sintered compacts. This is due to the fact that the surfaces of powder particles are uneven on a submicroscopic as well as on a coarse scale. Therefore, the powder particles are in contact with each other on a scale of atomic dimensions over only relatively small areas even after application of high pressure. Furthermore, the surfaces of powder particles are not clean, but are covered with oxide films and with adsorbed gases which interfere with adhesion between particles. The green density and the green strength of a compact depend on the mechanical properties, in particular yield strength, hardness and the extent of strain hardening of the powder material. Compacts from softer metal powders are denser in terms of relative density with respect to solid metal and have higher green strength at a given compacting pressure than those of harder metal powders, since the particles of softer metals deform more readily and produce larger areas of contact between them. The green density of compacts from a fine powder is generally lower than that from a coarser powder. The green strength of a compact depends upon the particle shape of the powder from which it is pressed. Powders with particles of irregular, more or less equiaxed shape and rough surface give compacts with higher green strength than those with particles which are either spherical or have a flake-like shape.

II. Effects of Powder Properties, Compacting and Sintering Conditions on Density and Dimensional Change in Sintering

Dimensional changes in compacts during sintering are closely related to changes in density. These parameters are important for both the theory and the technology of sintering. For fundamental studies, dimensional changes are determined on cylindrical specimens or sometimes specimens with rectangular cross sections as the difference in dimensions between the green and sintered compacts in the directions parallel and perpendicular to the direction of pressing, generally called radial and axial shrinkage or growth, respectively. They are expressed as percent of the green dimensions. Changes in dimensions in the radial and the axial direction during sintering are not identical, but the differences are usually small. A considerable literature exists on experimental data and on attempts to explain these differences. They will be briefly discussed in chapter 9. Green and sintered densities are, of course, the ratios of the masses of the compacts to their volumes. Although the changes in density between green and sintered compacts are primarily due to changes in dimensions, changes in mass, generally caused by the reduction of oxide skins on the powder particles by the sintering atmosphere, should not be neglected. Green and sintered densities are frequently expressed in percent with respect to the solid density of the metal. Total porosity in percent is equal to 100 - relative density in percent. Densification is occasionally expressed as a parameter:

$$\text{Densification parameter during sintering} = \frac{\text{sintered density - green density}}{\text{solid density - green density}} \times 100$$

This parameter eliminates the effect of small experimental variations in green density in the calculation of densification.

From a technological point of view, dimensional control during sintering centers around the changes in the radial direction between the dimensions of the die in which the compact is pressed and of the sintered compact. The expansion of compacts when they are ejected from the die—the springback—should be subtracted from any shrinkage or added to any growth measured as the difference between green and sintered compact dimensions.

In this section the effects of sintering temperature and time, of compacting pressure and of powder properties upon dimensional and density changes during sintering are presented on the basis of results of typical experiments. The simplest approach to studying these changes is to heat the compacts rapidly to the sintering temperature, hold them at this temperature for a given length of time, cool them rapidly to room temperature and determine dimensional and density differences between the green and sintered compacts. The data obtained are often taken as representing the "isothermal" shrinkage of the compacts, even though they include the shrinkage during heating and cooling.

The effect of sintering time and temperature upon the isothermal shrinkage of compacts from a −74 +43 μm sieve fraction of reduced copper powder compacted at 138 MPa (20,000 psi) and sintered in hydrogen are shown in Figure 8-1.[4] The compacts first shrink rapidly, but the rate of shrinkage slows down with increasing sintering time. The higher the sintering temperature, the sooner the decrease in shrinkage rate. On the other hand, the higher the sintering temperature, the larger the shrinkage of the compacts. The data in this figure clearly illustrate the fact that high sintered densities can be obtained much more rapidly by increasing the sintering temperature than by prolonging the sintering time. Since sintering temperature affects shrinkage much more than sintering time, data are commonly plotted as density changes as a function of sintering temperature at a given constant sintering time. Such plots for molybdenum powder loosely filled into a mold and for powder compacts pressed at 325 and 580 MPa (47 and 84,000 psi) and sintered for three hours in hydrogen are shown in Figure 8-2. The curves show that for molybdenum powder compacts the temperature at which shrinkage begins depends only slightly upon the compacting pressure. The percentage increase in density at a given temperature is the greater the lower the pressure. For compacts of the fine molybdenum powder investigated by Grube and Schlecht,[5]

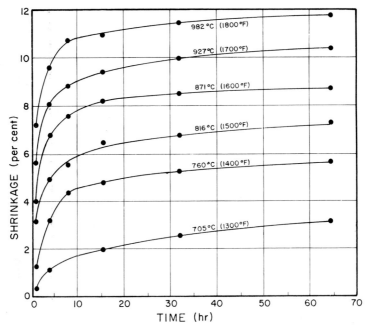

Figure 8-1 Shrinkage of copper powder compacts from −74+43 μm sieve fraction compacted at 138 MPa (20,000 psi) and sintered at various temperatures as a function of sintering time.

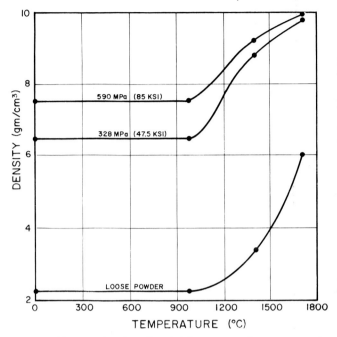

FIGURE 8-2 Sintered density of molybdenum powder compacts, loose, pressed at 325 and 580 MPa and sintered for three hours in hydrogen as a function of sintering temperature.

pressed at 590 MPa (85,000 psi), densities of 98% of theoretical are reached at a sintering temperature of 1600° C, which is less than 0.7 of the melting point of molybdenum in ° K.

Another approach to studying sintering is to determine the dimensional changes with a dilatometer. A compact is heated at a constant rate of heating to a given sintering temperature and then cooled at this rate from the maximum sintering temperature back to room temperature. This approach was pioneered by Pol Duwez[4] in the late 1940s. A dilatometer curve showing the dimensional changes in the axial direction during sintering of a compact from the same reduced copper powder as in Figure 8-1, again pressed at 138 MPa (20,000 psi) is shown in Figure 8-3. Curve 1 represents the dimensional changes of the compact heated at a rate of 3.9° C/ min (7° F/ min). The first portion of the curve up to 315° C (600° F) is identical with the curve for solid copper. When sintering begins, the shrinkage counteracts thermal expansion and the curve deviates from that for solid copper. As the temperature increases, shrinkage proceeds faster and the compact contracts. Immediately after reaching the maximum temperature of 927° C (1700° F) the compact is cooled, again at 3.9° C/ min

215

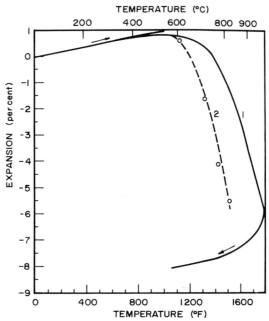

Figure 8-3 Expansion of compacts from −74 +43 μm copper powder pressed at 138 MPa (20,000 psi) and heated at a rate of 3.9° C/min to 925° C and then cooled at the same rate (curve 1). Curve 2 extrapolated for an infinitely slow rate of heating.

(7° F/min). During cooling, the contraction is first at a rate faster than the normal thermal contraction, since the compact still sinters, but eventually the slope becomes that for normal thermal contraction.

Since shrinkage during sintering is a function of time, curve 1 in Figure 8-3 would be displaced for different rates of heating and cooling. Curve 2 in the figure is an extrapolation showing dimensional changes for a compact heated infinitely slowly. It was obtained using the data for isothermal sintering shown in Figure 8-1. It is interesting to note that the temperature at which the dilatometer curve first deviates from the normal expansion curve is little affected by the rate of heating.

Dilatometric curves for the sintering of molybdenum powder compacts pressed at a series of compacting pressures and heated at a rate of 3.9° C/min (7° F/min) to a temperature of 1150° C (2100° F) are shown in Figure 8-4.[4] The temperature at which the dilatometer curve deviates from the normal expansion curve is only slightly influenced by the compacting pressure, which corresponds to the results shown in Figure 8-2. The total amount of shrinkage for compacts pressed at the low pressure of 276 MPa

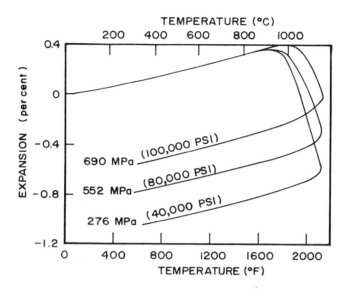

FIGURE 8-4 Expansion of compacts from molybdenum powder pressed at 40, 80 and 100,000 psi (276, 552 and 690 MPa) heated at a rate of 3.9° C/min to 1150° and cooled at the same rate.

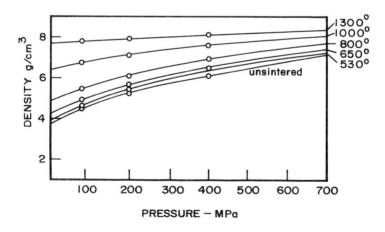

FIGURE 8-5 Density of carbonyl nickel powder compacts sintered at a series of temperatures for 2 hours as a function of compacting pressure.

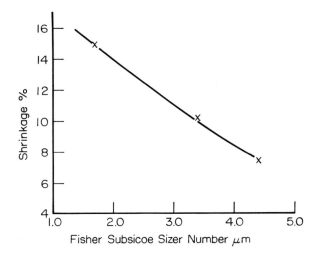

FIGURE 8-6 Shrinkage in percent for tungsten powder compacts sintered 5 hours at 1800° as a function of tungsten particle size, expressed as a Fisher Subsieve Sizer number.

(40,000 psi) is almost twice that for compacts pressed at the high pressure of 690 MPa (100,000 psi).

Another example showing the influence of compacting pressure upon the change in density is shown in Figure 8-5. The density of compacts from carbonyl nickel powder sintered at a series of temperatures for 2 hours is plotted as a function of the compacting pressure.[5] The higher the compacting pressure, the smaller is the shrinkage, i.e. the change from green density to sintered density. Compacts from this very fine powder, when sintered at 1300° C, i.e. 0.91 of the melting point of nickel in ° K, show considerable shrinkage and reach relative densities of 88% and higher regardless of compacting pressure.

In addition to compacting pressure and sintering temperature and time, the properties of the powder from which a compact is pressed, in particular its particle size, have a strong influence upon the dimensional changes during sintering. In Figure 8-6[6] the shrinkage in percent for tungsten powder compacts sintered 5 hours at 1800° C is shown as a function of the particle size of the powder given as a Fisher Subsieve Sizer Number. The finer the powder, the greater will be the shrinkage for a given sintering time. Data for the changes in density with increasing sintering time are shown in Figure 8-7 for compacts from two sieve fractions of electrolytic copper powder, a 75-105 μm (–150 +200 mesh) and a –44 μm (–325 mesh) fraction pressed at 276 MPa (40,000 psi), sintered at 865° C. The compacts

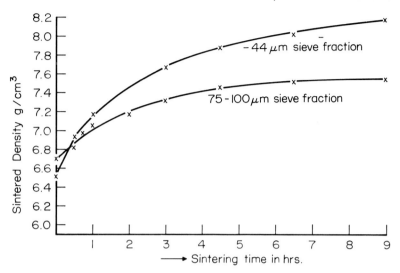

Figure 8-7 Density of compacts from electrolytic copper powder of two particle size fractions, 75-105 μm and –44 μm sintered at 865° C as a function of sintering time.

from the finer powder shrink considerably faster than those from the coarser powder.

The examples for changes in density and dimensions during sintering presented so far include both fcc metals (Cu and Ni) and bcc metals (Mo and W). From a qualitative point of view, there is no significant difference in the behavior of metals with the two types of space lattices, although compacts from bcc metals generally reach densities near that of the solid metal at lower relative temperatures than those from fcc metals. Of particular interest with regard to both the theoretical and the technological aspects of sintering is the behavior of iron powder compacts, since iron transforms from the bcc to the fcc modification at 910° C. The considerable difference in the rate of shrinkage of iron powder compacts depending on whether they are sintered below or above 910° C is illustrated in Figure 8-8, taken from work by Duwez.[4] Compacts from reduced iron powder were heated in hydrogen in a dilatometer at a rate of 3.9° C/min (7° F/min) to either 882° C (1620° F), just below the transformation temperature, or to 921° C (1690° F), just above the transformation temperature, and then held at these temperatures for 2 hours. Figure 8-8 shows both plots of temperature vs. time and of dimensional change in percent vs. time. As in the other dilatometric experiments, the compacts underwent normal thermal expansion, in this case to about 540° C (1000° F), then the expansion slowed

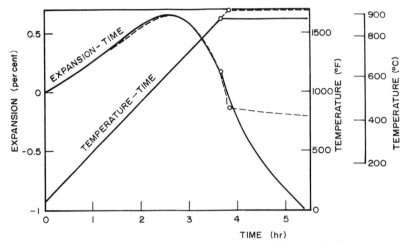

FIGURE 8-8 Expansion of compacts from reduced iron powder heated in hydrogen at a rate of 3.9°C/min to 882°C and 921°C respectively and held at these temperatures for two hours.

down, a maximum dimension was reached and contraction set in. For compacts sintered at 882°C (1620°F) the contraction continued during the isothermal stage of sintering at a rapid rate. For those heated to 921°C (1690°F) a rapid contraction is observed at the transformation temperature of 910°C (1670°F) because of the higher density of γ-iron, compared to α-iron, but during subsequent isothermal sintering the rate of contraction is much slower than for compacts sintered in the α range. One of the reasons for the slower rate of shrinkage in the γ-range compared with that in the α-range is the difference in the self-diffusivity of α and γ iron. At the transformation temperature the self-diffusivity of α-iron is 330 times that of γ-iron. However, experiments of Ciceron and Lacombe[7] have shown that there are other reasons for the difference in shrinkage rate besides the difference in self-diffusivity. Figure 8-9 shows curves for dimensional changes vs. time determined by dilatometry for two compacts from carbonyl iron powder. Curve (a) is that for a compact heated to 895°C (1643°F) in the α-range and held at this temperature. The black dot on the curve indicates the time when the isothermal sintering temperature was reached. During the subsequent 20 hours of isothermal sintering at 895°C (1643°F) the compacts continued to shrink although at a gradually decreasing rate. The linear shrinkage during the isothermal stage of sintering was approximately 8%. Curve (b) is for a compact heated to 920°C (1683°F) above the transformation temperature, held at this temperature for 40 minutes, then cooled to 870°C (1598°F) below the transformation

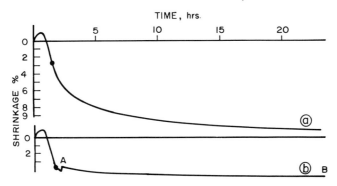

FIGURE 8-9 Dilatometric curves showing the shrinkage of compacts from carbonyl iron powder; curve (a) for a compact heated to 895°C and held at this temperature; curve (b) for a compact heated to 930°C, held at this temperature for 40 minutes, cooled to 895°C and held at this temperature for 20 hours (part A-B of the curve). The black dots indicate the end of the heating-up period.

temperature and held at this temperature for 20 hours. The sudden expansion due to the α to γ transformation at point A of the curve can be clearly seen. During subsequent sintering for 20 hours in the α-range, from point A to point B on the curve, the compact shrank less than 1% linearly, much less than the compact which had not undergone the intermediate excursion in the γ-range. This difference in the rate of shrinkage in the two experiments is related to changes in the microstructure and will be discussed in connection with the effect of sintering upon microstructure.

All of the data on dimensional changes in sintering, presented so far as examples, showed shrinkage of the compacts during sintering rather than growth. In the great majority of cases where growth during sintering of compacts from a single metal powder is observed, it is due to the effect of gases. Growth is more common when compacts from metal powders as soft as or softer than copper are sintered, particularly when they have been pressed at relatively high pressures. Under these conditions, gases are entrapped in closed-off pores of the compacts during compacting. Growth is generally not observed in compacts from iron or other harder metal powders at the usual compacting pressures up to 700 MPa (100,000 psi). Growth behavior of copper powder compacts is illustrated in Figures 8-10, 8-11 and 8-12. In Figure 8-10 the changes in density of copper powder compacts pressed at 520 MPa (75,000 psi) and sintered for 1 hour are plotted as a function of sintering temperature. At increasing temperatures the compacts first increase in density during sintering, but after reaching a maximum at 900°C, the density increase becomes smaller until at 1030°C

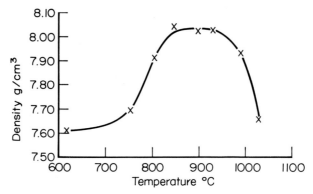

FIGURE 8-10 Sintered density of compacts from electrolytic copper powder pressed at 520 MPa (75,000 psi) and sintered for 1 hour at a series of sintering temperatures.

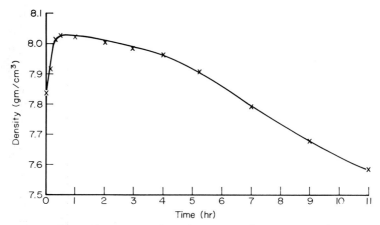

FIGURE 8-11 Sintered density of compacts from electrolytic copper powder pressed at 520 MPa (75,000 psi) and sintered at 900°C as a function of sintering time.

the sintered density is equal to the green density. In Figure 8-11 the changes in density as a function of time at a constant sintering temperature of 900°C are shown for compacts of copper powder pressed at a pressure of 520 MPa (75,000 psi). The density first increases, reaches a maximum at a sintering time of 30 minutes, then decreases below that of the green compacts with a gradually decreasing slope. An unusual case of expansion during sintering is shown in Figure 8-12 from the early work of Trzebia-towski.[8] An extremely fine copper powder (particle size about 1 μm) was

222

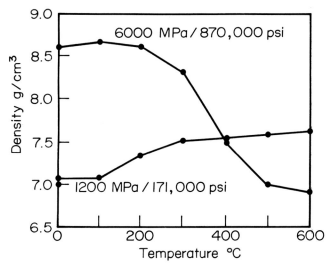

Figure 8-12 Density of compacts from a copper powder with 1 μm particle size, pressed at 1200 and 6000 MPa respectively, sintered for one hour in vacuum as a function of sintering temperature.

pressed at pressures of 1200 and 6000 MPa (171,000 and 870,000 psi). The compacts were sintered for one hour in vacuum at temperatures up to 600°C (1100°F). The compacts pressed at 6000 MPa (870,000 psi) decreased in density with increasing sintering temperatures from 8.65 to less than 7g/cm³. The growth of the copper powder compacts at certain sintering times and temperatures must be attributed to gas expansion. Gas entrapped in pores, closed-off during compacting, would consist of nitrogen and oxygen with the nitrogen expanding and oxygen forming copper oxide. For compacts sintered in hydrogen the expanding gas may also be water vapor formed by reaction of the hydrogen diffusing through the copper lattice and the oxygen or oxide entrapped in the pores. A semi-quantitative explanation of changes in density with sintering time in isothermally sintered copper powder compacts will be presented in chapter 9.

The examples illustrating the changes in dimensions and in density of compacts during sintering as a function of particle size of the powder, compacting pressure, and sintering time and temperature were chosen from papers in which the authors were primarily concerned with the theory of sintering. Many compacts exhibit large linear shrinkages during sintering, particularly those from fine powders, such as carbonyl iron or nickel powders, or reduced molybdenum or tungsten powders, those pressed at low pressures having low green densities and those sintered for long times at temperatures near the melting point of the metal. In the technology of P/M parts from single metal powders such high shrinkage and the corresponding increase in density during sintering may be desirable in certain cases, e.g. in tungsten powder compacts which, after sintering, are swaged and drawn into incandescent lamp wire. They are pressed from very fine tungsten powder, are sintered quite near the melting point of tungsten and shrink about 15% linearly during sintering, which corresponds to an increase in density from $11g/cm^3$ as pressed to $18g/cm^3$ as sintered.

In the fabrication of structural parts from single metal powders or from homogeneous solid solution alloy powders the aim is also to obtain high sintered densities. As will be discussed in more detail in a later section of this chapter, high sintered density generally means superior mechanical properties. However, in structural parts the control of final parts dimensions is also an important requirement. For this reason, it is not desirable to have the green compacts shrink large amounts during sintering. With such large changes in dimensions the final dimensions of the part are difficult to control and the parts may warp during sintering. In other words, the compacts should have high green density and should undergo little dimensional changes during sintering. Compacts with high green densities can be pressed by using powders with high compressibility and by pressing with high compacting pressures. This is the road generally taken in producing iron and steel structural parts. Gas entrapment is no problem when iron powder compacts are pressed at high pressures; when the compacts are sintered in the γ-range of iron, shrinkage is quite low, as illustrated in the shrinkage curves for iron. For this reason, it is often possible to sinter iron and steel structural parts under conditions where shrinkage during sintering is small, usually less than 0.5%. In structural parts from other powders, such as copper powders, brass and nickel silver powders and stainless steel powders, it is often not possible to obtain desirable mechanical properties in the finished parts without some shrinkage during sintering. Particularly, compacts from copper and some of the copper alloy powders cannot be pressed at high compacting pressures without encountering difficulties with growth of the compacts during sintering, as outlined above. They are therefore pressed at lower pressures and will shrink during sintering. In order to obtain satisfactory control of dimensions in the final parts, they are often sized or repressed after sintering.

III.   The Effect of Sintering Upon Microstructure

When a green metal powder compact is carefully sectioned, polished and etched and its microstructure examined, the original powder particles, with their boundaries outlined, and the pores in the structure can be observed. In compacts from powders with a wide particle size distribution, this distribution is also seen in the structure of the green compacts. Larger powder particles in the structure may appear flattened and distorted, the more so the softer the powder and the higher the compacting pressure. If the particles are polycrystalline, grain boundaries within the particles may be seen. An exception are compacts from tin powder pressed at 415 MPa (10,000 psi) at room temperature. Kuzmick[9] observed a recrystallized structure in which the original particle boundaries were obliterated. He postulated that tin with its low melting point recrystallizes at room temperature. This is not observed in compacts from powders of metals with higher melting points.

When a compact is sintered, there is a transition in structure. The original particle boundaries can no longer be observed. Instead, the structure becomes similar to that of the metal in the wrought and annealed condition, except that it contains pores. It consists of an array of equiaxed grains separated by grain boundaries. The transition in grain structure as a function of sintering temperature is illustrated in Figures 8-13, 8-14 and 8-15. They show micrographs at 180x of the structure of compacts from Ancorsteel 1000 (atomized iron powder) pressed to a density of 6.8 g/cm^3. Figure 8-13 represents the unetched structure of a compact sintered 30

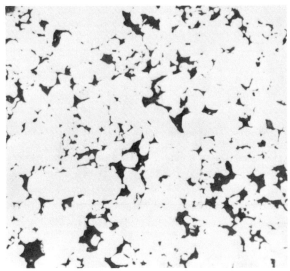

FIGURE 8-13 Microstructure of compact from atomized iron powder (Ancor 1000) pressed to a density of 6.8g. cm³, sintered 30 minutes at 1700°F (926°C) in dissociated ammonia, unetched, x180.

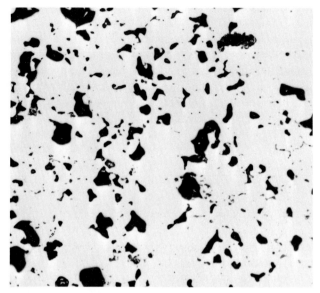

FIGURE 8-14 Microstructure of compact from atomized iron powder (Ancor 1000) pressed to a density of 6.8g/cm³, sintered 60 minutes at 2400°F (1315°C) in dissociated ammonia, unetched, x180.

FIGURE 8-15 Microstructure of the same compact as in Figure 8-14, etched with 2% nital, x180.

min. at 926°C (1700°F), Figure 8-14 that of a compact sintered 60 min. at
1315°C (2400°F), Figure 8-15 the structure of the same compact as in
Figure 8-14, but etched with 2% nital. The outline of the original iron
powder particles, particularly the small ones which are more or less spherical, is still clearly visible in Figure 8-13. They are no longer discernible in
Figure 8-14. Instead, as seen in Figure 8-15, a more or less continuous
structure of α-iron grains, formed during cooling from the sintering
temperature, has appeared.

The transition is, of course, also a function of sintering time as a
comparison of Figure 8-16 and 8-17 shows. Both represent the unetched
structure of compacts from the same powder pressed to the same density
and sintered at 1120°C (2050°F). The sintering time for the compact for
Figure 8-16 was 5 min. at temperature, that of the compact for Figure 8-17
40 min. at temperature. Many original particle boundaries can be seen in
Figure 8-16, while much fewer of these boundaries remain in Figure 8-17.

The transition in microstructure of compacts from a reduced copper
powder pressed at 550 MPa (80,000 psi) has been studied by Bier and
O'Keefe.[10] They found that in compacts sintered 30 minutes at 900°C
(1650°F) particle boundaries were still discernible, while during sintering
for 12 minutes at 955°C (1750°F) a typical sintered microstructure was
developed, in which by merging of particles the particle boundaries had
disappeared and a new grain structure developed by subsequent grain
growth. Sintering for 60 minutes at 955°C (1750°F) did not cause much
additional grain growth, but sintering at 1010 or 1055°C (1850 or 1930°F)
increased the grain size of the sintered microstructure.

In his early work on sintering, Sauerwald[11,12,13] contrasted the phenomena of recrystallization and grain growth in metals which are cold
worked and annealed and in metal powder compacts during sintering. In
cold worked and annealed metal the temperatures at which recrystallization and subsequent grain growth occur are the lower the larger the
amount of cold work. Sauerwald therefore expected that during sintering
of compacts recrystallization and grain growth should be observed at
lower temperatures for compacts pressed at higher pressures in which the
metal would be more severely strain hardened. In his metallographic
examination of sintered metal powder compacts he observed, however,
that the appearance of a sintered microstructure was more or less independent of the compacting pressure and occurred roughly at two thirds of
that of the melting point, generally well above the temperatures at which
recrystallization and grain growth are usually observed in cast and worked
metals.

The reason for the observation is that grain growth across powder
particle boundaries is restricted, until sintering has progressed to the point

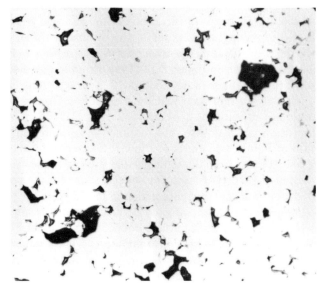

FIGURE 8-16 Microstructure of compact from atomized iron powder (Ancor 1000) pressed to a density of 6.8g/cm³, sintered 5 minutes at 1120°C (2050°F) in dissociated ammonia, unetched, x180.

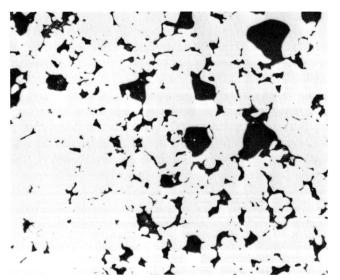

FIGURE 8-17 Microstructure of compact from atomized iron powder (Ancor 1000) pressed to a density of 6.8 g/cm³, sintered 40 minutes at 1120°C (2050°F) in dissociated ammonia, unetched, x180.

where a substantial increase in the contact area between particles has taken place. This increase in contact area is impeded by the network of voids. It is not responsive to those effects of cold work during compacting which are relieved at temperatures well below those at which the development of sintered microstructures is observed. The principal effect of higher compacting pressure upon grain growth during sintering is to facilitate the establishment of extensive contacts.

It is now quite clear that Sauerwald's original statement is much too sweeping. The establishment of a sintered microstructure is dependent in many cases upon particle size of the powder, the compacting pressure and the purity of the powder in addition to sintering temperature. In tungsten powder compacts Smithells et al.[14] observed an influence of compacting pressure, with compacts pressed at 440 MPa (64,000 psi) showing grain growth at 927°C, those pressed at 110 MPa (16,000 psi) showed incipient grain growth only at 1227°C.

The fact that the temperature at which a sintered microstructure is observed is well above that at which cold work is relieved by recovery and recrystallization can be readily shown by examining X-ray diffraction line profiles. Observations of this kind go back to the work of Trzebiatowski.[15] For compacts from tungsten powders having particles in the micron size range exaggerated grain growth (secondary recrystallization) during sintering in the temperature range from 2500 to 2800°C is commonly observed.[16] This exaggerated grain growth is much more pronounced in compacts from fine tungsten powder with particle sizes of 1 μm or less. Compacts from these powders readily develop grain sizes with several mm diameter. Compacts from coarser powder with particles several μm diameter end up with a structure having much smaller grain size after sintering under the same conditions.

The shape of the pores in compacts sintered at high temperatures or for long times show a characteristic change, which can be observed in many photomicrographs of compacts. In order to illustrate these changes a series of photomicrographs of artificial compacts taken by Alexander and Balluffi[17] are shown in Figure 8-18 and Figure 8-19. The compacts were produced by winding several layers of copper wire, 0.0128 cm in diameter, upon a copper rod and sintering the resulting spool at 1075°C, quite near the copper melting point, for times from 4 hours to more than 408 hours in hydrogen. When these artificial sintered compacts are sectioned, the pores first have a cusp-shaped cross section. In the direction perpendicular to the plane of sectioning the pores are, of course, continuous. The artificial compacts are therefore quite different from ordinary compacts produced by pressing, not only because of the peculiar shape of the pores, but also because the pores are originally all of the same diameter, compared with

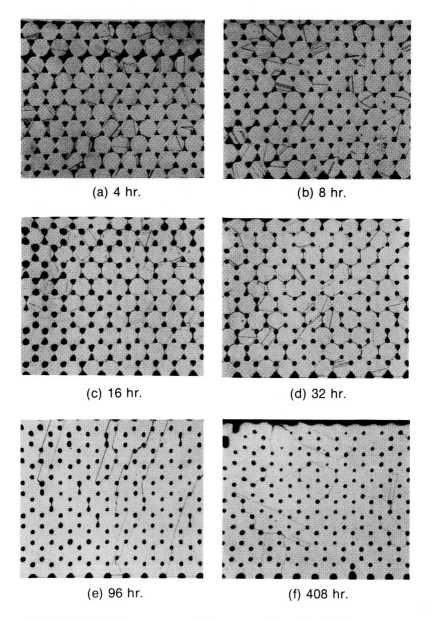

(a) 4 hr. (b) 8 hr.

(c) 16 hr. (d) 32 hr.

(e) 96 hr. (f) 408 hr.

FIGURE 8-18 Microstructure of the cross-sections of artificial compacts produced by winding 0.128 mm dia. copper wire on a copper spool and sintering at 1075° C for various lengths of time (a) 4 hr, (b) 8 hr, (c) 16 hr, (d) 32 hr, (e) 96 hr, (f) 408 hr, x50.

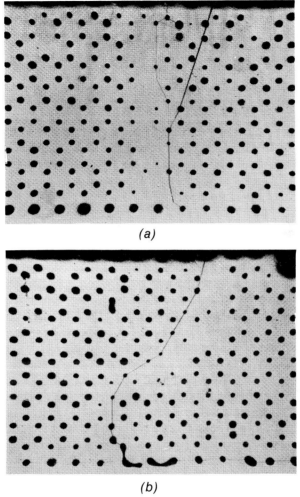

(a)

(b)

FIGURE 8-19 Enlarged photomicrographs of the artificial compact of
Figure 8-18 f (408 hr at 1075°C) showing elimination of voids in the vicinity
of isolated grain boundaries, x80.

the wide variation in pore size in normal compacts. However, because of
the original uniform size of the pores, the changes in pore geometry can be
particularly well observed in the artificial compacts. As the sintering time
is increased to 32 hours the pore cross sections become more and more
circular and, at the same time, diameters of the circles become smaller, the
compacts shrink. The pores with the circular cross sections are connected

with each other by grain boundaries. At sintering times longer than 32 hours, the grain boundaries are gradually swept out and the structure of the artificial compacts approaches that of a single crystal in which pores with circular cross sections are embedded. When grain boundaries are eliminated, the shrinkage of the pores slows down very markedly. As seen in Figure 8-19, only in areas, where a grain boundary was preserved by chance, do the pores continue to shrink and eventually disappear. This observation is of great importance in the understanding of the mechanism of sintering.

In the discussion of the work of Cizeron and Lacombe[7] on density changes during the sintering of carbonyl iron powder compacts, it was shown that compacts sintered at 870°C below the α-γ transformation temperature shrink much more than those which were first heated for 40 minutes to 920°C above the transformation temperature, then cooled below this temperature and held at 870°C in the α-range for many hours. The compacts heated to 920°C develop, by nucleation and growth of grains of the γ-phase, a much larger grain size than that of the α-iron structure before transformation from α to γ. When the compacts are cooled from 920°C to 870°C and an α-iron structure is reestablished, its grain size is also considerably larger than the original α-iron structure before the excursion in the γ-range. In the original α-iron structure most pores were located at grain boundaries. After the excursion in the γ-range there are many pores inside the grains in the α-iron structure. Much of the slow rate of shrinkage in the compacts which had been subjected to the excursion into the γ-range must, therefore, be attributed to the development of the coarser α-iron structure with pores within the grains rather than at grain boundaries.

IV. The Effect of Sintering Upon Mechanical Properties

In the early studies of sintering it was customary to determine the mechanical properties of sintered compacts as a function of compacting pressure, sintering temperature and sintering time together with the determination of sintered density as it depends upon processing parameters. Sauerwald and Kubik[18] and Bier and O'Keefe[10] presented data on compacts from copper powder, Eilender and Schwalbe[19] and Libsch et al.[20] on compacts from iron powder and Wulff[21] on compacts from stainless steel powder. In parallel with the increase in density with increasing compacting pressure, increasing sintering temperature and increasing sintering time, the mechanical properties, in particular tensile strength and elongation, also increase. However, when compacts are sintered for several hours, no further increases in strength and density are generally observed, but ductility continues to increase. With longer sintering times the pores in the compacts become more rounded, even when there is no further decrease in pore volume. The notch effect of irregular pores, which strongly affects ductility, is somewhat relieved when the pores become rounded.

The effect of compacting pressure upon mechanical properties was generally studied on the basis of plots of pressure vs tensile strength and other properties. Such a plot, taken from the work of Squire[22] for compacts from six different grades of iron powder, all of them sintered for 1 hour at 1100°C, is shown in Figure 8-20. For the same six powders Squire also plotted the density of the compacts, sintered under the same conditions, vs compacting pressure, as seen in Figure 8-21. A comparison of these plots shows considerable similarity. The higher the level of the curve for sintered density vs compacting pressure for a given powder, the higher is, generally, the level of the curve for tensile strength vs compacting pressure. Squire realized that, at least for compacts from powders which do not shrink very much during sintering, such as the iron powder investigated, two effects are combined in the tensile strength vs compacting pressure curves, that of the compressibility of the powder, which is not related to sintering, and that of sintered density upon mechanical properties. To eliminate the effect of compressibility, Squire proposed that tensile strength be plotted vs sintered density. This plot for the six grades of iron powder investigated is shown in Figure 8-22. As a first approximation, Squire suggested that the tensile strength of compacts from different iron powders sintered under the same conditions is primarily a function of sintered density and drew the curve shown in the figure. It is evident that there is still considerable scatter around this curve, although much less than in that for tensile strength vs compacting pressure. Repeated efforts have been made to establish mathematical relationships of the type

$$\sigma_p/\sigma_o = \exp(-kP)$$

between the ratio of the tensile strength of a porous compact, σ_p, that of the solid metal, σ_o, and the porosity of the compact P. The fact that no single value of the constant k describes the relationship for all powders or even for powders of one metal, such as iron, has been discussed by Exner and Pohl.[23] Squire's data (see Figure 8-22) show that the strength values for powder 4, which is finer than the other iron powders, lie above the average curve. Powders with rough particle surfaces also develop higher levels of sintered strength than those from other grades.[24] This is of practical importance when powders produced by different methods are compared with each other.[25] Exner and Pohl[23] developed equations, in which the notch effect of the pores was considered in the relationship between

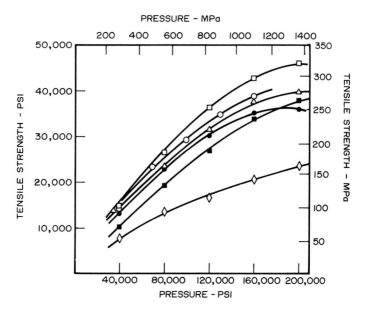

FIGURE 8-20 Tensile strength of compacts from six different grades of iron powder, sintered 1 hour at 1100°C as a function of compacting pressure.

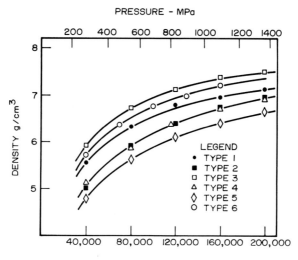

FIGURE 8-21 Density of compacts from the same six grades of iron powder as in Figure 8-20, sintered 1 hour at 1100°C as a function of compacting pressure.

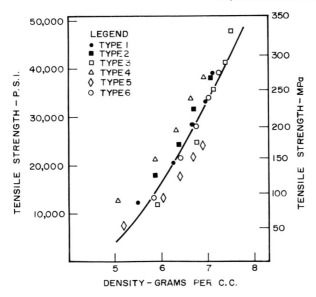

FIGURE 8-22 Tensile strength of compacts of the same six grades of iron powder as in Figure 8-20, sintered 1 hour at 1100°C as a function of sintered density.

relative tensile strength and porosity by including such parameters as the mean pore intercept length and obtained considerably better agreement between data and equation.

Plots of hardness, elongation and impact energy as determined on unnotched Charpy specimens vs sintered density for compacts from the same grades of iron powder sintered under the same conditions are shown in Figures 8-23, 8-24 and 8-25. The plots for tensile strength and hardness do not show the same knee as those for elongation and impact energy. Tensile strength and hardness increase steadily with increasing sintered density. Even at a relative density of only 90% of theoretical, i.e. 7.1 g/cm^3, the values for tensile strength, which range from 35 to 39,000 psi, are not too far below those of fully dense Armco iron or low carbon steel, i.e. 45,000 psi. On the other hand, much higher relative densities are needed to obtain values of elongation, near those of fully dense iron. R. Haynes[26] has recently derived a semiempirical equation for the relationship between relative ductility expressed as a fraction of fully dense material vs fractional porosity; it fits the experimentally found curves fairly well. Impact energy, when determined on unnotched Charpy specimens is affected even more drastically by small amounts of porosity than elongation. If the Charpy values had been determined on the usual notched rather than on unnotched specimens, densities of well over 99% of theoretical would have been necessary to obtain values approaching those of pore-free specimens.

The data in Figures 8-22 to 8-25 were determined on compacts sintered one hour at 1100° C. Raising the sintering temperatures or increasing the sintering time will have relatively little effect on tensile strength when compared on the basis of equal sintered density, but will noticeably increase the values for elongation.

Squire's method of comparing mechanical properties on the basis of sintered density is now widely used for the properties of compacts from different powders and sintered under different conditions. For the fabrication of parts it is, of course, of primary importance to know how a high sintered density is best obtained, e.g. by choosing highly compressible powders, using high compacting pressures, repressing and resintering after the first sintering step, or even by sintering under conditions at which the compacts shrink.

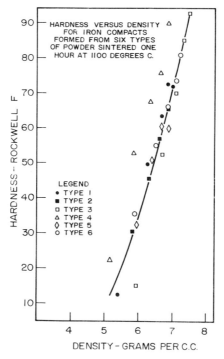

FIGURE 8-23 Rockwell F. hardness of compacts from the same six grades of iron powder as in Figure 8-20, sintered 1 hour at 1100° C as a function of sintered density.

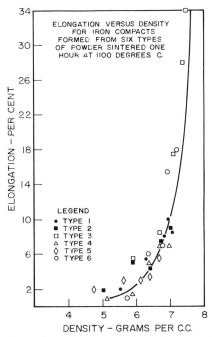

FIGURE 8-24 Elongation of compacts from the same six grades of iron powder as in Fiugre 8-20, sintered 1 hour at 1100°C as a function of sintered density.

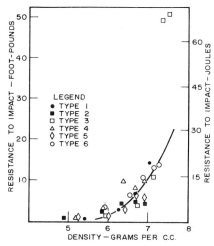

FIGURE 8-25 Unnotched Charpy impact energy of compacts from the same six grades of iron powder as in Figure 8-20, sintered 1 hour at 1100°C as a function of sintered density.

Conventional indentation hardness measures not only the resistance to indentation of the metal itself, but also of the pore volume, which does not resist. The Vickers, Rockwell or Brinell hardness of a compact is therefore affected by both the density of the compact and the inherent hardness of the structure of which the compact consists. The hardness of compacts pressed at low pressure will increase with increasing sintering temperatures because the compact shrinks and gets denser. Compacts pressed at high pressures will often first decrease in hardness with increasing sintering temperature because the cold work introduced during compacting will be relieved. At still higher temperatures the hardness will increase again. This is illustrated in Figure 8-26[5] in which the hardness of compacts from carbonyl nickel powder pressed at three different pressures is plotted as a function of sintering temperature for a sintering time of 2 hours.

When very light loads are used in indentation hardness measurements as in microhardness tests, the indentation is often located away from pores and the inherent hardness of the structure can therefore be determined, at least approximately. This is illustrated in Figure 8-27,[27] in which the hardness of iron powder compacts of varying relative densities, measured on the conventional Rockwell F scale and with a micro-hardness tester, are compared with each other. The fact that the inherent hardness of a porous structure can be determined by microhardness tests has been made use of by applying it in specifications, particularly those for surface hardened, i.e. carburized or carbonitrided porous materials. An example is an interna-

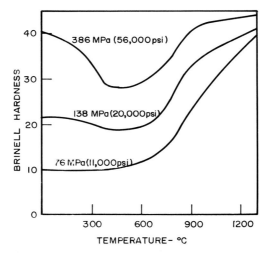

FIGURE 8-26   Brinell hardness of compacts from carbonyl iron powder pressed at 76, 138 and 386 MPa, sintered for 2 hours as a function of sintering temperature.

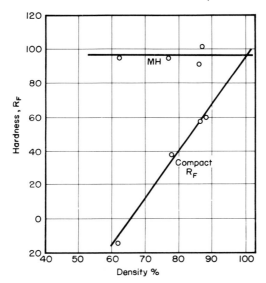

FIGURE 8-27 Rockwell F and microhardness of iron powder compacts as a function of density expressed as percent of solid density.

tional standard[28] for the determination and verification of effective case depth by the Vickers microhardness testing method for sintered ferrous materials, carburized or carbonitrided. According to this standard, the effective case depth and total case depth are determined using Vickers microhardness readings with a load of 0.9807 N.

References

1. F. N. Rhines, Seminar on the theory of sintering, Trans. AIME, vol. 166, p. 474-487, (1947).
2. W. D. Jones, Fundamental Principles of Powder Metallurgy, London, p. 242-260, (1960).
3. C. G. Goetzel, Treatise on Powder Metallurgy, vol. 1, p. 267, N.Y., (1949).
4. P. Duwez and H. Martens, A dilatometric study of the sintering of metal powder compacts, Trans. AIME, vol. 185, p. 571-577, (1949).
5. G. Grube and H. Schlecht, Sintering of metal powders and properties of metal compacts, Z. Elektrochemie, vol. 44, p. 367-374, (1938).
6. R. F. Cheney, G. T. E. Sylvania, Towanda, Pa., Lecture on sintering of tungsten and molybdenum, (Oct. (1970).
7. G. Cizeron and P. Lacombe, A dilatometric study of the isothermal sintering of carbonyl iron powder, Comptes Rendues, vol. 241, p. 409-410, (1955).
8. W. Trzebiatowski, Solidification in pressed metal powders, Z. physik. Ch., vol. B24, p. 87-97, (1934).

9. J. F. Kuzmick, Some experiments on the effect of pressure on metal powder compacts, Trans. AIME, vol. 161, p. 612-630, (1945).

10. J. Bier and J. F. O'Keefe, The sintering of metal powder - copper, Trans. AIME, vol. 161, p. 596-611, (1945).

11. F. Sauerwald, On synthetic metal bodies I, Z. f. Anorg. Ch., vol. 122, p. 277-294, (1922).

12. F. Sauerwald, Synthetic metal bodies III, Z. f. Metallkunde, vol. 16, p. 41-46, (1924).

13. F. Sauerwald, Lehrbuch der Metallkunde, Berlin, p. 21, (1929).

14. C. J. Smithells, W. R. Pitkin and J. W. Avery, Grain growth in compressed metal powder, J. Inst. of Metal, vol. 38, p. 85-97, (1927).

15. W. Trzebiatowski, Solidification of pressed metal powders, Z. physik. Ch., vol. B24, p. 75-86, (1934).

16. C. J. Smithells, Tungsten, 3rd ed., London, p. 131, (1952).

17. B. H. Alexander and R. W. Balluffi, The mechanism of sintering of copper, Acta Met., vol. 5, p. 666-677, (1957).

18. F. Sauerwald and S. Kubik, On synthetic metal bodies VI, Z. f. Elektro-chemie, vol. 38, p. 33-41, (1932).

19. W. Eilender and R. Schwalbe, Arch.f.d.Eisenhüttenwesen, vol. 13, p. 267-272, (1939).

20. J. Libsch, R. Volterra and J. Wulff, The sintering of iron powder, "Powder Metallurgy," ed. by J. Wulff, Cleveland, p. 379-394, (1942).

21. J. Wulff, Stainless steel powder, ibid., p. 137-144.

22. A. Squire, Density relationship of iron-powder compacts, Trans. AIME, vol. 171, p. 485-503, (1947).

23. H. E. Exner and D. Pohl, Fracture behavior of sintered iron, Powder Metal-lurgy Int., vol. 10, No. 4, p. 193, (1978).

24. F. V. Lenel, H. D. Ambs and E. Lomerson, "Symposium on Powder Metallurgy," The Iron and Steel Inst., Special Report, No. 58, London, p. 90-96, (1956).

25. R. B. Gold, Precision Metal,, p. 37, (June 1977).

26. R. Haynes, The study of the effect of porosity content on the ductility of sintered metals, Powder Metallurgy, vol. 20, No. 1, p. 17-20, (1977).

27. R. Steinitz, Metals and Alloys, vol. 17, p. 1183, (1943).

28. ISO International Standard 4507-1978E.

9

Sintering of a Single Phase Metal Powder — Mechanism of Sintering

The study of the mechanism of sintering of single phase metal powder aggregates and compacts[1,1A] has been concerned primarily with the geometrical changes occurring rather than changes in physical and mechanical properties. These geometrical changes can be divided into three stages of sintering. Such a division is helpful in this discussion even though the stages are not clearly separated from each other, but blend into each other.

In the first stage the initial relatively small contact areas (necks) between powder particles expand. This also involves changes in shape of the pores which become rounded. Simultaneously, the powder aggregate begins to densify which means a decrease in total void volume and a decrease in the center to center distance between particles. In spite of neck growth and initial shrinkage the particles in the original powder aggregate are still distinguishable.

In the second or intermediate stage the particles can no longer be distinguished, the pore channels in the powder aggregate gradually pinch off and close. In this stage migration of the grain-boundaries between the original particles by grain-growth becomes possible. On the other hand, the pores continue to form a more or less connected continuous phase throughout the aggregate.

In the third stage the pores become isolated and are no longer interconnected.

An at least semiquantitative delineation of the stages is possible on the basis of the topological approach to sintering by Rhines and his school.[2,3,4] This approach, which is briefly introduced here, is based on the concept of the genus of a complex surface. The genus is equal to the number of cuts which must be made to reduce the surface to a form which can be made spherical merely by distortion of the surface. Accordingly, the genus of a sphere is 0, but so is the genus of a table with four legs since it can be distorted into a sphere without making any cuts. The genus of a doughnut is 1 since one cut must be made in order to make it possible to distort the surface into a sphere. The genus of an aggregate of particles (assumed to be spherical for simplicity's sake) which touch each other is illustrated in Figure 9-1. In general, the genus of an aggregate containing P particles which make a total of C contacts is

$$G = C - P + 1$$

Based on the genus concept, the first stage is the one in which the genus stays constant or may even slightly increase because new necks are established by

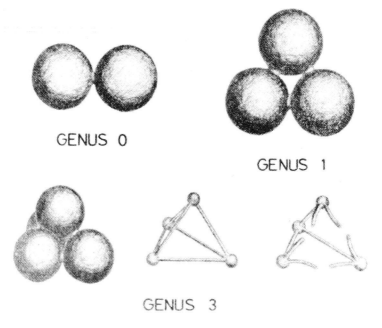

GENUS 0

GENUS 1

GENUS 3

FIGURE 9-1 Illustration of the genus of particle aggregates with genus zero, one and three.

rearrangement of particles. In the second or channel closing stage of sintering G decreases continually. In the third or isolated pore stage of sintering, the genus again stays more or less constant, except when a pore isolated by shrinkage disappears completely.

Rhines' topological approach is well suited to follow the extremely complex changes in the geometry of sintering aggregates from commercial powders which usually have irregular shapes and a wide particle size distribution. Up to the present, efforts to describe the mechanism of sintering using this approach have remained primarily qualitative. A few conclusions based on this approach which refer to the channel closure stage of sintering will be presented in connection with a discussion of this stage.

The initial stage of sintering has received the greatest attention of those who base their approach to sintering theory upon fundamental thermodynamic and atomistic ideas and try to describe quantitatively the driving forces and the material transport mechanisms involved. The discussion of the initial stage in section I of this chapter will therefore be based on this approach. The treatment of the later stages of sintering in section II will include not only the fundamental approach, but also some of the findings of Rhines' topological approach and the phenomenological approaches to sintering theory of the Finnish and Soviet groups in which quantitative relationships of the changes in porosity and volume are developed analytically. A third section will be concerned with a semiquantitative treatment of expansion due to entrapped gases during sintering.

I. The Initial Stage of Sintering

In the experimental study of the initial stage of sintering the principal research effort has been on model experiments, such as sintering a sphere to a sphere, a sphere to a plate, arrays of two or more wires, or wires wound on a spool rather than on sintering actual powder compacts.

There has been a tendency to generalize the results of such model experiments and to assume that a quantitative explanation of a model experiment, such as sintering together two spheres, also means a quantitative explanation of all sintering phenomena. This tendency to generalize must be guarded against. On the other hand, without the model experiments the advancement to the present stage of understanding would have been impossible. The initial stage of sintering involves, first of all, establishing contact between powder particles. What occurs during contact formation has been relatively little studied. It will be briefly presented in connection with the discussion of plastic flow as a transport mechanism during sintering. Also included in the initial stage of sintering are:

     (a) increase in contact area and pore rounding and

(b) shrinkage of the powder aggregates involving center to center approach of particles.

One could image that phenomenon (a) could occur alone without phenomenon (b). This would mean that the shape of the pores in the compact would change until all of them are isolated and completely spherical, but without any change in pore volume. Although in actual sintering, phenomena (a) and (b) generally occur simultaneously, it is sometimes worthwhile to consider the two phenomena independently.

The principal driving force for sintering is the lowering of free energy when particles grow together, voids shrink and the total surface area of the aggregate decreases. The specific energy associated with surface has dimensions of energy per unit area or of force per unit length and is commonly called surface tension.

A. Driving Forces

Surface tension forces, often also called capillary forces, are well known in liquids, including liquid metals (spherical shape of mercury droplets). Surface tension forces in solids cause stresses which are related to the curvature of the surface according to LaPlace's equation:

$$\sigma = \gamma \left(\frac{1}{r_1} + \frac{1}{r_2} \right)$$

in which σ is the stress, γ the surface tension and r_1 and r_2 the two radii of curvature of the surface. Under convex surfaces, for which the curvature is negative, the stress is tensile ($\sigma < 0$), for concave surfaces with a positive curvature the stress is compressive. When the LaPlace equation is applied to crystalline solids, it is assumed that the surface tension or specific surface energy is isotropic and insensitive to strain rate. For solids at high temperature and particularly crystals with cubic symmetry the anisotropy of surface tension can generally be neglected. The insensitivity to strain rate is justified by assuming that, at the sintering temperature, the atomic mobility is high enough so that perturbation in surface structure are rapidly relaxed by local adjustment.

Acceptable accurate measurements of the surface tension of solid metals were first made by Udin, Shaler and Wulff[5] on copper. They determined the dimensional changes in copper wires to which varying weights had been attached at temperatures near the melting point of copper. Applying the LaPlace equation to the thin copper wires the surface tension of solid copper can be calculated directly from the force applied by the attached weight, if the weight is so chosen that the copper wire neither shortens (due to surface tension) nor elongates (due to the applied tensile stress). The equilibrium force is given by

$$w = \gamma.r.\pi$$

where w is the force in newtons applied by the weight
   r is the diameter in meters
   γ is the surface tension in N/m

This equation is based on the assumption that the interfacial energy of the grain boundaries in the wire is negligibly small compared with the surface tension. Later research showed that this assumption is not tenable and that the interfacial tension is on the order of 1/3 of the surface tension. Fortunately, the approximate number and area of the grain boundaries in the wires of the Udin, Shaler and Wulff experiment were known. The wire had a "bamboo" structure, all grain boundaries were oriented in planes perpendicular to the axis of the wire. Therefore Udin was able to correct the earlier values taking into account the interfacial tensions. Udin's corrected equation is:

$$w = \pi r \gamma - n/l.\pi r^2 \gamma_{GB}$$

where n/l is the number of grains per unit length of wire and γ_{GB} is the interfacial tension between grains. Making the assumption that the interfacial tension is one third of the surface tension, Udin determined the surface tension of solid copper near its melting point as 1.65 N/m in place of the original value of 1.37 N/m.

In single phase systems the stress defined by the LaPlace equation causes a gradient in chemical potential $\Delta\mu$ between surfaces with different radii of curvature:

$$\Delta\mu = \sigma\Omega$$

where Ω is the atomic volume.

LaPlace's equation may be used to calculate the stresses and chemical potential gradients due to surface tension forces in model experiments designed to determine the mechanism of material transport in sintering. In the arrangement shown in Figure 9-2, in which two spheres are sintered together, the neck surface is characterized by 2 radii, the radius ρ with a negative sign because radii measured outside the condensed phase are taken negative and the radius x, positive because the radius is measured in the condensed phase. We, therefore, write for the stress

$$\sigma = \gamma \left(-\frac{1}{\rho} + \frac{1}{x} \right)$$

Since $x \gg \rho$ the term $1/x$ can be neglected compared to $1/\rho$ and the stress taken as $\sigma = -\gamma/\rho$ with the negative sign, indicating a tensile stress.

The corresponding chemical potential gradient may be written

$$\mu - \mu_o = \sigma\Omega = -\gamma\Omega/\rho$$

in which μ is the chemical potential over the convex surface with the curvature ρ which is under the tensile stress $\sigma = -\gamma/\rho$ and μ_o is the chemical potential at the adjacent flat surface which is unstressed.

In the early stages of sintering, forces other than the surface tension forces may contribute to atom movements in sintering. They are gravity forces and forces due to residual stresses in compacts. Lenel, Hausner, Roman and Ansell[7] attempted to determine the relative contributions of gravity forces compared to surface tension forces experimentally. They compared, for loose powder aggregates and for compacts from copper powder sintered at a series of temperatures, the shrinkage in the radial and axial directions, and the radial shrinkage at the top and bottom of the aggregates and the compacts. In an analysis of the shrinkage results on loose powder aggregates, Exner[8] showed that only a small part of the difference in shrinkage observed is actually due to gravity forces. When powder is compacted, residual stresses are induced at the surface of the compacts, which make a

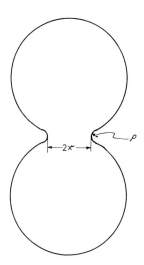

FIGURE 9-2 Two spheres sintering together, showing the diameter 2x of the neck between the particles and the radius ρ at the ends of the neck.

small contribution to the dimensional changes observed when the compacts are sintered at temperatures low enough so that the residual stresses are not yet relieved.[9] For most studies of dimensional changes in sintering, however, forces other than surface tension forces may be neglected.

B. Material Transport Mechanisms

Under the influence of the surface tension or capillary forces material is transported during sintering. Several mechanisms of material transport are customarily distinguished:

> Evaporation and condensation
> Viscous flow
> Diffusional flow
> Plastic flow

Before the mechanisms of diffusional flow, which is by far the most important in the sintering of metallic materials, is treated in detail, viscous flow and evaporation and condensation will be briefly discussed. Plastic flow is considered at the end of this section.

Evaporation and Condensation

For a model system consisting of two spheres (Figure 9-2) the gradient in chemical potential $\mu-\mu_0$ between the highly convex surface with the radius ρ and the adjacent flat surface with radius $r = \infty$ causes a gradient in vapor pressure given by the Gibbs-Thomson equation:

$$\mu - \mu_0 = - \frac{\gamma\Omega}{\rho} = RT \, (\ln p - \ln p_0)$$

in which p and p_0 are the vapor pressures over the stressed (curved) and the unstressed (flat) surface respectively, R the gas content and T the absolute temperature. Because of the difference in vapor pressure, material will evaporate from the flat surface and condense on the curved surface. The rate at which this material transfer takes place can be calculated by combining the Gibbs-Thomson equation for the difference in vapor pressure with the Langmuir equation for the rate of evaporation as a function of vapor pressure.

Evaporation and condensation is important for crystalline solids with a high vapor pressure near their melting point. It has been confirmed as the material transport mechanism in the sintering of NaCl.[10] For most metals, evaporation and condensation does not play an important role because of the low vapor pressure of most metals at their melting point.

Viscous Flow

This mechanism was suggested by the Russian physicist, J. Frenkel,[11] as occurring in both crystalline and amorphous solids. In crystalline solids he attributed this viscous flow to the presence of lattice vacancies. However, Nabarro[12] showed that, in a single crystal lattice, vacancies cannot give rise to a flow described by viscosity. Nevertheless, the mechanism is important in the sintering of glasses. An equation for the rate of growth of the neck radius x between two glass spheres of radius a derived by Frenkel reads:

$$\frac{x^2}{a} = \frac{3}{2} \frac{\gamma}{\eta} t$$

in which γ is the surface tension, η the viscosity of the glass and t the time. Experiments have shown reasonable agreement with this equation when glass spheres are sintered together.

Diffusional Flow

Diffusional flow as a transport mechanism is based on the concept that a certain concentration of vacancies exists in the lattice of a metal. The vacancy concentration is a function not only of temperature but also of the chemical potential or stress to which the surface of the metal is subjected. Again considering the model of two spheres (Figure 9-2) the gradient in chemical potential $\mu - \mu_o$ between the highly curved surface with radius ρ and the adjacent flat surface with radius $r = \infty$ causes a gradient in vacancy concentration between the material below the two surfaces. Again applying the Gibbs-Thomson equation:

$$\mu - \mu_o = -\frac{\gamma\Omega}{\rho} = RT \, (\ln c - \ln c_o)$$

in which c and c_o are the vacancy concentrations below a stressed (curved) and an unstressed (flat) surface.

For $\dfrac{\gamma\Omega}{RT} \ll 1$ this may be written

$$\mu - \mu_o = -RT\frac{c-c_o}{c} \quad \text{or} \quad c - c_o = -\frac{\gamma \, c_o \, \Omega}{RT} \cdot \frac{1}{\rho}$$

The difference in vacancy concentration $c - c_o$ causes a flux of vacancies away from the neck surface towards the flat surface which, of course, is equivalent to a flow of atoms in the opposite direction. In order to determine the material transport by diffusion, it is necessary to combine the Gibbs-Thomson equation for the difference in vacancy concentration with Fick's first law for the rate of diffusional flow under a concentration gradient. To make this calculation the relationship between the diffusivity of vacancies D_{vac} and the volume diffusivity D of the material is needed.

$$D_{vac} = D \cdot c_o$$

in which c_o is the equilibrium concentration of vacancies (under a flat surface). The transport of atoms from the flat portion of the neck to the volume underneath the surface with the small radius causes the neck to grow. Making simplifying approximations regarding the geometry of the neck and assuming that the material transport occurs through volume diffusion, Kuczynski[13] derived an equation for the rate of growth of the neck between two spheres:

$$\frac{x^5}{a^2} = \frac{40 \, \gamma \Omega D}{RT} \, t$$

in which x and a are the radii of the neck and of the sphere respectively, γ is the surface tension, D is the volume diffusivity, Ω is the atomic volume and t is the time. The equation can be readily adapted to the case of sintering spheres to a flat plate, which Kucynski studied experimentally. He used 100 μm diameter copper spheres sintered to a copper plate and found that the measured rate of growth of the neck was, indeed, proportional to the fifth root of time and fitted the calculated equation reasonably well. Kuczynski also derived an equation for the rate of neck growth when the material transport from the flat surface to the neck takes place by surface diffusion rather than volume diffusion. In this case the rate of neck growth should be proportional to the seventh root of time. Figure 9-3 shows schematically the three types of material transport discussed so far: path (a) through the gas phase (evaporation and condensation), path (b) by surface diffusion and path (c) by volume diffusion from the flat surface to the neck. Kuczynski's experiments and calculations were an important first step in the understanding of the mechanism of material transport in sintering. However, even for simple model experiments his approach had to be modified in two respects:

> Additional diffusion paths involving the grain boundary have to be considered

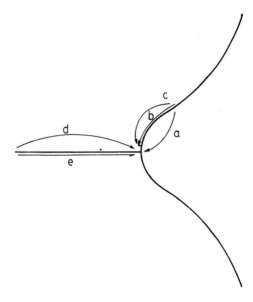

FIGURE 9-3 Types of material transport when two spheres sinter together: Path a) through the gas phase (evaporation and condensation); path b) surface diffusion from the flat surface to the neck; path c) volume diffusion from the flat surface to the neck; path d) volume diffusion from the grainboundary between the spheres to the neck; path e) grainboundary diffusion from the grainboundary to the neck.

The effect of the simplifying approximations in calculating rate equations has to be taken into account.

The idea that grain boundaries may act as vacancy sinks is based upon the work of F. R. N. Nabarro[12] and C. Herring.[14] They developed the concept that a shear stress applied to a polycrystalline solid causes a diffusion current from grain boundaries under a compressive stress to grain boundaries under a tensile stress as illustrated schematically in Figure 9-4. This is equivalent to a flux of vacancies proceeding in the opposite direction. The current or flux is proportional to the applied stress. The diffusional flow expected according to Nabarro and Herring is illustrated by flow path (d) in Figure 9-3. Vacancies flow by volume diffusion from the neck which is under the tensile stress of surface tension forces to the grain boundaries which are under compressive stress. There is a fourth diffusional flow path, in which vacancies flow from grain boundaries to the neck not by volume diffusion as in path (d), but by grain boundary diffusion, as in path (e). This last diffusional flow mechanism was first postulated by Coble.[15]

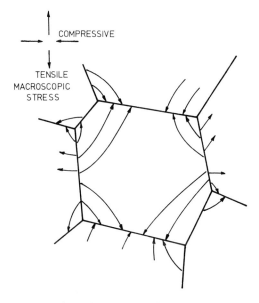

FIGURE 9-4 Diffusional flow from grainboundaries under a compressive stress to grainboundaries under a tensile stress as postulated by Nabarro and Herring.

Kuczynski's picture of diffusional flow from the flat particle surface to the neck can readily make neck growth plausible, but cannot explain how the centers of the particles approach each other and, therefore, how shrinkage of the compacts occurs. Compacts shrink by a mechanism in which vacancies flow from the pores as vacancy sources to the grain boundaries as vacancy sinks or, conversely, material from the grain boundaries is transported to the pores, which in this way are gradually filled. The validity of this concept is clearly demonstrated by the experiments discussed in chapter 8. Figures 8-13 and 8-14 taken from the work of Baluffi and Alexander[16] show that pores in artificial compacts made from copper wire shrink as long as they are connected with each other by grain boundaries but cease to shrink when the grain boundaries are swept out by grain growth. In the experiments of Cizeron and Lacombe[17] shrinkage in carbonyl iron compacts is very slow when the compacts have been heated into the gamma region. During the transformation from alpha to gamma iron a large-grained structure is formed with the pores inside the grains. Shrinkage is much faster for compacts in which the position of the pores at the boundaries of the alpha iron grains has been preserved through the sintering process.

The concept of grain boundaries as vacancy sources has played an important commercial role in the development of dense bodies of alumina ("Lucalox"). Only when the disappearance of grain boundaries by grain growth is inhibited by the addition of small amounts of dopants to the fine alumina powder compacts, can they be sintered to near theoretical density.

In view of the fact that there are four possible diffusional flow mechanisms:

> Vacancy flow from the neck to the flat surface on the sphere by volume diffusion
>
> Vacancy flow from the neck to the flat surface of the sphere by surface diffusion
>
> Vacancy flow from the neck to a grain boundary by volume diffusion
>
> Vacancy flow from the neck to a grain boundary by grain boundary diffusion,

it is clear that, in general, the rates of neck growth and of shrinkage in model experiments cannot be related to only one of the four mechanisms. This was clearly recognized by Rockland.[18,19] Johnson[20] was the first to show how the sintering of two spheres could be described quantitatively by taking into account simultaneously transport by surface, grain boundary and volume diffusion. This is possible by adding the individual contributions to the rate of reaction:

$$\frac{dx}{dt} = c_s \, f(x)_s + c_v \, f(x)_v + C_g \, f(x)_g$$

in which dx/dt is the rate of change of neck radius or of distance between the particle centers. In the constants c the physical constants (surface tension and diffusivities) and in the $f(x)$ the geometrical relationships for the three diffusional mechanisms, surface diffusion (subscript s), volume diffusion (subscript v) and grain boundary diffusion (subscript g) are comprised. A closed integration for all three kinds of diffusion has not yet been possible, the equations must be evaluated numerically. Ashby[21] has proposed another way to indicate the effect of the various transport mechanisms in his so-called sintering diagrams, illustrated by Figure 9-5. In these diagrams boundaries are shown as a function of sintering time, temperature and particle size along which two mechanisms contribute 50 percent each to material transport.

A second modification in developing sintering rate equations along the lines of Kuczynski's original approach concerns his simplifying geometrical approximations. One of these approximations, also used in the derivations of many later investigators of transport mechanisms, relates to the contour

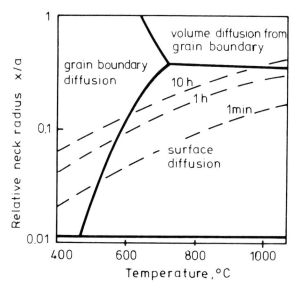

FIGURE 9-5 Sintering diagram according to Ashby for two copper spheres. The fully drawn lines are for sintering conditions for which the mechanisms stated in the neighboring fields contribute 50% each to material transport in the neck.

of the neck which is formed between two spheres. According to this approximation the neck contour consists of a convex surface having a circular arc section with radius ρ adjoining abruptly a concave surface having a circular arc section with the sphere radius a. This so-called "idealized geometry" cannot be stable from a thermodynamic point of view because of the sudden unsteady change from tensile stress in the convex neck contour to compressive stresses in the concave sphere contour. Instead, the contour should be determined by a step-wise calculation at small time intervals on a computer based on a numerical solution of the differential equations for material transport. Nichols[22,23] and Nichols and Mullins[24,25] first took this approach in deriving equations for surface diffusion and later Exner and Bross[26,27,28] expanded it. The method of determining material transport mechanisms by simply finding the exponents n and m in the equation relating neck radius x, sphere radius a and time t:

$$\frac{x^n}{a^m} = Ct$$

or similar exponents in the equations for the rate of shrinkage, is based on derivations using the idealized, but incorrect two circular arc contour of the neck. This also goes for the "scaling laws," a series of equations of the type

$$\Delta t_1 = \lambda^n \Delta t_2$$

which Herring[29] developed. In these equations Δt_1 and Δt_2 are the times required for a given change in shape to take place in two "clusters" which differ in linear dimensions by the ratio λ. "Clusters" refer to a particle or particles which have started to grow together. According to Herring, the exponent n is a function of the material transport mechanism. For viscous flow, n = 1, for evaporation and condensation, n = 2, for volume diffusion, n = 3 and for surface diffusion, n = 4. Derivations based on the true neck contour do not yield such characteristic constant exponents in the sintering equations. Determining material transport mechanisms on the basis of experimentally measured exponents is therefore wrong for two reasons:

1. The exponents measured are not constant, even if only one material transport mechanism contributes
2. Quite frequently more than one mechanism contributes.

As a matter of fact, the material transport mechanism in the original experiment of Kuczynski is probably surface diffusion rather than volume diffusion.[21] The calculated sintering rates for a two sphere model based on the true geometry of the model will be substantially different from those based on the "idealized geometry." These differences become much more important when an attempt is made to extend the calculations from a two sphere model to aggregates of larger numbers of particles, even spherical particles, where the presence of adjoining particles has constraining effects upon neck growth and shrinkage. Lenel and Eloff[30] called attention to these effects. They were studied quantitatively by Exner and associates[31] and are illustrated by the examples in Figures 9-6 and 9-7. Figure 9-6 shows a chain of three spherical copper particles approximately 50 μm in diameter sintered for 1 minute, 2 hours and 150 hours near the copper melting point. It can be clearly seen that, during the sintering, the chain of particles stretches, i.e. the angle between the lines connecting the centers of the particles becomes larger. For other arrangements, the angle may become smaller. In theoretical work it has been proven that such changes in angles are to be expected due to the asymmetry of the arrangement of the particles surrounding the necks. Figure 9-7 shows what this change in angles will do to a planar array of spherical copper powder particles when it is sintered for 2 and 12 hours at 1020° C. The area of the small pores decreases. Most of the triangular pores disappear but the area of the large pores grows steadily. Centers of densification are formed where rapid shrinkage takes place. This

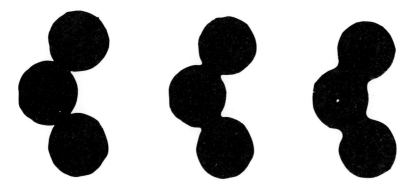

FIGURE 9-6 Sintering effect on an array of three spherical copper particles sintered at 1020°C, 300x (a) 1 minute, (b) 2 hours, (c) 50 hours. The angle between the particles, initially larger than 90°C, increases with increasing sintering time.

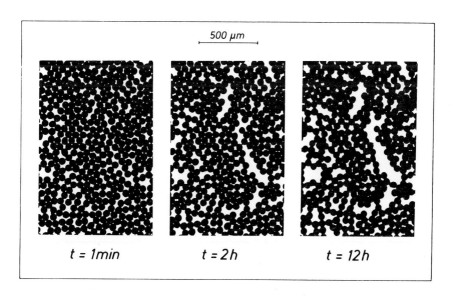

FIGURE 9-7 Sintering of planar layers of spherical copper powder at 1020°C in hydrogen, 65x (a) Original arrangement, (b) Sintered 2 hr., (c) Sintered 12 hr.

255

shrinkage is partially compensated by the opening of large pores. The constraints which surrounding particles exert upon sintering in a planar array of loose spherical powders are, of course, different from the constraints exerted by surrounding particles in a three-dimensional pressed compact of irregular particles, but there is clear evidence that these constraints exist and have an important effect upon sintering. Because of these constraints it seems rather doubtful that a complete calculation of neck growth and shrinkage in powder aggregates or compacts on the basis of known surface tension, diffusivities and initial geometry will ever be possible.

Plastic Flow

The question whether plastic flow, i.e. the transport of relatively large numbers of atoms accompanying dislocation motion as in slip, makes an important contribution to material transport in sintering, has been very controversial. For this reason, a discussion, at this point, of the basic mechanism of plastic flow in creep may be helpful. The phenomenon is different from the diffusional flow postulated by Nabarro[12] and Herring[14] which causes the shortening and lengthening of the wires in the Udin-Shaler-Wulff[5] experiment, although both "plastic flow" due to the motion of dislocations and Herring-Nabarro diffusional flow may cause creep, i.e. the plastic deformation of a solid under a constant applied stress. Which type of creep predominates will depend on the temperature and stress range of the experiment. Models for creep by dislocation motion are based on the concept of dislocation climb, i.e. the motion of dislocations in a direction perpendicular to its slip plane. Dislocation climb is a recovery process. During creep, recovery by climb produces a substructure which is in a state of dynamic equilibrium, i.e. at any instant the substructure is equivalent to the structure at any other period during steady state creep. According to a model developed by Weertman,[32,33] dislocations are generated under the action of an applied stress from sources in the crystal. The dislocations glide on their slip planes out from the dislocation sources. Eventually they interact with dislocations generated from sources on other slip planes and are blocked from further glide. They then climb towards one another and annihilate. As soon as a dislocation is annihilated, a new dislocation can be generated. Hence the concept of dynamic structural equilibrium, since the number of dislocations moving in the solid is constant. This mechanism involves the operation in sequence of several processes, dislocation generation, motion in the slip plane and dislocation climb.

In order to have appreciable material transport by a dislocation mechanism, it is not sufficient to show that dislocations which happen to exist in the neck area move under the influence of shear stresses. Any transport due to motion of existing dislocations would be minimal. It must also be shown that new dislocations can be generated by surface tension forces.

The first possibility would be that there are dislocation sources in the neck, e.g. the so-called Frank-Reed sources which consist of dislocations lying on a plane containing its Burgers vector pinned at two points by dislocation segments out of the slip plane. When a shear stress is applied to such a source, the dislocation line bows, expands, wraps around the pinning points and joins forming a new dislocation. A calculation shows that, even if the Frank-Reed sources happen to be located quite near the neck surface, the shear stresses available are smaller than the critical stress necessary to activate the source. This means that the dislocations necessary to provide material transport must be generated at a free surface.

The rate of dislocation generation at a free surface as a function of the stress due to surface tension has been calculated.[34] According to this calculation only at surfaces with sufficiently small radii of curvature, e.g. 40 nm, would the stresses be high enough so that dislocations are generated at a high enough rate to cause plastic deformation.[35] Experiments to determine whether dislocation generation and motion takes place during the early stages of sintering have not been entirely conclusive.[36] In the absence of applied external forces, plastic flow may be expected to contribute to material transport in sintering only in the very early stages of sintering when the radii of curvature at the cusps of the pores are small. When the radii of curvature become larger, the stresses due to surface tension forces become too small for dislocation generation. However, when pressure is applied during sintering, as in hot-pressing, plastic flow involving dislocation generation and motion may become the predominant mechanism of material transport.

An interesting question which up to now has been given relatively little attention concerns the phenomena which occur when contact between particles is first established. On the basis of the discussion of plastic flow as a transport mechanism during sintering, it is clear that the stresses arising during contact formation may be very large. Easterling and Thölén,[37] who investigated by electron microscopy contact formation between small metal spheres, showed that they are flattened elastically against each other producing strain field fringe contrasts in the micrographs. They calculate that the bonding in the necks between metal particles must be metallic, as in metallic grain boundaries, rather than Van der Waals bonding. In a later paper. Hansson and Thölén[37A] showed that in the adhesion between gold particles 40 to 60 nm in diameter plastic deformation by twinning to relieve stresses in the contact zone is important.

Contact formation must play an important role not only in bonding between small but also between larger particles, where actual contact, i.e. approach to within atomic dimensions, is believed to occur over only quite small areas even when considerable pressure is applied.

II.   The Intermediate and Final Stages of Sintering

In the later stages of sintering, the intermediate or channel closure stage and the final or isolated pore stage, the same driving forces, i.e. surface tension forces and the same material transport mechanisms, i.e. primarily diffusional flow, control the sintering process as in the initial stage of sintering. Using this driving force and these material transport mechanisms a number of models have been constructed to calculate the rate of shrinkage during the later stages of sintering. The oldest and probably most discussed of these models was developed by Coble.[38,39] In it the pores in the sintering aggregate have cylindrical shape for the intermediate stage and spherical shape for the final stage. They are of uniform size and located at the edges and corners, respectively, of the grains which have the shape of regular polyhedrons. The calculation based on this model shows that the number of vacancies which diffuse away from the pores per unit time is independent of pore size. The decrease in porosity depends only on the number of pores and this number, in turn, depends on the volume of the grains. During the intermediate and final stages of sintering the number of grains in the aggregates decreases through grain growth. In order to calculate the decrease in porosity the change in the number of grains, i.e. the rate of grain growth in the aggregate, must therefore be known.

In applying his model to the sintering of Al_2O_3 Coble determined, in independent experiments, the rate of shrinkage of compacts and the rate of grain growth in the compacts. He found that grain size in the Al_2O_3 compacts was proportional to the third root of sintering time. By combining the equation for the rate of porosity decrease as a function of the number of pores, which in his model means as a function of grain size, with the equation for the rate of grain growth Coble[40] arrived at a logarithmic relationship between porosity P and time t:

$$P = P_o - C \cdot \ln t / t_o$$

in which P_o is the porosity at the time t_o, i.e. at the beginning of the second stage of sintering and C is a constant which contains, besides surface tension, diffusivity and atomic volume, a grain growth parameter. This relationship has been found to apply not only to Al_2O_3 compacts, but also to those of several metal powders. Coble's approach has two principal drawbacks. The geometrical model is unrealistic with its cylindrical or spherical pores of uniform size, since in most compacts, certainly those of commercial powders, the pores have a variety of sizes. Under the circumstances, one would expect not only shrinkage of pores, but also interaction of pores with each other, with larger pores growing at the expense of the smaller ones. Secondly, grain growth and shrinkage are treated as independent phenomena, while it is intuitively clear that grain growth must depend directly on the changes in shape, size and number of pores. To overcome these drawbacks, Kuczynski[41,42] developed a statistical theory of sintering. In doing so he used as a model for the intermediate stage of sintering a body perforated by a cylindrical pore of total length L per unit volume. The radius r of this pore varies along its length. Due to the instability for cylindrical surfaces the continuous pore breaks up into segments and finally into chains of discrete pores of more or less spherical symmetry. Thus the intermediate stage changes into the final stage. The model for this later stage is a body containing N spherical pores per unit volume of varying size. A distribution function for pore radii in the intermediate and final stage is assumed, so that one can write for the porosity P in the intermediate stage:

$$P = \pi \, L_v \, \overline{r^2}$$

and in the final stage

$$P = {}^4/_3 \, \pi \, N_v \, \overline{r^3}$$

where $\overline{r^2}$ is the average of the square of the radius in the intermediate and $\overline{r^3}$ the average of the cube of the radius in the final stage.

To develop an equation for the kinetics of pore shrinkage two phenomena mentioned above must be taken into account in the derivation. One is the change in pore diameter. In models which assume uniform pore diameter, such as the one of Coble, only the decrease in the volume of the pores which manifests itself in density increase must be considered. However, in a model which assumes a statistical distribution of pore diameters, an exchange of volume among neighboring pores, very similar to Ostwald ripening, will also occur. The second is grain growth. For this purpose Kuczynski adopts Zener's equation for the change in grain size as a function of the average pore radius $\overline{r}$ and of the volume fraction of porosity P which reads:

$$\overline{a} = K \, \overline{r} / P$$

259

in which a is the average grain size and K is a parameter, different for the two stages of sintering here considered, which depends only on the width of the distribution function of the pore radii.

Taking these two phenomena into account Kuczynski arrives at two final equations, one each for the intermediate and the final stage of sintering, which relate the kinetics of decrease in porosity with that of changes in average grain size and average pore size. The equation for the intermediate stage reads

$$\left(\frac{P_o}{P}\right)^{3/2\, x_1} = \left(\frac{\bar{a}}{\bar{a}_o}\right)^{\frac{3x_1}{1+x_1}} = \left(\frac{\bar{r}}{\bar{r}_o}\right)^{\frac{3x_1}{x_1-1}} = 1 + B_1 t$$

That for the final stage reads

$$\left(\frac{P_o}{P}\right)^{x_2\,+\,2/3} = \left(\frac{\bar{a}}{\bar{a}_o}\right)^{\frac{3x_2+2}{x_2+2}} = \left(\frac{\bar{r}}{\bar{r}_o}\right)^{\frac{3x_1+2}{x_2-1}} = 1 + B_2 t$$

In these equations P_o are the volume fractions of porosity, $\bar{a}_o$ the average grain size and $\bar{r}_o$ the average pore radius at the beginning of the intermediate and the final stage of sintering, respectively, while P, $\bar{a}$ and $\bar{r}$ are volume fraction of porosity, average grain size and average pore radius at any given time t during sintering. The parameters x_1 and x_2 depend only on the characteristics of the distribution functions for grain size and pore size, not on the material. The parameters B_1 and B_2 are functions of temperature through the coefficient of volume diffusion, which is the controlling mechanism for shrinkage. They also depend on the initial grain and pore size, as well as the width of the pore size distribution. The available shrinkage data fit the equations rather well. Not such good agreement is obtained for grain sizes. The data on average pore radii which increase with time are too inaccurate to be fitted into the equations. In general, the equations agree with the observation made by Coble that the pore shrinkage and grain growth during sintering are controlled by the same mechanism.

As pointed out in the introduction, the later stages of sintering have been treated not only by the "fundamental" approach based on thermodynamic and atomistic data, but also by Rhines' topological approach and by phenomenological approaches.

To further the understanding of the intermediate stage of sintering, the topological approach is concerned with the changes in the curvature of pore surfaces.[3] In an unsintered powder aggregate or compact the curvature of pore surfaces is primarily concave. During the initial stage of sintering the pore surface curvature changes, so that, when the intermediate sintering stage is reached, the configuration of the interior pore surfaces is characterized by saddle surfaces with radii of curvature of opposite signs, partially concave, partially convex. This configuration of the pore surfaces is preserved throughout the second stage of sintering. It insures that, as the density increases and the surface area decreases through closing off of pore channels (loss of connectivity), a minimal surface corresponding to the level of densification is maintained. As a consequence, a linear relationship between density and specific surface (expressed as area per unit volume) prevails throughout the second stage of sintering:

$$\delta = - m \, S_v + \delta_t$$

in which δ is the density of a compact of specific surface S_v, δ_t is the theoretical (bulk) density and m is a proportionality constant. Experiments showed that this relationship applies for powder aggregates of different particle sizes, for loosely stacked powder aggregates and those settled by vibration and for compacts pressed at various pressures. A graph of specific surface vs. density for aggregates from an electrolytic granular copper powder with particles in the range from 125 to 150 μm is shown in Figure 9-8.[3] The slope m in the relationship between density and specific surface depends upon the topological character, i.e. the connectivity or genus per unit volume, of the original aggregate or compact. The slope is shallower for aggregates from coarse powder than for those from fine powder, for aggregates settled by vibration than for those loosely stacked and for compacts pressed at higher pressure than for those pressed at lower pressure or for looose powder aggregates. A shallow slope would indicate that the increase in density of the aggregate or compact during second stage sintering is relatively small even though the interior surface area decreases rapidly.

In the phenomenological approaches efforts are made to express, by way of relatively simple equations, the results of isothermal sintering experiments in which the dependence of the changes in compact volume or pore volume, sintering time and sintering temperature are determined. In the cases where the empirical equations fit the experimental data well, attempts are made by the authors of the equations to draw conclusions from the character of the equations regarding the fundamentals of the sintering process. As was indicated in the first section of this chapter, the sintering process, in particular for compacts from commercial powders, is extremely complex. It therefore appears unlikely that valid connections between simple sintering equations for such compacts and the fundamen-

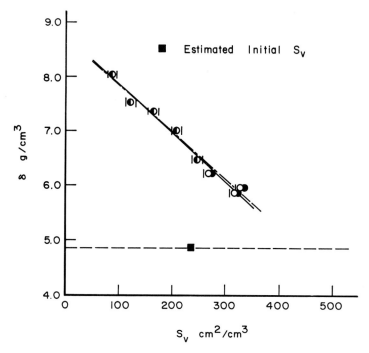

FIGURE 9-8 Quantitative metallographic determination of the variation of surface area with density for granular electrolytic copper powder, -100 +200 mesh, sintered in dry hydrogen at 1050°C for various times.

tals of the process can be established. On the other hand, some of these equations for the shrinkage behavior of compacts from commercial powders have practical use and are therefore presented here.

An equation for the densification of compacts as a function of time, first established by Makipirtti[43] for compacts of W, Ni and Cu, which sinter with a liquid phase, was later found by Tikkanen and coworkers[44] to be also valid for compacts from many simple metal powders. The equation reads:

$$\frac{V_o - V_s}{V_s - V_E} = (Ct)^n$$

in which V_o, V_s and V_E are the initial volume of the compact, its volume after sintering for the time t and its volume after complete densification, C and n are constants. The equation fits surprisingly well for compacts from electrolytic copper powder, carbonyl and sponge iron powders in the α, γ and δ regions, carbonyl and atomized nickel powders, cobalt powder and tungsten powder. Depending on the nature of the material the exponent n has values between 0.5 and 1.

Ivensen[45] in the Soviet Union first established a simple equation for the relationship between the pore volume of a compact before sintering, V_p, and its pore volume after sintering, V_s, for a given length of time at a given temperature. For an ideal isothermal sintering process without changes in composition and without the effect of trapped gasses:

$$\frac{V_s}{V_p} = \text{constant}$$

for compacts pressed to a fairly wide range of densities. As an example, carbonyl powder compacts pressed to densities from 3.43 to 5.42 g/cm^3 and sintered for 60 min. at 850° C had sintered densities from 6.04 to 7.46 g/cm^3 with ratios of pore volume equal to 0.285 ± 0.006 regardless of pressed densities.

Ivensen also established equations for the change in the pore volume as a function of sintering time. In differential form the equation reads

$$\frac{1}{v}\frac{dv}{dt} = -q\left(\frac{v}{v_{in}}\right)^m$$

in which v is the pore volume at any time t, v_{in} is the volume at the beginning of isothermal sintering, $q = dv/dt\, v_{in}$ is the rate of reduction in pore volume at $t = 0$ and m is a constant. After integrating between $t = 0$ and $t = t$ and setting the pore volume at time t_o equal to 1 one obtains

$$v = v_{in}\,(qmt + 1)^{-1/m}$$

This equation contains two constants q and m, like Tikkanen's equation. However, in order to derive the constants from experimental results the pore volume at the beginning of isothermal sintering and the time when isothermal sintering actually starts must be known. They can be extrapolated from data on experiments in which the compacts are rapidly heated to the isothermal sintering temperature, but the extrapolation is somewhat uncertain.

III. Expansion Due to Extrapped Gases During Sintering

In the discussion of the geometrical changes occurring during sintering, only shrinkage of compacts or loose powder aggregates were considered.

As was pointed out in chapter 8, when growth is observed during sintering of compacts from a homogeneous metal powder, it is attributed to the effect of gases. In Figure 8-11 the changes in density of copper powder compacts during isothermal sintering were shown as a function of sintering time, with densification during the early stage of sintering and expansion in a later stage. This behavior can be explained at least semiquantitatively by assuming that stresses due to surface tension forces and those due to gas expansion interact. The explanation is based on the assumption that during compacting a gas is entrapped in the pores of the compact, and that during heating of the compact to the sintering temperature, the pressure of the gas increases because it is heated at a constant volume. The gas must be of such a nature that it neither diffuses through the compact (such as hydrogen) nor reacts with the metal of the compact (such as oxygen). Compacts pressed in air will contain nitrogen and, if they are heated in hydrogen, also water vapor, which is formed by the reaction of the hydrogen diffusing into the pore with the oxide on the pore wall. Neither nitrogen nor water vapor will diffuse through or react with the metal.

A further assumption is that, due to the variation in size of the powder particles in the compact, the closed off pores also have a variety of sizes, say from 1 μm diameter to 40 μm diameter and that the pressure of the gas entrapped in the pores is atmospheric at room temperature at which the compact is pressed. When the compact is heated to the sintering temperature, the pressure inside the pores will rise. The pressure differential between the entrapped gas and the outside will be the same for all pores regardless of size. For a sintering temperature of 1200° K (927° C) the pressure differential will be (1200/300 x 0.1) - 0.1 or approximately 0.3 MPa. On the other hand, the shrinkage stress due to the surface tension forces will depend on pore size. For spherical pores it will be $2\gamma/r$ where γ is the surface tension (1.65 N/m) and r is the radius of curvature, i.e. the radius of the spherical pore. For a pore 22 μm in diameter ($r = 11$ x 10^{-6}m), the stress due to the surface tension force will be $3.3/11$ x $10^{-6} = 0.3$ MPa. For this pore size the expansion and the contraction stress will therefore just cancel. For pores smaller than 22 μm the contraction stress will be higher than the expansion stress and the pores will shrink. The pore volume will decrease with the cube of the pore radius. Therefore, the expansion stress will increase and will be inversely proportional to the cube of the pore radius. The contraction stress due to the surface tension forces will also increase, but it will be inversely proportional to only the first power of the pore radius. Therefore a pore size will eventually be reached where the stresses balance.

For pores larger than 22 μm the expansion stress will be larger than the contraction stress and the pores will grow. The expansion stress decreases and will again be inversely proportional to the cube of the pore radius, the contraction stress also decreases, but will be inversely proportional to the first power of the pore radius. Again, a pore size will be reached where the stresses balance. For very small pores the contraction stress due to surface tension forces will be several times the expansion stress (e.g. for a 1 μm diameter it will be $3.3/0.5 \times 10^{-6} = 6.6$ MPa) and the net contraction stress ($6.6-0.3 = 6.3$ MPa for the 1 μm diameter pore) will also be several times the expansion stress. On the other hand, the net expansion stress for the large pores can never be larger than 0.3 MPa. The rate at which the pores change their volume (the strain rate) will be a function of the stress, the larger the stress, the greater the rate of volume change.

In general, the volume contained in small (below equilibrium size) pores will be considerably smaller than the volume contained in large (above equilibrium size) pores. However, since the contraction stress in the very small pores is very large, their rate of shrinkage will be very rapid compared with the rate of growth of large pores. Even though the total volume of the small pores is small, the net change in volume during the early stages of sintering will be shrinkage. As soon as the rate of small pore shrinkage slows down, the net change in volume will become expansion because the influence of the large pores upon net volume change will predominate due to their relatively large volume. This is what is observed in the volume change of copper powder compacts.

References

1. H. E. Exner, Grundlagen von Sintervorgängen, Stuttgart (1978).
1A. English translation: Principles of Single Phase Sintering, Reviews on Powder Metallurgy and Physical Ceramics, Vol. 1, No. 1-4 (1979).
2. R. T. DeHoff, R. A. Rummel and F. N. Rhines, The role of interparticle contacts in sintering, in "Powder Metallurgy," ed. by W. Leszynski, N.Y., p. 31-51 (1961).
3. R. T. DeHoff, R. A. Rummel, H. P. LeBuff, and F. N. Rhines, The relationship between surface area and density in the second stage sintering of metals, in Mod. Dev. in Powder Metallurgy, Vol. 1, ed. by H. H. Hausner, N.Y., p. 310-331 (1966).
4. F. N. Rhines, and R. T. DeHoff, A topological approach to the study of sintering, in Mod. Dev. in Powder Metallurgy, Vol. 4, ed. by H. H. Hausner, N.Y., p. 173-188 (1971).
5. H. Udin, A. J. Shaler, and J. Wulff, The surface tension of solid copper, Trans. AIME, Vol. 185, p. 186-190 (1949).
6. H. Udin, Grain boundary effect in surface tension measurement, Trans. AIME, Vol. 191, p. 63 (1951).
7. F. V. Lenel, H. H. Hausner, O. V. Roman, and G. S. Ansell, The influence of gravity in sintering, Trans. AIME, Vol. 227, p. 640-644 (1963).

8. Reference 1A, p. 210-226.
9. F. V. Lenel, H. H. Hausner, I. A. El Shanshoury, J. G. Early, and G. S. Ansell, The driving force for shrinkage in copper powder compacts during the early stages of sintering, Powder Metallurgy, No. 10, p. 190-198 (1962).
10. W. D. Kingery, and M. Berg, Study of the initial stages of sintering solids by viscous flow, evaporation-condensation and self diffusion, J. Appl. Phys., Vol. 26, p. 1205-1212 (1955).
11. J. Frenkel, J. Phys. USSR, Vol. 9, p. 385 (1945).
12. F. R. N. Nabarro, Report on a conference on the strength of materials, The Physical Soc., London, p. 75 (1948).
13. G. C. Kuczynski, Self-diffusion in sintering of metallic particles, Trans. AIME, Vol. 185, p. 169-178 (1949).
14. C. Herring, Diffusional viscosity of a polycrystalline solid, J. Appl. Phys., Vol. 21, p. 437-445 (1950).
15. R. L. Coble, Diffusion models for hot pressing with surface energy and pressure effects as driving forces, J. Appl. Phys., Vol. 41, p. 4798-4805 (1970).
16. B. H. Alexander and R. W. Baluffi, The mechanism of sintering of copper, Acta Met., Vol. 5, p. 666-677 (1957).
17. G. Cizéron and P. Lacombe, Influence of self-diffusion phenomena in the sintering of pure carbonyl iron above and below the transformation point, Revue de Métallurgie, Vol. 52, p. 771-783 (1955).
18. J. G. R. Rockland, On the rate equation for sintering by surface diffusion, Acta Met., Vol. 14, p. 1273-1279 (1966).
19. J. G. R. Rockland, The determination of the mechanism of sintering, Acta Met., Vol. 15, p. 277-286 (1967).
20. D. L. Johnson, New method of obtaining volume, grain boundary and surface diffusion coefficients from sintering data, J. Appl. Phys., Vol. 40, p. 192-200 (1969).
21. M. F. Ashby, A first report on sintering diagrams, Acta Met., Vol. 22, p. 275-289 (1974).
22. F. A. Nichols, Coalescence of two spheres by surface diffusion, Acta Met., Vol. 37, p. 2805-2808 (1966).
23. F. A. Nichols, Theory of sintering of wires by surface diffusion, Acta Met., Vol. 16, p. 103-113 (1968).
24. F. A. Nichols, and W. W. Mullins, Surface, (Interface-) and volume-diffusion contributions to morphological changes driven by capillarity, Trans. AIME, Vol. 233, p. 1840-1847 (1965).
25. F. A. Nichols, and W. W. Mullins, Morphological changes of a surface of revolution due to capillarity-induced surface diffusion, J. Appl. Phys., Vol. 36, p. 1826-1835 (1965).
26. Reference 1A, p. 32-39 and 205-210.
27. H. E. Exner and P. Bross, Material transport rate and stress distribution during grain boundary diffusion driven by surface tension, Acta Met., Vol. 27, p. 1007-1012 (1979).
28. H. E. Exner and P. Bross, Computer simulation of sintering processes, ibid., p. 1013-1020.

29. C. Herring, Effect of change of scale on sintering phenomena, J. Appl. Physics, Vol. 21, p. 301-303 (1950).

30. P. C. Eloff and F. V. Lenel, The effect of mechanical constraints upon the early stages of sintering, Mod. Dev. in Powder Metallurgy, Vol. 4, ed. by H. H. Hausner, N.Y., p. 291-302 (1971).

31. H. E. Exner, G. Petzow and P. Wellner, Problems in the extension of sintering theories to real systems, in "Sintering and Related Phenomena," ed. by G. C. Kuczynski, N.Y., p. 351-362 (1973).

32. J. Weertman, Steady state creep of crystals, J. Appl. Phys., Vol. 28, p. 1185-1189 (1957).

33. J. Weertman, Steady state creep through dislocation climb, ibid., p. 362-364.

34. J. P. Hirth, The influence of surface structure on dislocation nucleation, in "Proceedings of the conference on the relation between structure and mechanical properties of metals," H. M. Stationary Office, London, p. 218 (1963).

35. J. E. Sheehan, F. V. Lenel, and G. S. Ansell, Investigation of the early stages of sintering by transmission electron micrography, "Sintering and Related Phenomena," ed. by G. C. Kuczynski, Plenum Press, N.Y., p. 201-208 (1973).

37. K. E. Easterling and A. R. Thölén, Surface energy and adhesion at metal contacts, Acta Met., Vol. 20, p. 1001-1008 (1972).

37A. I. Hansson and A. Thölén, Adhesion between fine gold particles, Phil. Mag. A, vol. 37, No. 4, p. 535-559 (1978).

38. R. L. Coble, Sintering crystalline solids, I, intermediate and final stage diffusion models, J. Appl. Phys., Vol. 32, p. 787-792 (1961).

39. R. L. Coble, Sintering crystalline solids II, Experimental tests of diffusion models in powder compacts, ibid., p. 793-799.

40. R. L. Coble and J. E. Burke, Sintering in crystalline solids, in "Reactivity of Solids," ed. by J. H. DeBoer, Amsterdam, p. 38-51 (1961).

41. G. C. Kuczynski, Statistical approach to the theory of sintering, in "Sintering and Catalysis," ed. by G. C. Kuczynski, N.Y., p. 325-337 (1975).

42. G. C. Kuczynski, Statistical theory of sintering, Z. f. Metallkunde, Vol. 67, p. 606-610 (1976).

43. S. Makipirtti, On the sintering of W-Ni-Cu heavy metal, in "Powder Metallurgy," ed. by W. Leszynski, N.Y., p. 97-111 (1961).

44. M. H. Tikkanen, and S. Ylassari, On the mechanisms of sintering, Mod. Dev. in "Powder Metallurgy," Vol. 1, ed. by H. H. Hausner, N.Y., p. 297-309 (1966).

45. V. A. Ivensen, Densification of metal powders during sintering, Consultants Bureau, N.Y. (1973).

10

Sintering of Compacts from Mixtures of Metal Powders Without a Liquid Phase

Sintering of compacts from mixtures of metal powders is of interest in powder metallurgy technology because it is an important method of producing alloy materials from metal powders. From the point of view of sintering theory, the material transport during interdiffusion between different metals in the solid state is the principal phenomenon of interest. In interdiffusion processes a movement of vacancies, i.e. the exchange of metal atoms with vacancies, takes place. The driving force is the chemical potential gradient which is caused by concentration differences. In the discussion of sintering of homogeneous metal powders it was shown that the most important transport mechanism is self-diffusion, which is also a flow of vacancies, but in this case under the influence of a chemical potential gradient caused by surface and interfacial tension (capillary) forces. In the sintering of compacts or aggregates from mixtures of metal powders self-diffusion caused by capillary forces and interdiffusion caused by concentration gradients are superimposed. This complicates the analysis of the process.

In this chapter only sintering without the appearance of a liquid phase is discussed. In the first section, production of alloys from mixtures of metal powders is compared with other methods of powder metallurgy alloy production. In the second section, the effect of the parameters governing the interdiffusion process is discussed in a qualitative way. It is shown how the knowledge of these effects may be applied to practical problems of alloy production from metal powder mixtures. In the third section, an example for the quantitative analysis of interdiffusion in compacts from mixtures of metal powders is presented. The complications involved in cases of simultaneous interdiffusion and neck growth by self-diffusional processes are discussed in the fourth section. In the last section, one of the so-called "activated sintering" processes is presented, in which rapid densification based on interdiffusion is observed.

I. Alloys by Powder Metallurgy

Alloy compacts may be produced by powder metallurgy from prealloyed powders. In prealloyed powders each powder particle has the same composition as the final compact. Prealloyed powders are often atomized from alloy melts. Commercially important prealloyed atomized powders are brasses, nickel silvers, low alloy steels and austenitic stainless steels. These prealloyed powders of irregular particle shape are compacted and sintered. The raw materials for the recently developed hot consolidation processes for tool steels, superalloys and titanium alloys are also atomized prealloyed powders, in these cases generally of spherical shape. A much less common method of producing prealloyed powders is the reduction of solid solution compound powders, e.g. a prealloyed iron-nickel powder by reduction of a precipitated iron oxide-nickel oxide solid solution powder. Producing alloys by powder metallurgy from prealloyed powders has the advantage that alloy formation does not depend on homogenization during the sintering step. When an alloy is to be obtained by sintering compacts from a mixture of the powders of the alloying ingredients, the interdiffusion between the powder particles upon which homogenization depends may require high sintering temperatures or long sintering times. On the other hand, because of the effect of solid solution hardening, the strength and hardness of alloys is generally higher than that of pure metals. Therefore, mixtures of elemental powders can often be pressed at a given compacting pressure to a higher density than prealloyed powders. An outstanding example of the use of mixtures of powders because of their better compressibility is that of iron powder and graphite powder or in some cases iron powder and carbide powder. Iron powder is much softer than powders of iron-carbon steels. Mixed with graphite powder, iron powders are therefore the raw material for structural parts with steel compositions. In this case the diffusion of carbon from the graphite into the iron particles is rapid enough so that completely homogeneous iron-carbon austenite is formed at the sintering times and temperatures commonly used for structural parts.

II. Parameters Affecting Interdiffusion

The parameters which control the kinetics of homogenization are particle size, sintering time and sintering temperature. Particle size is important because homogenization is related directly to the square of the diffusion distance. Powder particle size determines this distance and plays an important role. If the compacts are made of a mixture of two powders, the particle size of the minor constituent is more important in determining diffusion distances, since this minor constituent will be dispersed in a more or less continuous matrix of the major constituent. As near as possible ideal statistical distribution of the constituents contributes to more rapid homogenization, which means that the constituents should be well blended. A high compacted density, which means a high compacting pressure, will produce better contact between particles which must diffuse into each other and will accelerate the rate of homogenization. However, during homogenization at lower temperatures where surface diffusion may play an important role, a less dense compact with more internal surface may be advantageous.

Temperature affects rates of homogenization because of the temperature dependence of the diffusivity D between the constituents of the powder mixture. It is given by

$$D = D_o \exp(-Q/RT)$$

in which D_o is a constant independent of temperature, Q the activation energy of diffusion, R the gas content and T the absolute temperature. This means the diffusivity increases rapidly with temperature. At the usual sintering temperatures the diffusivity may be high enough so that complete homogenization occurs in mixtures of powders with the usual particle size distribution during commercial sintering times, as in the case of iron-graphite mixtures for structural parts. At a temperature of 1100°C, the diffusivity of carbon into austenite is on the order of 7×10^{-9} m^2/sec. On the other hand, when homogeneous iron-nickel or iron-manganese alloys are to be produced from mixtures of elemental iron with nickel or manganese powders, either very fine powders or long sintering times and high sintering temperatures are necessary. At 1100°C the diffusivity of iron into nickel is on the order of 8×10^{-14} m^2/sec, about 10^5 times lower than that of carbon into austenite at the same temperature. If a completely homogenized alloy steel composition from powders is desired, a prealloyed powder, containing the metallic alloying elements but no carbon, may be mixed with graphite powder, compacted and sintered. The compressibility of the alloy powder-graphite powder mixture will, however, be lower than that of an elemental powder mixture.

This does not mean that complete homogenization is always necessary in producing materials from mixtures of metal powders. The study of the properties of partially homogenized alloys is only in its beginning. In one investigation,[1] a "partially prealloyed powder", produced by diffusion bonding commercial nickel, copper and molybdenum powders to commercial iron powder was blended with graphite powder and conventionally pressed and sintered. When the resulting, only partially homogenized alloy was heat treated, excellent mechanical properties useful in structural parts were obtained.

The rates of homogenization in compacts from mixtures of iron powder and such alloying elements as chromium, silicon and molybdenum, having compositions in the ferritic region outside the gamma loop of the respective binary equilibrium diagrams, will be faster than in those from iron-nickel or iron-manganese powder mixtures. The reason for the more rapid homogenization is that the solid solutions formed of iron with chromium, silicon and molybdenum are body-centered cubic and the diffusivities between iron and alloying elements in which ferritic alloys are formed are about 100 times higher than those of iron and alloying elements in which austenitic solid solutions are formed. A practical example of the formation by interdiffusion of alloys ferritic at the sintering temperature is iron silicon alloys for magnetic applications. Three percent silicon alloys with homogeneous microstructure and satisfactory magnetic properties are produced by compacting mixtures of iron and ferrosilicon powder and sintering near 1300° C for one hour.

III. Quantitative Analysis of Interdiffusion

Homogenization during sintering of compacts of mixtures of metal powders was first studied semi-quantitatively by Rhines.[2] A more quantitative approach was taken by Rudman and coworkers.[3] Heckel[4] and his school developed and compared with each other models for the interdiffusion between metal powder particles and brought the knowledge of this field to its present more or less definitive stage. In this section, only one example for the quantitative analysis of interdiffusion is presented, rather than an exhaustive treatment, and much of the mathematical apparatus is omitted. The example relates to the interdiffusion in compacts from a binary mixture of metal powders, in which a complete series of solid solutions is formed, such as in copper and nickel. The model is the so-called concentric sphere diffusion model. It consists of a sphere of the minor constituent of the

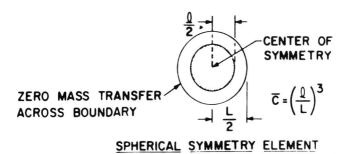

ZERO MASS TRANSFER
ACROSS BOUNDARY

CENTER OF
SYMMETRY

$$\bar{c} = \left(\frac{l}{L}\right)^3$$

SPHERICAL SYMMETRY ELEMENT

FIGURE 10-1 Symmetry elements for concentric sphere model.

mixture, e.g. nickel, having a diameter l, representative of the particle size of the nickel. The nickel sphere is embedded in a sphere of the major constituent, e.g. copper. The diameter of the copper sphere, L, is equal to that of the nickel sphere divided by the third root of the mean composition, $\bar{c}$, of the alloy:

$$L = l/\sqrt[3]{\bar{c}}$$

For a copper-nickel mixture containing 20 volume percent of nickel powder with a diameter of the nickel powder sphere equal to 100 μm, the diameter of the copper sphere would be $100/\sqrt[3]{0.2} = 171$ μm. The spherical symmetry elements of the model are shown in Figure 10-1. A solution to Fick's equation for unsteady state diffusion is developed for this single sphere-shell model and then applied to the entire mass of the compact. The solutions are complex and involve numerical methods and computer techniques. The results of the analysis for a single phase system, such as Cu-Ni, are shown in Figures 10-2, 10-3 and 10-4.

In Figure 10-2 schematic distance concentration profiles are shown for various times t from the center of symmetry along the radius r to the outer boundary of the shell at L/2. The initial particle compositions are c_A and c_B, the final composition is $\bar{c}$.

Figure 10-3 shows the degree of homogenization, F, as a function of the parameter Dt/l^2, in which D is the diffusivity between the constituents, l is the diameter of the sphere of the minor constituent (e.g. nickel) and t is the time. The degree of homogenization, F, is the ratio of the mass transferred through the surface at r = 1/2 (see Figure 10-2) in time t to the mass transferred in infinite time. Curves for F as a function of Dt/l^2 are shown for different mean compositions, $\bar{c}$, of the composite, e.g. $\bar{c} = 0.05$ (5% Ni, 95% Cu), $\bar{c} = 0.20$ and $\bar{c} = 0.50$.

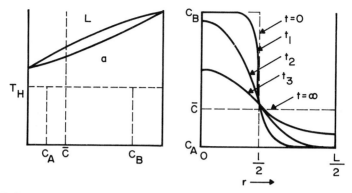

FIGURE 10-2   One-phase, concentric sphere homogenization model.

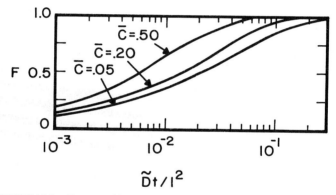

FIGURE 10-3   Degree of interdiffusion F calculated for the one phase concentric sphere model as a function of reduced time (Dt/l^2) using various mean compositions $(\bar{C})$.

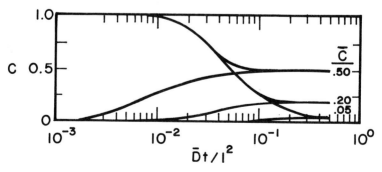

FIGURE 10-4   Range of compositions (C_{min} and C_{max}) calculated for the one phase concentric sphere model as a function of reduced time (Dt/l^2) using various mean compositions $(\bar{C})$.

Figure 10-4 is a plot of the compositions existing at the positions r = 0 (center of symmetry) and r = L/2 (boundary of the outer shell) again as a function of Dt/l² for different mean compositions.

Other geometrical models have been proposed, such as an ordered cube model produced by three dimensional stacking of cubes. Heckel[4] has shown that the concentric sphere model is not only easier to handle analytically, but also gives results which are closer to those found experimentally. This does not mean that the concentric sphere geometry is ideal; it assumes spherical minor constituents particles, perfectly distributed and with perfect contact at the sphere-shell interfaces. It also assumes space filling of the spherical composites which, of course, is impossible.

In order to assess experimentally the progress of homogenization in sintered compacts of single phase binary systems, such as copper and nickel, the changes in X-ray diffraction peak profiles are determined. A typical set of such diffraction peaks as a function of sintering treatment is shown in Figure 10-5. In the initial stage of homogenization the complete range of solid solutions from pure copper to pure nickel is formed; subsequently, the

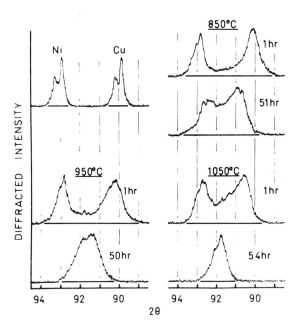

FIGURE 10-5 X-ray diffraction peak profiles for Cu-Ni compacts which have been given various thermal treatments. The (311) peaks using Cu Kα radiation are shown for mean compact positions of C̄ = 0.52 atomic fraction nickel. Both powders –100 +140 mesh.

pattern sharpens into a single peak characteristic of the mean composition of the compact, $\bar{c}$. From the X-ray data, the degree of homogenization, F, and the minimum and maximum concentrations present at any time are determined and compared with the mathematically developed analysis for homogenization. Such a comparison is shown in Figures 10-6 and 10-7 for degree of homogenization and for maximum and minimum concentrations. It is seen that

1. the parameter Dt/l^2 normalizes the data to a band which gets narrower with increasing homogenization,
2. the initial stages of homogenization proceed faster than predicted quite probably because of the contribution of surface diffusion to homogenization,
3. the later stages of homogenization proceed slower than predicted, mainly because the model is based on ideal mixing and very narrow particle size distribution.

In addition to determining homogenization by measuring X-ray diffraction peak profiles, electron micro-beam techniques have been developed.

Heckel[4] and his coworkers have developed mathematical models not only for single phase binary systems, such as Ni and Cu, but also for binary systems with more than one phase. In these multi-phase systems the sintering treatment results in structures which are generally not homogeneous, but, at the end point of the treatment, consist of the phases present in the system having compositions in thermodynamic equilibrium with each other

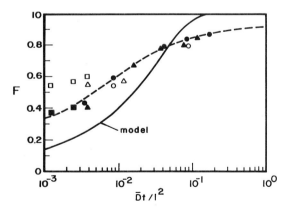

FIGURE 10-6 Comparison of experimental F data for compacted blends of Cu and Ni powders ($\bar{C} = 0.20$ atomic fraction Cu) with concentric sphere model predictions. □ ■ 850°C, △ ▲ 950°C, O ● 1050°C. Cu –200 +270 mesh ($l = 65\ \mu$m). Closed points - coarse, open points - fine Ni.

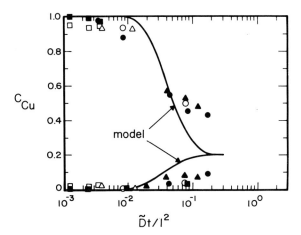

FIGURE 10-7   Comparison of experimental composition data (C_{min} and C_{max}) for compacted blends of Cu and Ni powders ($\bar{C}$ = 0.20 atomic fraction Cu) with concentric sphere model predictions. □ ■ 850°C, △ ▲ 950°C, O ● 1050°C. Cu –200 + 270 mesh (1 = 65 μm). Closed points - coarse, open points - fine Ni.

at the treatment temperature. Examples of two-phase systems investigated are Ni and W, and Cu and Ag. A three-phase system studied is that of Ni and Mo. At 1200°C, it consists of terminal solutions of fcc Ni-rich α, bcc Mo-rich γ and the Mo-Ni intermediate phase β. In these more complex systems the movement of the phase boundary or boundaries is of particular importance for the kinetics of the process. It can be analyzed by quantitative metallography. The analysis of these complex systems will not be presented.

IV.   Simultaneous Interdiffusion and Neck Growth

As pointed out in the introduction, the mechanism of interdiffusion between two metals is generally based on the interchange of atoms with vacancies in the lattice. During interdiffusion between the two constituents of a diffusion couple, the individual rates of atom-vacancy interchange for the two constituents will not be identical. One of the constituents will diffuse faster than the other. This phenomenon is called the Kirkendall effect.

As a result of the uneven rate of diffusion, excess vacancies will accumulate on the side of the diffusion couple with the faster diffusing constituent. The question arises what happens with these excess vacancies. In many cases they are taken care of at grain boundaries or at edge dislocations (by dislocation climb), since in many diffusion experiments no precipitation of excess vacancies in the diffusion couple is observed. In those cases where the concentration gradients involved are very steep—and this applies particularly to metal powder compacts—the excess vacancies are taken care of in other ways.

The effects produced by vacancy precipitation have been studied in detail by Kuczynski and Stablein,[5,6] whose work is used to illustrate them. One method of vacancy precipitation is shown in Figure 10-8. It is a section through an artificial compact produced by winding alternate layers of gold wire and nickel wire upon a nickel rod and sintering the assembly two hours at 900°C in vacuum. A cross section through the neck which is formed between the nickel rod and a gold wire of the first layer is shown. If a similar artificial compact from a single metal were sintered, the cross section of the neck between rod and wire would appear straight. In contrast, the neck between gold wire and nickel rod has grooves extending into the gold wire at the end of the contact surfaces. These grooves are formed by the precipitation of excess vacancies formed in the gold when it diffuses into nickel faster than the nickel diffuses into the gold. Another way of relieving an excess vacancy concentration is by nucleation and growth of pores inside the faster

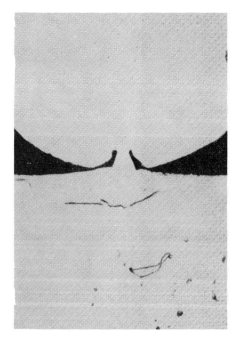

FIGURE 10-8   Cross section through a .005 inch gold wire sintered to a nickel cylinder for 2 hours at 900°C in vacuum, x500.

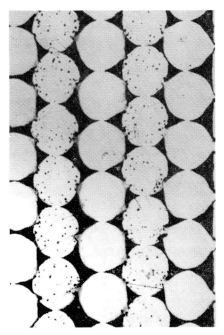

FIGURE 10-9   Section through an array of copper (99.9% pure) and nickel
wires originally .005 inch diameter, wound on a copper core. Sintered for 50
minutes at 1070°C in helium. Wires with porosity are copper wires. Etched
with potassium dichromate solution, 100x.

diffusing component. This is illustrated in Figure 10-9 showing an artificial
compact produced by winding alternate layers of 125 μm diameter copper
and nickel wires upon a copper core and sintering 3 hours at 995°C in
vacuum.

The nucleation of pores in the copper wires requires the presence of
minute amounts of impurities in the copper. However, even when no pores
are nucleated inside the copper wires, the geometry of artificial compacts of
copper and nickel wires will change drastically with increasing sintering
time, as is shown in the series of micrographs of Figure 10-10. It shows
sections through an array of 125 μm diameter copper and nickel wires
wound on a copper core and sintered for 20 minutes, 1 hour, 2 hours, 3
hours and 18 hours at 1070°C. The more rapid diffusion of copper into
nickel than of nickel into copper causes growth of the nickel wires and
gradual disappearance of the copper wires.

As Figure 10-9 and 10-10 indicate, the Kirkendall effect may cause pores
to appear and grow when compacts of mixtures of metal powders are

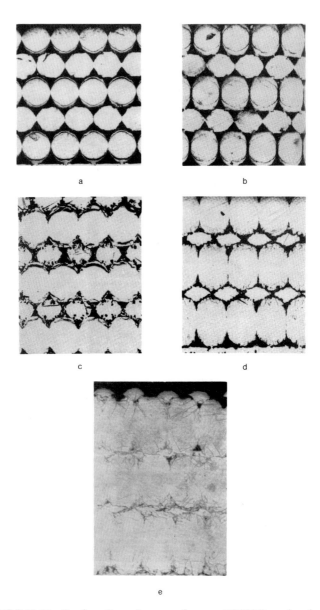

a   b

c   d

e

FIGURE 10-10 Sections through arrays of copper (99.999% pure) and nickel wires, originally .005 inch diameter, wound on a copper core and sintered in vacuum at 1070° C. First, third and fifth row from bottom are nickel. Etchant potassium dichromate, 63x.

a) 20 min. b) 1 hour c) 2 hours d) 3 hours e) 18 hrs. 10 min.

sintered, which means that the compacts themselves grow. Growth of compacts from mixtures of metal powders during sintering due to the Kirkendall effect is important in powder metallurgy technology. An example of growth during sintering, where the Kirkendall effect plays a role, will be discussed in chapter 11. On the other hand, in many practical cases of sintering where the size of the powder particles and therefore the size of the pores is small, the pore shrinkage due to the normal self-diffusional flow will overshadow any pore growth caused by the Kirkendall effect. The relative contribution of the Kirkendall effect and of normal pore shrinkage to dimensional changes during sintering of compacts of mixtures of metal powders has not yet been treated quantitatively.

V. Activated Sintering

For certain metals the addition of a small amount of an alloying element ("doping") may cause the rate of densification to increase as much as 100 times compared with undoped compacts. This phenomenon is one of those which have been called "activated sintering." It has been observed for the refractory metals tungsten, molybdenum, hafnium, tantalum, niobium and rhenium.[7] The best activators are palladium and nickel, but activation has also been observed with other metals in group VIII of the periodic table.

The best known example, first investigated by Agte[8] and Vacek,[9] is the sintering of tungsten powder compacts to which small amounts of nickel have been added. The nickel is often added as a solution of its salt which is reduced to the metal and forms a layer several monatomic layers thick on the surface of the tungsten powder particles.[10] Four to ten monatomic layers appear to produce optimum activation. Gessinger and Fischmeister[11] made experiments with compacts of tungsten powder with a particle size of 0.5 μm pressed at 97 MPa (14,000 psi) and sintered at 1000 and 1100° C without and with the addition of 0.5% nickel. Figure 10-11 shows the linear shrinkage of the compacts as a function of time. The strong effect of the nickel addition upon the rate of sintering is evident. They explain the effect by postulating that the nickel diffuses over the surface of the tungsten particles and accumulates at the necks between the particles, as shown schematically in Figure 10-12. From here the nickel penetrates the grain boundaries between tungsten particles. Grain boundary self-diffusion in tungsten is greatly enhanced by the presence of nickel in the grain boundaries. The rate at which tungsten atoms are transported by grain boundary diffusion through the grain boundaries and deposited at the necks of the tungsten particles is much greater for nickel-doped than for undoped tungsten and accounts for the more rapid densification. The enhancement of grain boundary diffusion in activated sintering has been related to the electronic structure of the metals involved.[12]

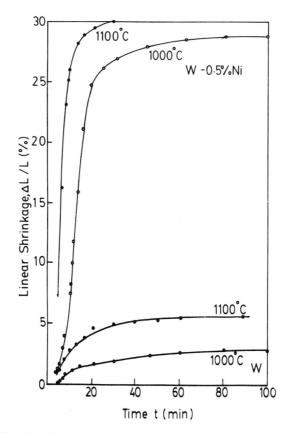

FIGURE 10-11 Comparison of the linear shrinkage as a function of time during sintering at 1000 and 1100°C of compacts of tungsten powder (0.5 μm particle size) without and with the addition of 0.5 weight % nickel.

Although the densification of tungsten can be greatly accelerated by small additions of nickel during solid state sintering, the effect is not generally made use of in producing high density tungsten because the alloys so produced are too brittle to be of commercial importance. On the other hand, alloys of tungsten with additions of nickel and copper or nickel and iron sintered with a liquid phase, the so-called "heavy alloy," are important applications of powder metallurgy. Their sintering will be discussed in chapter 11. When porous skeletons are to be prepared for infiltration with copper to produce tungsten-copper contact materials, it is customary to add very small amounts of nickel powder to the tungsten powder from which the porous skeletons are made. This facilitates the sintering of the skeletons and later the wetting of the skeleton when it is infiltrated with copper.

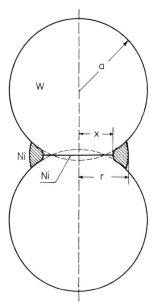

FIGURE 10-12 Geometrical model for sintering of doped tungsten.

A similar effect of activated sintering has been observed in compacts from refractory carbide powders.[13] The rate of densification of carbide compacts during solid state sintering is greatly enhanced when small amounts of the powders in the iron group are added to the carbide powders. In the production of cemented carbides, compacts of mixtures of tungsten carbide and cobalt powders are sintered at temperatures where a liquid phase is present, but the rapid densification is observed in the sintering of these compacts already at temperatures well below those where a liquid phase is formed.

References

1. P. Lindskog and O. Thornblad, Ways to improve the strength of sintered compacts made from partially prealloyed steel powders, Powder Metallurgy Int., vol. 11, No. 1, p. 10-11, (1979).
2. F. N. Rhines and R. H. Colton, Trans. ASM, vol. 30, p. 166 (1942).
3. B. Fisher and P. S. Rudman, J. Appl. Phys., vol. 32, p. 1604 (1961).
4. R. W. Heckel, Diffusional homogenization of compacted blends of powders in "Powder Metallurgy Processing," ed. by H. A. Kuhn and A. Lawley, Academic Press, N.Y., p. 51-97 (1978).
5. G. C. Kuczynski and P. F. Stablein, Jr., Sintering in multi-component systems in "Reactivity of Solids," ed. by J. H. DeBoer, Amsterdam, (1961), p. 91-103.

6. P. F. Stablein and G. C. Kuczynski, Sintering in multi-component metallic systems, Acta Met., vol. 11, p. 1327-1337 (1963).
7. R. M. German, An enhanced diffusion model of refractory metal activated sintering, presented at 4th International Round Table Conference on Sintering, Dubrovnik, Sept. (1977).
8. C. Agte, Hutnicke Listy, vol. 8, p. 227 (1953).
9. V. J. Vacek, Planseeberichte für Pulvermetallurgie, vol. 7, p. 6 (1959).
10. J. H. Brophy, L. A. Shephard and J. Wulff, The nickel-activated sintering of tungsten, in "Powder Metallurgy," ed. by W. Leszynski, N.Y., p. 113-134 (1961).
11. G. H. Gessinger and H. F. Fischmeister, J. Less-common Metals, vol. 27, p. 129 (1972).
12. R. M. German and Z. A. Munir, Systematic trends in the chemically activated sintering of tungsten, High Temp. Science, vol. 8, p. 267-280 (1976).
13. E. J. Sandford and E. M. Trent, Physical metallurgy of cemented carbides, "Symp. on Powder Metallurgy," Iron and Steel Institute, London, Special Report, No. 38, p. 84-91 (1947).

CHAPTER
11

Liquid Phase Sintering

When compacts from a single phase metal powder are sintered, the possible material transport mechanisms can be enumerated and their contributions to the sintering process assessed. This is not yet possible for liquid phase sintering which comprises at least two different processes which have in common only that a liquid phase appears at one stage during the process of sintering. In one of these processes the liquid is present during the entire time while the compacts are at the sintering temperature; they are sintered between the solidus and the liquidus of the alloy systems and are heterogeneous during the entire sintering cycle. This is the process by which the so-called heavy alloy consisting mainly of tungsten with 10% or less of nickel and copper or nickel and iron and the one by which cemented carbides are sintered. The mechanism is called the "heavy alloy mechanism". Because of its technological importance it has been investigated more thoroughly than other liquid phase sintering processes with new contributions to its understanding having been made quite recently.

Another liquid phase sintering process may be called transient liquid phase sintering. The liquid is formed while the compact is heated to the sintering temperature, but disappears by interdiffusion while the compact is at the sintering temperature. The solid alloy present at the end of the period during which the compact is held at the sintering temperature may be a homogeneous solid or a heterogeneous alloy consisting of two or more solid phases. An example where a homogeneous solid alloy is formed during sintering is the self-lubricating bearing from a mixture of 90% elemental copper and 10% elemental tin powder sintered below the solidus temperature of the alloy. Other examples are alloys of aluminum with copper, magnesium and silicon, certain alnico magnets and rare-earth-cobalt magnets. A heterogeneous alloy consisting of several solid phases is formed in dental amalgams of silver, tin and mercury which are sintered at room temperature. In spite of its technological importance sintering with a transient liquid phase is not yet well understood.

Whether the sintering of a compact from a mixture of metal powders should be included under the heavy alloy mechanism, under sintering with a transient liquid phase or even under solid state sintering with interdiffusion will sometimes depend upon the exact composition of the powder mixture, the temperature at which the compact is sintered or even the rate at which it is heated to the sintering temperature. When compacts from a mixture of 90% copper and 10% tin powder are sintered at a temperature between the solidus and the liquidus rather than below the solidus temperature of the alloy, a solid copper-rich and a liquid tin-rich phase are formed. If the resulting alloy is rapidly cooled, its structure will closely resemble that of the heavy alloy of tungsten, nickel and copper and its sintering should be described by the heavy alloy mechanism.

Compacts from a mixture of iron and copper powder with less copper than the limit of solid solubility of copper in austenite at a sintering temperature above the melting point of copper, e.g. an alloy of 93% Fe, 7% Cu sintered at 1150° C may be heated to the sintering temperature at a slow enough rate so that a homogeneous solid solution of copper in iron and therefore no liquid phase is formed, particularly if the particle size of the powder and therefore the diffusion distances are small. This would be solid phase sintering with interdiffusion. If the compacts are heated more rapidly, the conditions of sintering with a transient liquid phase may apply. Iron-copper alloys with higher copper contents above the limit of solid solubility of copper in austenite at the sintering temperature (approximately 7.5% at 1100° C) have been used as classical examples for the heavy alloy mechanism.

In the first section of this chapter the heavy alloy mechanism will be presented. In the second section model experiments on sintering compacts from mixtures of iron and copper powder are discussed. In these experiments powders with a particle size of 100 μm were used. This is a much coarser particle size than that for which rapid densification of compacts in accordance with the heavy alloy mechanism is observed in iron-copper powder mixtures. Compacts from these mixtures of relatively coarse powders show expansion during sintering which is an important phenomenon in the technology of iron-copper structural parts. On the basis of the experiments, it has been possible to elucidate at least partially the mechanisms contributing to the dimensional changes in compacts during liquid phase sintering. The same phenomena play a role in the sintering of compacts from mixtures of copper and tin powder, which is a principal example of sintering with a transient liquid phase, technologically important for self-lubricating bearings. However, no quantitative analysis of the phenomena has yet been possible in this case. Nevertheless, in view of the importance of sintering with a transient liquid phase, it was felt necessary to present, in the third section of this chapter, the experimental evidence available on sintering copper-tin compacts and attempt a qualitative interpretation of this evidence. Sintering with a transient phase in other alloy systems has been studied even less than in the copper-tin system. The sintering of some of those alloy systems which are of technological interest will be discussed in later chapters of this book.

I. The Heavy Alloy Mechanism

The most important features of sintering with a liquid phase characterized by the heavy alloy mechanism are presented by referring to sintering of tungsten-nickel-copper alloys as described in the classical paper by Price, Smithells and Williams.[1] Modifications of these features for other systems sintered with the heavy alloy mechanism are also indicated. The features are:

1.   The composition is so chosen that the material which forms the solid phase during sintering, e.g. tungsten, is soluble in the liquid phase, e.g. the nickel-copper alloy. The amount of liquid phase, which exists in the example chosen of a nickel-copper alloy with some tungsten in solution, must be small enough so that the compacts keep their shape during sintering. Since the goal of Price, Smithells and Williams' investigation was to produce alloys of high density, at least 90% of tungsten was used.

2.   The sintering temperature must be high enough so that an appreciable amount of liquid phase is present. Above this temperature rapid densification takes place. Figure 11-1 shows the density after 1 hour of sintering for compacts with 93% W, 5% Ni, and 2% Cu, pressed at 80 MPa to a green density of 10.5 g/cm^3 as a function of sintering temperature. The tungsten powder contained 71% of particles between 1 and 5 μm. The very rapid increase in density in the temperature range where a liquid phase appears is evident. At 1375° C a density of 17.2 g/cm^3 (97% of theoretical) is obtained. Theoretical density (17.8 g/cm^3) is reached after sintering for 6 hours at 1400° C.

3.   The particle size of the powder which forms the solid phase is important. The finer the powder, the more rapid and complete is the densification. For a tungsten powder with 98% of particles below 1 μm complete densification of a 90W-6Ni-4Cu alloy is obtained after 1 hour at 1400° C. Compacts from a powder mixture with 87% of the tungsten

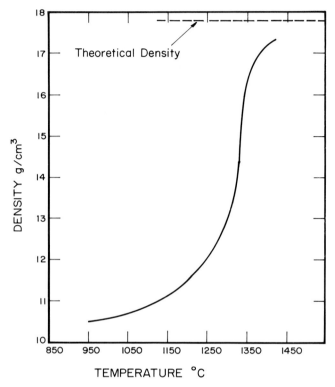

FIGURE 11-1 Effect of sintering temperature upon the density of compacts from 93% W, 5% Ni and 2% Cu powder sintered for 1 hour.

particles between 1 and 5 μm had to be sintered 3 hours at 1500° C to achieve this density. Recent work[2] has shown that near complete densification can also be obtained with powder mixtures in which as much as 50% of the tungsten is coarse (250 μm), if the rest of the powder is fine.

    4.   Final density is independent of the compacting pressure. Compacts pressed at low pressure will shrink correspondingly more than those pressed at high pressure in order to reach theoretical density.

    5.   Heavy alloy has a characteristic microstructure which is illustrated in Figures 11-2 and 11-3. It consists of single crystal grains of tungsten in a matrix of the nickel-copper phase which at the sintering temperature was liquid. The grains are rounded, but not perfect spheres. After sintering for one hour at 1400° C (Figure 11-2) the tungsten grains appear to be 20-40 μm in diameter, at least 10 times the diameter of the original tungsten particles. After sintering for an additional 5 hours (Figure

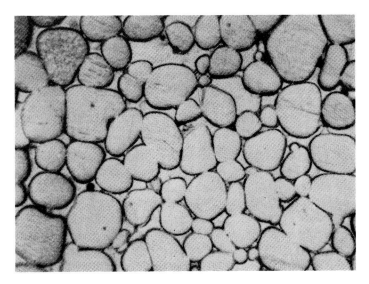

FIGURE 11-2 Microstructure of heavy alloy with 90% W, 7.5% Ni, 2.5% Cu, sintered 1 hour at 1450°C, x500.

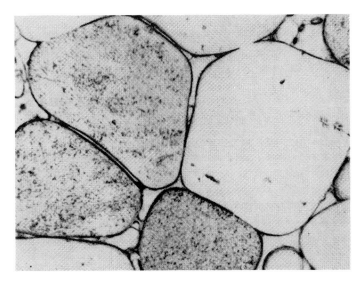

FIGURE 11-3 Microstructure of the same alloy sintered 6 hours at 1450°C, x500.

11-3) they have further doubled their diameter. Neither the rounded grain shape nor the large increase in the size of the grains of the solid phase is characteristic for all systems sintered with the heavy alloy mechanism. In micrographs of cemented tungsten carbides (see Figures 16-2 to 16-10) the WC grains appear triangular or rectangular rather then rounded and their size is generally only slightly larger than that of the WC particles from which the cemented carbides were produced.

Densification and growth of the solid particles in the heavy alloy mechanism has been customarily analyzed by dividing it into three stages:[3]

    1.   The liquid flow or rearrangement stage

In this stage densification, after the liquid melts, is brought about under the action of capillary pressure by collapse of melt bridges between particles and by rearrangement in which solid particles slide over one another.

    2.   The solution-reprecipitation or accomodation stage.

In this stage further densification and growth of the particles of the solid phase is achieved by solution and reprecipitation and by coalescence processes.

    3.   The solid state sintering stage.

In many cases of liquid phase sintering with the heavy alloy mechanism, particularly in the technologically important ones, such as heavy alloy and cemented carbides, complete densification is achieved during the first two sintering stages. Prolonged holding of the compacts at the sintering temperature may, in these cases, lead to microstructural changes in the dense compacts, including further growth of the particles of the solid phase and formation of a skeleton of the solid phase in cemented carbides. In other cases, a rigid skeleton of the solid phase is formed before complete densification has been obtained. This skeleton will interfere with rapid densification by rearrangement. A further slow increase in density is possible only by a third stage: solid sintering.[3]

First Stage:

Recent investigations[4] have shown that in the first and second stages of the sintering process similar material transport mechanisms prevail and that a sharp distinction between stages is not justified. Nevertheless, it is possible to study systems in which the solid phase is not soluble in the liquid phase and in which therefore no solution and reprecipitation can occur. The principal system of this kind which has been studied is the W-Cu system. Huppmann and Riegger[5] made model experiments on two and three dimensional arrays of spherical tungsten particles which had been coated with uniform layers of copper. The experiments showed that the length of time during which shrinkage takes place, after a liquid is formed, is primarily a function of the rate at which the melting process occurs. In the copper coated tungsten particle arrays shrinkage takes only seconds. The total amount of shrinkage also depends on the rate of melting, the faster the melting rate up to a limiting value, the greater the shrinkage. What is most significant, however, is that the amount of shrinkage depends upon the packing density of the particles and, in the case of mixtures of tungsten and copper powder, upon the quality of mixing. The closer the packing and the more ideal the mixing, the greater is the shrinkage. The total shrinkage is always less than that which the amount of liquid phase present would permit, if it flowed into all the pores. Geometrical factors are, therefore, primarily responsible for the rate and the extent of shrinkage in systems in which the solid phase is insoluble in the liquid. Huppmann and Riegger's results show that the simple relationship for the kinetics of densification in the first sintering stage which was developed by Kingery[6] must be oversimplified. To derive this relationship Kingery postulated that the driving force for densification, i.e. surface tension, is balanced by an intrinsic frictional force, i.e. the viscosity of the melt. He wrote for the rate of shrinkage:

$$\Delta L / L_o = 1/3 \ \Delta V / V_o = kt^{1+x}$$

in which ΔL and ΔV are changes in length and volume, L_o and V_o are the original length and volume of the compact, k is a constant, t is time and x is a small fraction. The strain occurring during viscous flow would be directly proportional to time, if the stress due to surface tension stayed constant. The fact that the capillary pore size must decrease and therefore the surface tension increase is taken care of by setting the strain rate proportional to t^{1+x} instead of t.

Second Stage:

In many cases of liquid phase sintering, particularly in heavy alloy, densification is accompanied by growth of the particles of the solid phase. The interpretation of the mechanisms of growth and of densification are controversial and a quantitative analysis of these processes has not been developed.

When Price, Smithells and Williams did their work on heavy alloy, they assumed that the rapid growth of the solid phase particles takes place by a process of solution of the smaller particles and reprecipitation on the larger ones due to the difference in solubility of small and large particles in the liquid phase. This difference is governed by another form of the Gibbs-Thomson equation:

$$c - c_o = \frac{2\gamma_{s,l} \, c_o \, \Omega}{RT} \cdot \frac{1}{\rho}$$

in which c is the solubility of a particle of the solid phase with radius ρ, c_o that of a flat surface of the solid and $\gamma_{s,l}$ the interfacial tension between solid and liquid. Analyses of the kinetics which govern the solution-reprecipitation process, commonly called "Ostwald ripening" have been given by Lifshitz and Slyozov[7] and by Carl Wagner.[8] Wagner showed that the kinetics are different for systems in which solution and reprecipitation are governed by diffusion of the solid through the liquid and those in which they are governed by reaction rate at the solid-liquid interface. The equations derived apply to systems in which the distance between solid particles is equal to or larger than the diameter of the particles. Efforts have been made to modify the equations[9] for the rate of growth to fit them for systems such as the heavy alloys in which the distances between particles are much smaller than their diameters. In postulating Ostwald ripening as the mechanism for growth of the solid particles it is generally assumed that particles with different orientations do not coalesce. This means that the dihedral angle, i.e. the angle which the liquid makes at the boundary between two grains of the solid phase is 0 or, at any rate, very small. The dihedral angle, as shown in Figure 11-4, is defined as

$$\theta = \text{arc cos} \, \frac{\gamma_{s,s}}{2 \, \gamma_{s,l}}$$

where $\gamma_{s,s}$ is the grain boundary tension between two grains of the solid phase. This equation assumes that the grain boundary tension is a constant, independent of the relative orientation of the solid grains. This is not strictly true, but applies primarily to large angle grain boundaries. A zero dihedral angle means that two solid-liquid interfaces can be maintained at a lower energy than one solid-solid interface or grain boundary, which makes coalescence of particles with different orientations impossible.

Growth by coalescence in the heavy alloy mechanism has, on the other hand, repeatedly been postulated. The process has also been treated analytically.[10,11] If the dihedral angle in the sintering system is positive, grain boundaries between coalescent particles will be formed and aggregates of two, three or more grains will be established. These aggregates may lead to the formation of a rigid skeleton. If at the time of skeleton formation complete densification has not yet been accomplished, the skeleton should

LIQUID

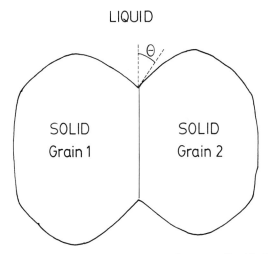

FIGURE 11-4 Schematic of dihedral angle between a liquid phase and two grains of a solid phase.

interfere with further densification. This will be discussed in connection with the solid state sintering stage. In the structure of sintered heavy alloy no polycrystalline aggregates, but only tungsten single crystals have been observed. The question has therefore been asked, how grain boundaries between coalescent particles may be eliminated. Elimination by grain boundary motion would require lengthening of the grain boundaries and therefore an increase in grain boundary energy. It has also been suggested that grain boundaries may be eliminated by grain boundary rotation; but this process appears to be too slow to be effective. Only in the case where small particles coalesce with large ones with formation of a small dihedral angle can grain boundaries be readily eliminated as was shown by Buist, Jackson, Stephenson, Ford and White.[12]

There is, however, another type of coalescence which was found by Huppmann and Yoon[13] when they sintered mixtures of large single crystal tungsten particles, 200-250 μm in diameter, and 2% nickel at 1640°C for times from a few minutes to an hour. As shown in Figure 11-5 the particles grew into their neighbors during sintering pushing a layer of liquid ahead of them. The migration of the layer of liquid evidently takes place by solution and reprecipitation across the boundary. Through etching and through microprobe tests Huppmann and Yoon were able to show that the composition of the particles on either side of the advancing liquid layer differed. The receding particle is pure tungsten, while the advancing particle is the equilibrium solid solution of 0.15% Ni in W. The advance of the boundary which

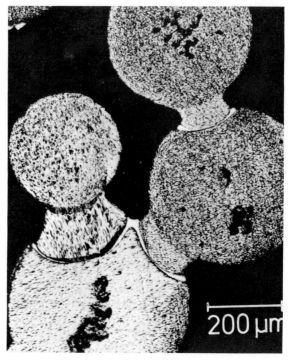

FIGURE 11-5 Microstructure of tungsten spheres of 200-250 μm diameter sintered with 2 weight % nickel for 60 minutes at 1640° C, etched in Murakami's solution.

eventually leads to coalescence into a single crystal is therefore driven by a chemical potential gradient due to the concentration differences. Up to now this type of coalescence has been observed only on aggregates of large tungsten particles and nickel, but it appears probable that it plays a major role in the initial rapid growth of the fine particles observed during sintering with the heavy alloy mechanism. On the other hand, there appears little doubt that particle growth during the later stages of the heavy alloy mechanism takes place by Ostwald ripening, as was shown by detailed studies of growth kinetics.[14]

Not only the mechanism of particle growth, but also that of densification has been the subject of controversy. In general, densification is a process of rearrangement of particles. It will be more pronounced when the conditions are conducive to grain growth. Grain growth by Ostwald ripening will lead to spherical grains of the solid phase. However, as already shown in the experiments of Price, Smithells and Williams,[1] when the quantity of liquid is small, complete densification is impossible if the spherical shape is maintained. Instead the grains form polyhedra with rounded corners which accomodate themselves to each other. Hence the term "accomodation stage" for the second stage of sintering. Kingery[6] has suggested a theoretical model for this process of accomodation between grains on the basis of which he derived equations for the rate of densification. He reasoned that the densification forces arising from the presence of pores produce very high stresses at the grain contacts due to which the solubility of the solid phase in the melt is markedly increased there. Because in other regions the solubility is lower, material is transported away from the contact regions and reprecipitated in other regions. However, in this explanation of "contact flattening" Kingery ignored grain growth. His mathematical analysis of the process assumed uniform and constant grain size. In recent experiments Huppmann and Yoon[2] sintered mixtures of large (250 μm) and small tungsten powder particles with nickel additions. During sintering the percentage of small particles decreased by solution and reprecipitation on the larger particles. As shown in Figure 11-6 the growth of these particles

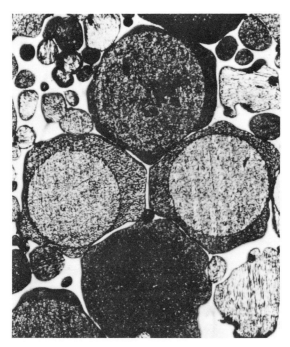

FIGURE 11-6 Microstructure of a mixture of 48 weight % large W spheres, 48 weight % fine W powder and 4 weight % Ni, sintered at 1670°C for 120 minutes, 125x.

occurred preferentially away from the areas where the grains are in close contact and thus shape accomodation occurred during growth and enabled complete densification. The contact flattening process postulated by Kingery does not appear to be important when the typical heavy alloy structure of large tungsten grains having polyhedral shapes is formed. This conclusion was confirmed by an order of magnitude calculation of the rates of densification by the two mechanisms.

Third Stage:

Cannon and Lenel[3] have postulated that in liquid phase sintering of systems with a positive dihedral angle the multigrain aggregates formed from coalescing solid particles will lead to rigid skeletons which will interfere with further densification. The formation of such a skeleton was made probable in experiments with compacts from 75% carbonyl iron powder and 25% copper powder sintered at 1100° C. Some of these compacts attained near theoretical density after sintering for a few hours, while others reached a limiting level of density short of complete densification even after sintering for many hours. Samples from both types of compacts were made anodes in an experiment, in which the copper was dissolved in a cyanide plating bath by electrolysis. Those of near theoretical density, in which the individual iron particles were presumably separated by copper, disintegrated rapidly during electrolyzing, while those with a limiting lower density having a rigid iron skeleton could not be disintegrated by electrolyzing. Further densification after a rigid skeleton has been formed would be a third stage of sintering. Its rate would be much smaller than densification during the second stage, because the process would be similar to that in solid state sintering.

II. Sintering of Compacts from Mixtures of Iron and Copper Powder

Compacts from a mixture of iron powder with the usual commercial particle size distribution, i.e. particle sizes mostly between 44 and 150 μm, and of up to 20% copper powder generally expand when they are sintered at temperatures where a liquid phase is present. A liquid phase is present at temperatures above 1094° C, the peritectic temperature in the iron-copper system. This expansion was first attributed to the fact that the liquid copper diffuses into the iron and that thereby the volume of the iron increases. This phenomenon is the Kirkendall effect which was discussed in chapter 10. The fact that diffusion takes place from a liquid solution of iron in copper into iron rather than between two solids does not change the picture. A dilatometric study by Dautzenberg[15] showed, however, conclusively that the diffusion of copper into iron cannot be the only cause of expansion. Dautzenberg studied the dimensional changes of compacts pressed to a density of 6.75 g/cm³. He compared the behavior of compacts from pure iron powder with that of compacts with 3% copper, which were heated to 1200° C at a rate of 300° C/hour. As seen in Figure 11-7, the behavior of the two types of compacts is practically identical up to the temperature where a liquid phase appears. At this temperature, however, the iron-copper powder compacts expand almost instantaneously by about 1%. Such instantaneous swelling cannot be due to diffusion which is relatively slow, but must be due to penetration of the liquid phase into the interstices between iron particles. The contribution of diffusion, on the one hand, and of penetration, on the other hand, were determined in model experiments by Kaysser.[16] He compacted mixtures of iron powder and 10% copper powder at a pressure of 785 MPa. Both iron and copper powder particles

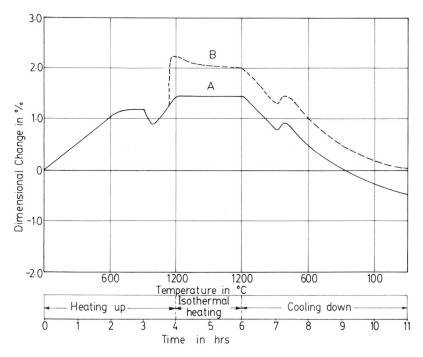

FIGURE 11-7 Expansion of compacts from pure iron powder (A) and from a mixture of iron powder with 3% Cu (B), heated at a rate of 300° C/hr to 1200° C, held 2 hours at 1200° C and cooled at the same rate.

were spherical and had diameters of 100 μm. The compacts were sintered isothermally at 1165° C. Photomicrographs of a green compact and of one sintered for 8 minutes are shown in Figures 11-8 and 11-9. It is seen that the liquid, a saturated solution of iron in copper, has penetrated first into the interstices between iron particles and then into the grain boundaries within the iron particles. Because of the high green density of the compacts no rearrangement of iron particles has occurred during the first 8 minutes of sintering. The space originally occupied by copper particles are pores. With no rearrangement of iron particles, no shrinkage counteracting the observed swelling takes place. How much of the swelling of the compacts is due to penetration of the liquid phase first into the interstices between and then into the grain boundaries within the iron particles and how much is due to diffusion of copper into iron was determined by quantitative metallographic analysis. It was assumed that the width of the liquid copper seams between particles stays constant and that the iron dissolved at the grain boundaries is reprecipitated at the particle interfaces. The total increase in volume of the compacts in the model experiments in the first 8 minutes of sintering amounts to 6%. The analysis shows that 65% of this increase is due

FIGURE 11-8   Micrograph of green compact pressed at 785 MPa from a mixture of 90% spherical iron powder particles and 10% spherical copper particles, both of them 100 μm in diameter.

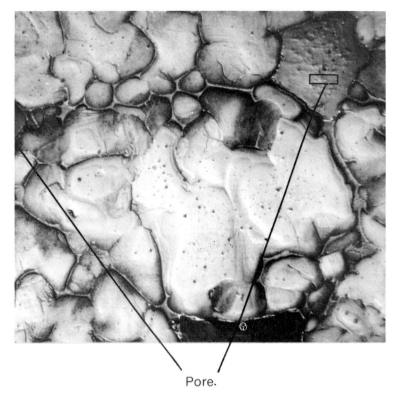

Pore.

FIGURE 11-9 Micrograph of the same compact as in Figure 11-8 after sintering for 8 minutes at 1165°C.

to penetration of copper into the interstices between particles and into grain boundaries and 35% is due to diffusion of copper into iron. It should be noted that the penetration of the liquid solution of iron in copper into the iron grain boundaries indicates that the dihedral angle between this liquid and iron is near zero at the sintering temperature of 1165°C. Other measurements[17] have shown that the dihedral angle is positive at temperatures below as well as above 1165°C and that it is also positive when the liquid solution of iron in copper is in contact with a saturated solid solution of copper in iron rather than with more or less pure iron.

While during the first 8 minutes of sintering in this model experiment rearrangement of the iron particles can be excluded, this is no longer the case for longer sintering times. A photomicrograph of a 90% Fe - 10% Cu compact after 23 minutes of sintering is shown in Figure 11-10. As a consequence of the progressive penetration of the melt along the grain

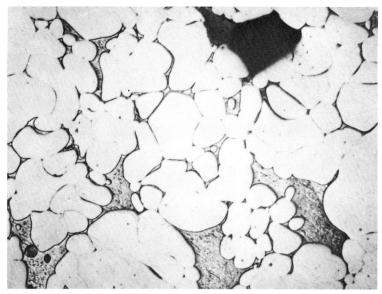

FIGURE 11-10 Micrograph of the same compact as in Figure 11-8 and 11-9 after sintering 23 minutes at 1165°C.

boundaries within the iron particles, individual grains have been separated from the particles. Where these separated grains have migrated into the neighboring melt, they have been rounded through solution and reprecipitation. The rearrangement of the iron particles occurring in this stage of sintering will cause shrinkage of the compacts counteracting the expansion due to the penetration of copper and its diffusion into the iron. During continued sintering a new phenomenon is observed as seen in Figure 11-11, which shows the structure of a loose powder aggregate of 60% spherical iron and 40% spherical copper particles sintered for 20 hours. Small bridges are formed between the neighboring grains of the iron-copper solid solution which originally were separated by a thin layer of melt. This is the beginning of the skeleton formation discussed in the section on the heavy alloy mechanism.

The model experiments on compacts and loose powder aggregates of mixtures of iron and copper powders show the following phenomena occurring during sintering:

> Movement of the liquid phase by capillary pressure
> Interdiffusion between phases
> Rearrangement of solid particles
> Solution and reprecipitation

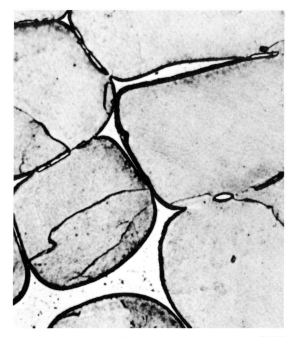

FIGURE 11-11   Micrograph of a loose powder aggregate of 60% spherical iron and 40% spherical copper powder particles sintered 20 hours at 1165°C, 750x.

> Particle coalescence
> Grain growth
> Skeleton formation

For the early stages of sintering in highly pressed 90% Fe - 10% Cu compacts the contributions to the dimensional change (swelling) of the compact by the movement of the liquid phase and by interdiffusion could be determined quantitatively. The effects of the other phenomena could be shown qualitatively by photomicrography.

III.   Sintering of Compacts from Mixtures of Copper and Tin Powders

The analysis of the sintering process for compacts from a mixture of 90% copper and 10% tin powder at a temperature below the solidus temperature of the alloy is made difficult by the complexity of the copper-tin equilibrium diagram which is shown in Figure 11-12. The final product of the sintering

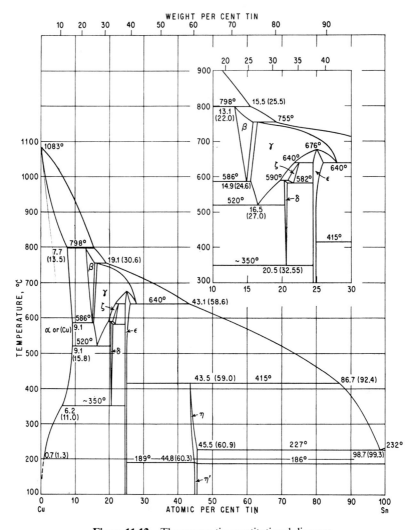

Figure 11-12 The copper-tin constitutional diagram.

process is a homogeneous face-centered cubic solid solution of tin in copper. However, on the way to this structure, several intermediate solid or intermetallic compounds of copper and tin and liquid phases of varying compositions may be expected to appear, since the powders in the green compacts are pure copper and tin. Some of these intermediate phases are the hexagonal η, $Cu_6 Sn_5$, the pseudohexagonal ϵ, $Cu_3 Sn$, the cubic γ, $Cu_3 Sn$, the body centered cubic β and the cubic δ, Cu_{31} Sn_8. Of these phases, Duwez[18] has identified by X-ray diffraction the η and the ϵ phase in compacts which were cooled to room temperature, after being heated at a rate of 3.5° C/min to temperatures between 180 and 370° C for the η phase and between 180 and 480° C for the ϵ phase. The γ and β phases which are stable only at elevated temperatures could be identified only by quenching the partially sintered alloys or by high temperature X-ray diffraction analysis. Such experiments have not yet been made. The appearance of the β phase during sintering will depend upon the size and shape of the powder particles, the compacting pressure and therefore the greeen density of the compacts and particularly the rate of heating the compacts to the sintering temperature. On the basis of photomicrographs of partially sintered compacts and of moving pictures of the changes in structure during sintering, two scenarios of the sintering process have been postulated. According to the first scenario, suggested by Drapeau,[19] Hall[20] and Sauerwald,[21] the tin, when it melts, is drawn by capillary force into the voids between the copper particles forming a cementing medium whereupon the interdiffusion of tin and copper, i.e. diffusion of tin into the copper, and, at the same time, solution of the copper in the liquid tin-rich phase follows. This scenerio describes, in a general way, what happens during sintering of self-lubricating bearings of a copper-tin powder mixture which are heated quite rapidly to the final sintering temperature between 800 and 830° C. Metallographic evidence of the formation of pores in the place of tin particles in the green compacts, which have melted and been drawn into voids, has been given by K. Watanabe.[22]

Another scenario has been suggested by W. L. Wain[23] who investigated the sintering of compacts of 88.2% Cu, 9.8% Sn and 2% graphite using compacting pressures of 35, 70, 105, 165, 280, 490 and 700 MPa and sintering temperatures of 300, 500, 700, 800 and 830° C. On the basis of his micrographs, particularly those of compacts sintered at 300 and 500° C, Wain describes the sintering process as follows:

"When the tin particles melt, they rapidly dissolve sufficient copper to form the higher melting point η solid solution—the compound Cu_3Sn with additional tin in solid solution according to Hoyt's diagram—which is solid up to temperatures above 600° C. Thus the tin rich areas at this stage consist of a solid shell of η possibly containing a molten core of higher tin content depending on the size of the original tin particles. This core gradually absorbs copper by solid diffusion through the η shell. The formation of a relatively high melting point envelope explains the lack of evidence of any flowing of molten tin or tin rich solution into the pores and cavities of the copper matrix. On further heating to 500° C the diffusion trend is for tin to diffuse out into the copper matrix; thus one gets α bronze fringes around the tin areas which at this stage are duplex η plus δ. With continued heating at 500° C the α bronze area increases in extent at the expense first of η then of δ".

This description may have validity for compacts pressed at low pressure from fine tin powder and heated quite gradually to the final sintering temperature.

In the industrial production of self-lubricating bearings the dimensional changes during sintering—generally an expansion—and the control of these changes are very important. Several dilatometric experiments have been made, in which compacts from a mixture of 90% Cu and 10% Sn powders were heated at a uniform rate to the sintering temperature and the dimensional changes as a function of temperature measured. Certain features are common to these curves, but details vary widely again depending upon the size and size distribution of both the copper and tin powders, the compacting pressure and the rate of heating. Also, any additions of graphite powder and of lubricant to the mixture of copper and tin powders alter the character of the curves markedly. Up to temperatures of about 600° C small changes in slope of the dilatometric curves are observed at temperatures where a liquid phase is formed, but the amounts of expansion and shrinkage in the sintered compacts corresponding to these slope changes are minor compared to what happens at higher temperature. They are, however, of interest because, from these slope changes, conclusions may be drawn as to the type of phenomena occurring in this temperature range. Esper and Zeller,[24] e.g., found in the temperature range just above the melting point of tin, that the slope became steeper for compacts pressed at 400 MPa and shallower for compacts pressed at 100 MPa. The shallower slope must be interpreted as due to shrinkage, when particles rearrange themselves upon the appearance of a liquid phase. In compacts pressed at higher pressure shrinkage due to rearrangement does not occur. The steeper slope above the tin melting point should be attributed to the liquid phase penetrating between the solid particles, to the faster diffusion of the tin or the tin-rich liquid phase into the solid copper-rich particles rather than vice versa (Kirkendall effect) or to both of these effects. The dilatometric curves of Duwez,[18] who pressed his compacts at relatively high pressures of 280, 420 and 560 MPa and heated them at the slow rate of 3.5° C/min, have a steeper slope than the one corresponding to thermal expansion in the temperature ranges just above the melting point of tin and from 415 to 480° C where the reaction $\eta \rightarrow \epsilon + L$ would produce a new liquid phase. Again this swelling may be attributed to liquid phase penetrating between the particles, to the Kirkendall effect or to both phenomena. Many of the dilatometric curves, e.g. those of Duwez,[18] of Esper and Zeller[24] (for 8 μm diameter copper powder) and of Peissker[25] show rapid shrinkage in the temperature range between 600 and 750° C, which would be tentatively attributed to rearrangement in the presence of a liquid phase. At a temperature of 798° C Mitani[26] observed a sudden expansion, which he called "abnormal expansion". The observation was confirmed by Peissker.[25] This expansion is probably the reason why self-lubricating bearings grow during sintering. It is directly related to the appearance of a liquid phase due to the peritectic reaction $\beta \rightarrow L + \alpha$, which occurs at 798° C. If much of the tin has already diffused into the α solid solution before the compacts have reached 798° C, the expansion at this temperature is small; if much of the tin is still present as β solid solution or tin-rich liquid, the expansion is large. The degree of expansion can therefore be reduced by

1. heating the compacts slowly, i.e. giving interdiffusion time to occur,
2. using copper powder which is fine and has a large specific surface and tin powder which is also fine,
3. using compacts with a low green density.

Mitani's investigation indicates that the expansion at 798° C is to some extent due to the release of gas, e.g. hydrogen, from the sintering atmosphere, when a liquid is formed during the peritectic reaction $\beta \rightarrow L + \alpha$. Finally, a homogeneous structure of grains of the α face-centered cubic solid solution of tin in copper is formed by interdiffusion. In producing self-lubricating bearings the sintering temperature is generally above 800° C but below the solidus temperature of the solid solution of 10% Sn in Cu, because in this temperature range the rate of homogenization is considerably faster than at lower temperatures and is accompanied by grain growth. A coarse-grained structure is desirable for the bearings.

In sintering compacts from a mixture of copper and tin powder the same phenomena appear to play a role which are observed in sintering with the heavy alloy mechanism and in the sintering of iron-copper powder compacts. They are primarily interdiffusion between phases, rearrangement of the solid particles in the presence of a liquid phase, movement of the liquid phase by capillary pressure and grain growth. However, the details of the processes and their rates, as they depend upon particle size, green density and rate of heating, are not yet well understood.

References

1. G. H. S. Price, C. J. Smithells and S. V. Williams, Sintered alloys. Part I: Copper-nickel-tungsten alloys sintered with a liquid phase present, J. of the Inst. of Metals, vol. 62, p. 239-264, (1938).

2. D. N. Yoon and W. J. Huppmann, Grain growth and densification during liquid phase sintering W-Ni, Acta Met., vol. 27, No. 4, p. 693-698, (1979).

3. H. S. Cannon and F. V. Lenel, Some observations on the mechanism of liquid phase sintering, Proc. 1st Plansee Seminar, ed. by F. Benesovsky, Reutte, Austria, p. 106-121, (1953).

4. W. J. Huppmann, Sintering in the presence of a liquid phase, in "Sintering and Catalysis", ed. by G. C. Kuczynski, N.Y., (1975), p. 359-378.

5. W. J. Huppmann and H. Riegger, Modelling of rearrangement processes in liquid phase sintering, Acta Met., vol. 23, No. 8, p. 965-971, 1975.

6. W. D. Kingery, Densification during sintering in the presence of a liquid phase, I, Theory, J. Appl. Phys., vol. 30, p. 301-306, (1959).

7. I. M. Lifshitz and V. V. Slyozov, J. Phys. Chem. Solids, vol. 19, p. 35 (1961).

8. Carl Wagner, Theory of the aging of precipitates by solutions and reprecipitation, Z. f. Elektrochemie, vol. 65, No. 7/8, p. 581-591, (1961).

9. A. J. Ardell, The effect of volume fraction on particle coarsening: theoretical considerations, Acta Met., vol. 20, No. 1, p. 61-71, (1972).

10. T. H. Courtney, Microstructural evolution during liquid phase sintering, Part I, Development of microstructure, Met. Trans., vol. 8A, No. 5, p. 679-684 (1977).

11. T. H. Courtney, Part II, Microstructural coarsening, ibid., p. 685-689.

12. D. S. Buist, G. Jackson, I. M. Stephenson, W. F. Ford, J. White, Trans. Brit. Ceram. Soc., vol. 64, p. 173 (1965).

13. D. N. Yoon and W. J. Huppmann, Chemically driven growth of tungsten grains during sintering in liquid phase, Acta Met., vol. 27, No. 6, p. 973-977 (1979).

14. T. K. Kang and D. N. Yoon, Coarsening of tungsten grains in liquid nickel-tungsten matrix, Met. Trans., vol. 9A, No. 3, p. 433-438 (1978).

15. N. Dautzenberg, Observation of sintering processes in the system Fe/Cu by dilatometry and hot-stage microscopy, Arch. f. das Eisenhüttenwesen, vol. 41, p. 1005-1010, (1970).

16. W. Kaysser, Comparative investigations of liquid phase sintering in the systems Fe-Cu, Fe-Sn, Cu-Bi, Cu-Sn, Ni-Bi, W-Cu, Ph.D. thesis, Univ. of Stuttgart, (1978).

17. D. Berner, H. E. Exner, and G. Petzow, Swelling of iron-copper mixtures during sintering and infiltration, Mod. Dev. in Powder Metallurgy, vol. 6, ed. by H. H. Hausner and W. E. Smith, Princeton (1974), p. 237-250.

18. P. Duwez, The formation of alloys by diffusion in powder metallurgy, Powder Metallurgy Bulletin, vol. 4, No. 5, p. 144-156, and vol. 4, No. 6, p. 168-174 (1948).

19. J. E. Drapeau, Sintering of powdered copper-tin mixtures, Ch. 32, p. 332, in "Powder Metallurgy", ed. by J. Wulff, Cleveland, (1942).

20. H. E. Hall, Sintering of copper and tin powder, Metals and Alloys, vol. 10, p. 297 (1939).

21. F. Sauerwald, Present status of powder metallurgy, Metallwirtschaft, vol. 20, p. 649 and 671, (1941).

22. K. Watanabe, J. Japan Soc. Powder Met., vol. 12, p. 219, (1965).

23. H. L. Wain, Powder metallurgy, influence of some processing variables on the

properties of sintered bronze, ACA-25, Australia Council for Aeronautics, Melbourne, (1946).

24. F. J. Esper and R. Zeller, The sintering process of 90% copper-10% tin compositions, Powder Metallurgy Int., vol. 1, No. 3, p. 1-5, (1970).

25. E. Peissker, Pressing and sintering characteristics of powder mixtures for sintered bronze 90/10 containing different amounts of free tin, Mod. Dev. in Powder Metallurgy, vol. 6, ed. by H. H. Hausner and W. E. Smith, p. 597-614 (1974).

26. H. Mitani, Abnormal expansion of Cu-Sn powder compacts during sintering, Trans. Japan J.I.M., vol. 3, p. 244-251 (1952).

12

Loose Powder Sintering, Slip Casting and Infiltration

In this chapter three subjects are treated:

> The technology of sintering loose powders
> The slip casting process for metal powders
> Infiltrating porous compacts with a lower melting
>   point metal.

I. Loose Powder Sintering

From the discussion of the mechanism of sintering in chapter 9 it is clear that contacts between powder particles are established and grow by the action of capillary or surface tension forces when the powder particles are heated in contact with each other. Application of pressure as in compacting is not necessary for this purpose. An early classical example of loose powder sintering is the investigation of Schlecht et al.,[1] in which spherical carbonyl iron powder with particle sizes in the range between 1 and 10 μm was filled into porous ceramic tubes and sintered without the application of pressure. Sintering for 24 hours at 800° C in hydrogen resulted in sintered rods with a density of 7.12 g/cm^3. Nevertheless, in most technological applications of powder metallurgy a compacting step precedes sintering. This permits the production of metal powder parts with relatively low porosity and close control of the part dimensions. When semifinished products such as wire, rod, plate, strip, sheet or tubing are to be produced, densities close to theoretical can be achieved more readily for most metal powders by starting with sintered pressed compacts. The principal technological applications for loose powder sintering are in porous metallic materials, where porosities from 40 volume percent to as high as 90 volume percent are required. Two examples of such porous materials are bronze filters and porous nickel membranes used as electrodes for alkaline storage batteries and fuel cells.

The first step in producing bronze filters is to fill spherical bronze powder of a closely controlled particle size range into a mold. The mold is leveled and vibrated. The powder may be atomized from a liquid 90% copper-10% tin alloy or may be a spherical copper powder coated with a layer of very fine tin powder which has been partially alloyed with the copper. The particle size range is adjusted according to the size of the particles which the filter is to remove and may be as coarse as 20-30 mesh or as fine as 80-125 mesh. Graphite molds and molds from stainless steel are widely used. Their shape reproduces that of the filter to be sintered. Mold and powder are sintered in a reducing atmosphere at a temperature near the solidus temperature of the alloy for times on the order of 30 minutes. The powder particles sinter together with the filter having dimensions slightly smaller than those of the mold in which they are sintered and porosities in the range from 40 to 50%. The resulting product has quite adequate strength for a filter because of the strong bonds between the spherical powder particles.

The fabrication of porous nickel membranes[2] by loose powder sintering is similar to that of bronze filters. Two types of carbonyl nickel powder are used. The fibrous type has apparent densities below 1 g/cm^3 and particle sizes in the range from 2.6 to 3.6 μm. The equiaxed type has apparent densities in the range from 2.0 to 2.5 g/cm^3 and particle sizes from 4 to 7 μm. After the powder has been filled into the mold, leveled and vibrated, mold and powder are sintered for times on the order of 15 minutes in hydrogen at temperatures in the range from 700-1000° C. Most of the porosity in the loose powder bed is retained in the final sintered plaque. For membranes from the fibrous carbonyl nickel powders the porosities are in the range from 80 to 90%, with pore sizes in the neighborhood of 10-15 μm, for those from equiaxed powder the porosities are in the range from 50 to 60% with finer pores sizes near 5 μm. In order to increase the electrical conductivity of porous nickel membranes they are often fabricated with a central support of perforated strip or woven wire mesh. Details will be described in chapter 16.

A variation of loose powder sintering for porous nickel membranes is slurry sintering.[2] Woven wire mesh or perforated strip is coated by immersing it in a slurry of nickel powder in a suitable vehicle, such as a 3% solution of methyl cellulose in water, which also assists in bonding the slurry to the strip. The coated strip is dried and sintered. Slurry sintered membranes are slightly less porous and have finer pores than loose powder sintered membranes, since the powder packs somewhat more closely in the liquid slurry.

An interesting example in which loose powder sintering has been proposed to produce a product with near theoretical density is the pressureless sintering of loose beryllium powder to densities of 95-98% of theoretical.[3] This technique of sintering beryllium powder is an alternative to the usual hot pressing of the powder in vacuum. In order to achieve the desired shrinkage of the powder aggregates on the order of 20% linearly, the particle size of the powder and the sintering conditions must be controlled. Particles with average diameters of 20-35 μm are used, which are obtained by milling beryllium turnings. The sintering temperature is 1200 to 1220° C and the sintering time 4 to 12 hours. The powder is filled into graphite molds, hand tapped and mechanically vibrated. Molds and powder are sintered by induction heating in a diffusion pump vacuum. The advantages claimed for the loose powder sintering technique are its simplicity, versatility and economy of operation. Shapes of high length to diameter ratio can be readily produced. Nuclear fuel elements may be produced directly by sintering the beryllium powder around a uranium-ceramic core.

Loose powder sintering is occasionally used as an intermediate step in producing powder metallurgy products. Two examples are found in the fabrication of steel-backed precision automotive main and connecting rod bearings. For the production of one type of these bearings a layer of copper and nickel powder having a given thickness is applied loose to a cold rolled steel strip. The strip with the applied loose powder layer is run continuously through a roller hearth furnace and heated near the melting point of copper, thereby forming a copper-nickel sponge layer on the strip. This sponge layer is later vacuum infiltrated with a lead base babbit. The purpose of the sponge is to provide an excellent bond between steel and lead base babbit.

In the production of another type of bearing a layer of atomized copper-lead powder is applied loose to cold drawn steel strip to a certain thickness. The strip with the loose powder layer is sintered at a temperature where the powder bonds to the steel. The porous copper lead alloy layer is later made dense by passing the strip through a rolling mill and then resintering it. Details for both procedures are presented in chapter 22.

II. Slip Casting

Metal powders may be slip cast into a desired shape by a technique similar to slip casting of ceramics.[4] A porous mold is prepared from plaster of Paris,

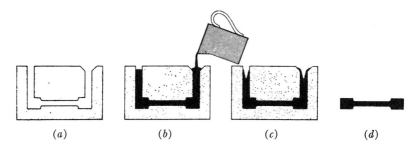

FIGURE 12-1 Schematic of slip casting of metal powders (a) assembled mold, (b) filling the mold, (c) absorbing water from the slip, (d) finished piece, removed from the mold and trimmed.

the compound Ca SO_4 . $\frac{1}{2}$ H_2O. The mold has the negative of the shape to be slip cast. The slip to be cast consists of a suspension of the metal powder in water. The slip should have a low viscosity so that it can be readily poured, it should be stable during standing and have a low rate of settling; the slip cast shape should easily release from the mold, it should have a low shrinkage on drying and high strength after drying. Using sufficiently fine powder with particles under 5 μm these properties of the slip may be achieved in many ceramics by the control of the p_H of the slip. Through this control electrical charges are built up on the surface of the powder particles which make the particles repel each other and prevent deflocculation. A plot of the p_H of a slip vs its viscosity often shows a minimum viscosity at a given p_H value, which is the optimum for slip casting. In metal powder slips it is generally necessary to use both a deflocculant and control of the p_H of the slip to obtain optimum viscosity. For very fine molybdenum powder a slip can be prepared by suspending the powder in a 5% aqueous polyvinyl alcohol solution with a minimum viscosity at a p_H value of 7. For a coarser type 316 spherical stainless steel powder (80% –325 mesh powder) Hausner[5] prepared slips by mixing 80.7% metal powder, 19% water and 0.3% of sodium alginate as a deflocculant and adjusting the p_H to a value near 10.0. The actual process of slip casting is shown schematically in Figure 12-1, which shows the assembled plaster mold, filling the mold, absorbing water from the slip into the porous mold and the finished piece removed from the mold and trimmed. For successful slip casting the time for letting the slip set in the mold and the time for letting the slip cast shape dry in air at room temperature before drying it further in an oven and then sintering it are critical. The density of a slip casting is considerably higher than the tap density of the powder from which the casting is made, since suspension in the slip decreases the friction between powder particles. Hausner gave values for the density of his stainless steel powder slip castings as

4.88 g/cm^3 as cast

4.62 g/cm^3 as dried

and

6.94 g/cm^3 as sintered for 2 hours at $1300°$ C in dry hydrogen.

W. D. Jones[6] enumerated the advantages and disadvantages of slip casting as a process for producing shapes from metal powder:

1. articles can be made with shapes or in large sizes that could not possibly be pressed;
2. no expensive equipment is required;
3. it works best with the finest powders, which are at the same time most suited to sintering;
4. consequently the finished products have excellent physical properties.

The principal disadvantage is that the process at present is a craftsman's job, and therefore slow and costly in man hours. At the present time, slip casting of metal powder has very limited commercial applications.

III. Infiltration

Infiltration is the process in which the pores of a sintered solid are filled with a liquid metal or alloy. The solid must have a melting point considerably higher than that of the infiltrating metal or alloy. Kieffer and Benesovsky[7] have given an extensive table of possible binary systems in which a skeleton of the higher melting phase may be infiltrated by the lower melting metal. The number of systems where infiltration is used commercially, in particular those of W and Mo infiltrated with Cu and Ag and that of Fe infiltrated with Cu, is quite limited, but the quantity of infiltrated parts produced in this limited number of systems is large and infiltration is therefore an important powder metallurgical process. The theoretical aspects of the infiltration of powder metallurgy products have been discussed in several papers,[8,9,10] but will be treated here only in a qualitative manner. In order to infiltrate a porous skeleton of a solid phase with a liquid phase it is necessary that the total surface free energy of the system after the infiltration, including the surface free energies of the solid and liquid phases by themselves and the interfacial energy between solid and liquid, be lower than the total surface free energy before infiltration. The relationship between the surface free energies of solid and liquid and the interfacial solid and liquid free energy is given by the relation

$$\gamma_{s,l} = \gamma_s - \gamma_l \cos \theta$$

in which $\gamma_{s,l}$ is the solid-liquid interfacial free energies, γ_s and γ_l the surface free energies of solid and liquid and θ the liquid-solid contact or wetting angle. A low wetting angle, i.e. ready wetting of the solid by the liquid, therefore, promotes infiltration. Infiltration may be further aided by establishing a pressure gradient, e.g. putting the porous skeleton in a vacuum and applying pressure to the liquid which is to infiltrate the porous skeleton. As Shaler and coworkers[11] have shown, wetting may also be promoted by additions, which may be gaseous, liquid or solid, to the systems to be infiltrated. Shaler has called these additions "detergents". The rate of infiltration has been related to the rate of flow of liquid through a capillary tube, i.e. Poiseuille's law. On the basis of this analogy Semlak and Rhines[12] derived an equation for the rate of rise of a liquid metal through a porous skeleton which is perfectly wetted

$$h = \frac{2}{\pi} \left(\frac{R_c \, \gamma \, t}{2\eta} \right)^{1/2}$$

in which h is the height of rise in the time t for a liquid of viscosity η and surface energy γ in a porous skeleton with an "average capillary radius" R_c. The parabolic rate of rise during the early stage of infiltration could be clearly demonstrated by the authors, but they found it difficult to assign an "average capillary radius" to any dimension measured on the porous skeleton. The infiltration of a porous skeleton with a liquid metal comprises the actual penetration stage and a stage, after the pore volume has been filled, in which the infiltrant while still liquid will react with the skeleton. For an understanding of these stages a number of points must be considered in addition to the basic requirement of lowering the total surface free energy:

1. The liquid and the solid must not react to form a solid compound or alloy having a specific volume as great as or greater than their combined preinfiltration specific volumes; otherwise the reaction product blocks the entry of additional liquid before infiltration is complete.

2. The porosity in the porous skeleton should be interconnected since the infiltrant cannot penetrate into closed-off pores in the skeleton, except under the conditions shown under 4.

3. Ideally the skeleton material should be insoluble in the liquid infiltrant, as is the case in the systems W-Cu and W-Ag. The solubility of γ-iron in liquid copper at and above the melting point of copper, which amounts to several percent, complicates the infiltration of iron and steel by copper as will be discussed later.

4. The dihedral angle at the intersection of a boundary of two grains of the solid phase and the liquid phase plays an important role in infiltration. If the dihedral angle is zero, the liquid will penetrate along grain boundaries in the solid skeleton. This may be advantageous because the liquid may in this way reach closed-off pores in the skeleton. On the other hand, the final properties of the infiltrated material may depend upon the good mechanical properties of the porous skeleton in the as-sintered condition. Penetration of the liquid along the grain boundaries of the skeleton may considerably weaken the original porous skeleton structure. It may also make the control of dimensional changes during infiltration more difficult.

The aim of infiltration is to obtain a more or less pore-free structure. The question may be asked whether this aim may not be reached more readily by liquid phase sintering, in which the materials forming the skeleton and the infiltrant are mixed as powders, compacted and sintered at a temperature above the melting point or eutectic temperature of the liquid. As was outlined in chapter 11, such a pore-free structure may indeed be obtained in systems which can be sintered with the "heavy alloy mechanism". For this reason most of the development work on cemented carbides, in which a porous carbide skeleton was infiltrated by the binder metal,[13] has not led to extensive commercial applications, since liquid phase sintering using the "heavy alloy mechanism" proved a simpler approach. On the other hand, in such systems as tungsten-copper in which the liquid is insoluble in the skeleton material, simple liquid phase sintering does not lead to pore-free structures for the reasons shown in chapter 11 and infiltration must be resorted to. In chapter 11 the sintering of compacts from mixtures of iron and copper is also discussed. Liquid phase sintering may lead to pore-free structures in compacts from a very fine iron powder, but this approach is generally more expensive than infiltration.

In the following section applications of the infiltration process including the methods used for infiltrating the porous skeletons will be discussed. Electrical contacts[14] are probably the oldest powder metallurgical material produced by infiltration. They are most commonly compound materials with a tungsten skeleton infiltrated with copper. Combinations of tungsten and silver, of molybdenum and silver and of tungsten carbide and copper are also used. The tungsten-copper compound materials have also found an important application for resistance welding electrodes. When a tungsten-copper material with a maximum copper content of 25% is to be produced, a compact is pressed from a tungsten powder coarse enough so that it can be readily infiltrated. The compacting pressure is chosen to give the desired porosity in the compact. It is sintered at about 1200° C and infiltrated by dipping it in molten copper. For materials with higher copper contents, loose tungsten powder is filled in a graphite container and a body of copper of the appropriate mass positioned on top of the tungsten powder. The assembly is heated to a temperature at which the molten copper penetrates the porous tungsten body. The properties of these materials will be discussed in chapter 24 on electrical applications of P/M materials.

An interesting example of an infiltrated tungsten-silver compound material[15] was developed for use as billets for rocket nozzles weighing as much as 200 lbs with dimensions of 10″ in diameter and 6″ long. The fabrication of these nozzles, in which each step has to be very carefully controlled, has been described in some detail. Tungsten powder of a particle size of 4-5 μm is isostatically pressed at 10 to 20 tsi into a hollow or solid cylinder with 60% of theoretical density. The compact is sintered in a resistance or an induction furnace in dry hydrogen according to a carefully controlled sintering schedule to a density of 75-83% of theoretical. The sintered compact is infiltrated in hydrogen with pure liquid silver which is made to penetrate the compact either from the top down or from the bottom up. The hydrogen which is dissolved in the liquid silver is released during solidification. It takes up the void space, also produced during solidification of the silver. The uniformly distributed voids in the infiltrated material serve to avoid local areas of stress concentration when the silver expands during use of the nozzle. The composite material must have a minimum tensile strength of 228 MPa (33,000 psi) at 1500° F (825° C) and a sufficiently high thermal conductivity to minimize thermal stresses when used as a rocket nozzle.

In the 1950's the infiltration process was applied to the production of turbine blades with complex shapes,[16] in which a skeleton of titanium carbide was infiltrated with a liquid nickel or cobalt base superalloy in vacuo. The process was similar to investment casting in inert molds. It made it possible to produce graded products, in which the metal content changed from nearly 100% in the root tip and the leading and trailing edge, to below 50% in the center of the airfoil. Such a graded product was to accomodate a wide range of thermal and mechanical stresses along the length and width of the product.

A similar infiltration process of a titanium carbide skeleton with a variety of liquid steel alloys was the original basis of the Ferro-tic cemented carbide alloys. However, these alloys are now produced by liquid phase sintering of a mixture of titanium carbide and metal powders.

The most widely used infiltration process is the one in which the porous skeleton has an iron or steel composition and the infiltrant is copper or copper alloy. This process is used for ferrous structural parts which have densities of 7.4 g/cm^3 and higher and better mechanical properties than parts which have been only compacted and sintered. Depending upon the application, the porous skeleton may be only partially or almost completely infiltrated. The usual method of infiltrating iron and steel skeletons is to place a compact pressed from the powder of the infiltrant material next to the skeleton. The compact of infiltrant powder may be positioned on top or underneath the skeleton compact or two infiltrant compacts may be used, one on top, the other underneath the skeleton compact. The exact amount of infiltrant needed may be compacted in the same die in which the powder for the porous skeleton is pressed. After the skeleton has been sintered, the green compact or compacts of infiltrant powder are positioned next to the skeleton and the assembly heated to the infiltration temperature. Sintering of the skeleton and infiltration may be combined into one operation, in which the green compact or compacts of the infiltrant is positioned next to the green compact of the skeleton. By controlling the rate of heating, the skeleton compact will be adequately sintered by the time the melting point of the infiltrant is reached. This operation has been called "sintrating". If only part of the porous skeleton is to be infiltrated, e.g. the teeth of an infiltrated gear, the skeleton may be positioned in a graphite container in which space is provided adjacent to the gear teeth to be preferentially infiltrated. This space is then filled with the appropriate amount of infiltrant in powder form. Because iron has some solubility in copper, the liquid infiltrant will attack the surface of the skeleton when it first comes in contact with it and severe erosion may take place at this point. One method to minimize erosion is to use a copper alloy, e.g. an 80% Cu, 20% Zn brass, as infiltrant. Because the brass does not melt at one temperature, but over a range of temperatures, less erosion takes place. Another method is to use an alloy of copper with iron as an infiltrant. However, it is difficult to specify the exact amount of iron since too little iron may cause erosion and too much iron an undesirable adherent deposit on the infiltrated part. A third method[15] is to use an alloy of copper, iron and a third alloying constitutent, such as manganese, which oxidizes in the atmosphere in which infiltration takes place. In this case a crust containing an oxide of the oxidizing alloying ingredient of the infiltrant is formed which does not adhere to the skeleton, but can be more or less readily removed. The control of dimensions during infiltration may cause problems. It depends upon the composition of the steel with carbon free iron showing the greatest growth during infiltration. It also depends upon the exact temperature of infiltration and the time during which the infiltrated compact is above the liquidus temperature of the infiltrant. The problems are somewhat similar to those in liquid phase sintering of iron-copper compacts which were discussed in chapter 11.

A last example of the commercial use of the infiltration process is in the fabrication of one type of steel backed precision automotive main and connecting rod bearings. A steel strip to which a sponge layer of a nickel-copper or an iron-copper alloy has been applied is infiltrated in vacuo with a lead base babbit alloy. In this case an excess of the infiltrant is provided on top of the infiltrated sponge layer. This will be discussed in chapter 22.

References

1. L. Schlecht, W. Schubardt and F. Duftschmidt, On consolidation of carbonyl iron powder by thermal and pressure treatment, Z. f. Elektrochemie, vol. 37, p. 485-491, (1931).
2. V. A. Tracey, Production of porous nickel for alkaline battery and fuel cell electrodes, practical and economic considerations, Powder Metallurgy, vol. 8, No. 16, p. 241-255 (1965).
3. T. R. Barrett, G. C. Ellis and R. A. Knight, The pressureless sintering of loose beryllium powder, Powder Metallurgy, No. 1/2, p. 122-132 (1958). See also W. W. Beaver and H. F. Larson, The powder metallurgy of beryllium, "Powder Metallurgy," ed. by W. Leszynski, N.Y., (1961), p. 760.
4. H. H. Hausner and A. R. Poster, Slip casting of metal powders and metal-ceramic combinations, in "Powder Metallurgy," ed. by W. Leszynski, N.Y., (1961), p. 461-506.
5. H. H. Hausner, Slip casting of metal powders, Proc. Metal Powder Association, 14th meeting, (1958), p. 79-90.
6. W. D. Jones, Fundamental principles of powder metallurgy, London, (1960), p. 376-382.
7. R. Kieffer and F. Benesovsky, The production and properties of novel sintered alloys (infiltrated alloys), Berg-und Hüttenmännische Monatshefte, vol. 94, No. 8-9, p. 284-294, (1949).
8. P. Schwarzkopf, The mechanism of infiltration, in "Symposium on Powder Metallurgy", Special Report No. 58, The Iron and Steel Inst., London, (1956), p. 55-58.
9. C. G. Goetzel and A. J. Shaler, Mechanism of infiltration of porous powder metallurgy parts, J. of Metals, vol. 16, p. 901 (1964).
10. A. J. Shaler, Theoretical aspects of the infiltration of powder metallurgy products, Int. J. of Powder Metallurgy, vol. 1, No. 1, p. 3, (1965).
11. M. Kimura, J. C. Kosco and A. J. Shaler, Detergency during infiltration in powder metallurgy, Proc. Metal Powder Industries Fed., 15th annual meeting, p. 56 (1959).
12. K. A. Semlak and F. N. Rhines, The rate of infiltration of metals, Trans. AIME, vol. 212, p. 325-331, (1958).
13. R. Kieffer and F. Kölbl, About the production of cemented carbides by infiltration, Berg - und Hüttenmännische Monatshefte, vol. 95, No. 3, p. 49-58, (1950).

14. H. Schreiner, Pulvermetallurgie elektrischer Kontakte, Berlin, (1964).
15. C. G. Goetzel and J. B. Rittenhouse, The influence of processing conditions on the properties of silver infiltrated tungsten, in "Symposium sur la Metallurgie des Poudres", Edition Metaux, Paris, (1964), p. 279-288.
16. U. S. Patents 2,827,874; 2,828,226; 2,843,501; 2,922,721.
17. P. W. Taubenblat, W. E. Smith and R. Lewis, An infiltration system achieving high strength in P/M parts, Mod. Dev. in Powder Metallurgy, vol. 8, ed. by H. H. Hausner and W. E. Smith, p. 149-162 (1974).

Hot Consolidation
of Metal Powders

This chapter is concerned with the hot consolidation of metal powders to fully dense compacts, an operation in which the step of applying pressure to the powder occurs at elevated temperature and is combined with the sintering step. The chapter includes hot pressing, hot extrusion, hot isostatic pressing and hot forging of metal powders. An additional method of hot consolidation, used for aluminum powder, i.e. hot powder rolling has already been discussed in chapter 6 in connection with other methods of powder rolling. Hot forging of preforms, a process in which metal powder compacts after having been pressed and usually also sintered are hot forged into parts of high density and closely controlled dimensions, is not included in this chapter. The process is of greatest importance for structural parts from metal powders and will be presented in chapter 18 following the discussion of structural parts in chapter 17. At first glance, one would expect that an operation, in which pressure application and sintering are combined in one step, should have economic advantages over two step processing, in which the powder is first cold pressed and then sintered. However, because hot consolidation will, in general, have to be done under conditions in which the powder, before it is consolidated to near theoretical density, must be protected from reaction with air, makes the operation more complex and costly. Hot pressing in dies requires that the powders are heated, held under pressure and cooled in a protective atmosphere. In hot extrusion, hot isostatic pressing and hot forging, the powder is usually "canned", i.e. encapsulated in a container, which prevents reaction of the powder with the atmosphere; can and powder are then hot consolidated together. Hot consolidation of metal powders is, therefore, reserved for expensive materials with special properties. The special properties are due to microstructural features which can be obtained in certain materials when they are produced by hot consolidation of metal powders rather than by conventional means. In beryllium and in magnesium alloys, powder consolidation leads to a finer grain size. In super alloys, homogeneous microstructures free of segregation can be produced. In high speed steel, the size and distribution of the carbide phase and in aluminum alloys the size and distribution of intermetallic compound phases can be better controlled. Most dispersion strengthened alloys containing a very fine dispersion of an insoluble phase, usually an oxide, in a metallic matrix can be produced only by hot consolidation of powders. The question often arises whether the improved mechanical and chemical properties of the materials which are based on these microstructural features make hot consolidation of metal powders competitive with more conventional methods of production, even though fabrication from powder is generally high in cost. This question is particularly important when hot consolidation of powders leads to a compact of unfinished dimensions and shape, often called a billet or even an ingot, which has to be finished by forging, machining, grinding, etc. For this reason, recent developements in hot consolidation of powders have been aimed at producing directly more complex shapes than simple billets. In particular, in hot isostatic pressing shapes close to that of the final shape of the product, what is called "near net shape," can now be produced. Jet engine turbine disks from nickel base superalloy powders are an example. In this case, production by hot isostatic pressing is economically justified not only because of the improved microstructure and properties of the component, but also because the amount of raw material needed in hot isostatic pressing is less than that required for a casting which has to be forged and machined.

The following methods of hot consolidation:

> hot pressing including "spark sintering"
> hot extrusion
> hot isostatic pressing and
> hot forging of metal powders

are discussed individually. The principle of each method and its characteristics in comparison with other methods of hot consolidation are described. Examples of how the method is applied to particular types of powders are given and differences in processing when the method is used for other powders are shown. A discussion of individual materials, their properties and how these properties compare with conventionally processed materials, are reserved for later chapters.

I. Hot Pressing

During hot pressing, the total amount of deformation of the compact is relatively limited in contrast to hot extrusion and hot forging, but complete densification is generally achieved.

One of the principal problems in hot pressing of metal powders is the choice of a suitable mold material. It must be strong enough at the hot-pressing temperature to withstand the applied pressure without plastic deformation and it must not react with the powder to be hot-pressed. A widely used material is graphite, which is used in hot pressing beryllium and cemented carbides.

Hot pressing is the most commonly used method for consolidating beryllium powder.[1,2] The powder is filled in a hot pressing unit made of graphite, which is often quite large with diameters as much as 1875 mm (75"). The punches are also of graphite. The unit may be heated with resistance heating elements or by supplying current directly to the mold by direct resistance or by induction heating. Beryllium powder is hot pressed at temperatures of 1050 - 1100°C under pressures of 1.4-4 MPa (200 - 600 psi). Beryllium is always hot pressed in a vacuum of less than 100 mm Hg pressure. Rather than producing only rectangular or cylindrical blocks by vacuum hot pressing, pressing methods for shapes have been investigated in which a green preform is first pressed, then hot pressed in a closed die configuration.[2] Since the powder has to be heated to the hot-pressing temperature, held at this temperature and then cooled back to room temperature in a vacuum, hot pressing of beryllium is rather slow.

High temperatures of the order of 1400°C and somewhat higher pressures in the range up to 17 MPa (2500 psi) are used for hot-pressing cemented carbide powder in graphite molds which are heated by resistance or by induction. When hot consolidation of nickel supperalloys was first investigated, the powders were hot-pressed in molds made of molybdenum alloy TZM in an inert atmosphere. Hot isostatic pressing has now completely superseded hot pressing as a technique for hot consolidating nickel and cobalt base superalloy powders.

In producing aluminum alloy powder compacts by hot consolidation, the hot pressing step is usually preceded by cold pressing of the powder and followed by a hot working step, such as extrusion or forging. Aluminum alloy powder compacts can be readily hot pressed in dies made of hot die steel.

Certain types of metallic friction materials are pressed in the shape of thin disks from a mixture of metal powders and friction producing ingredients, such as silica. They are sintered and simultaneously bonded to a steel backing with pressure application at the sintering temperature by a rather specialized process. It will be discussed in chapter 20 in the section on friction materials.

II. "Spark Sintering"

A method of hot consolidating metal powders closely related to hot pressing is electrical resistance sintering under pressure, which was investigated in the 1950's.[3] Powder or a green powder preform is placed between two punches which also serve as electrodes for conducting a controlled amount of low voltage, high amperage current through the powder. The powder is heated to the hot-pressing temperature by the current and simultaneously pressed. The die stays relatively cold during this type of hot-pressing. A variation of this method of consolidation, in which both direct current to establish contact and alternating current are used, has been promoted during the last few years under the name of "spark sintering". Its principal use has been in pressing beryllium powder compacts in graphite molds[4] under conditions in which a dense compact is produced in a very short time. Other applications of "spark sintering" for consolidating metal powders, e.g. titanium alloy powders,[5] have also been advocated. Boron fiber reinforced aluminum has been successfully produced by resistance sintering a mixture of atomized aluminum powder and boron fibers, 200 μm in diameter for periods of 1.2 to 1.5 seconds.[6]

III. Hot Extrusion

Hot extrusion[7] combines hot compacting and hot mechanical working, yielding a fully dense product. The type of deformation occurring in hot extrusion is compared with that in hot isostatic compacting and in hot powder rolling in the schematic drawing, Figure 13-1. In hydrostatic pressing, forces are equal in all directions and little deformation takes place beyond that necessary for consolidation. In hot powder rolling, compressive forces are applied locally, resulting in densification as well as plastic elongation. In extrusion, large hydrostatic compressive forces occur and a unidirectional force component makes the compact flow through the die. Frictional forces produce a shear component which results in a characteristic shear pattern in the extruded metal. The energy expended in shear represents almost one half of the total energy needed for extrusion. The total amount of deformation in extrusion is much larger than in any other single metal working step.

The three basic methods of hot extrusion of powder are shown in Figure 13-2. In the first method loose powder is placed into the heated extrusion container and extruded directly through the die. This method has been

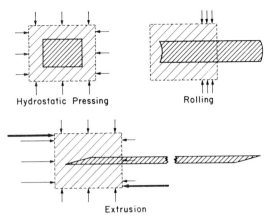

FIGURE 13-1   Forces and deformation of powders in hydrostatic pressing, in rolling and in extrusion.

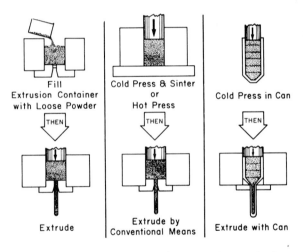

FIGURE 13-2   Three hot extrusion methods.

developed for the extrusion of certain magnesium alloy powders[8,9] or magnesium alloy pellets, as they have been termed, because of their relatively large particle size in the range from 70 - 450 μm. No atmospheric protection is provided and the heat of the container is used to raise the temperature of the powder sufficiently to allow extrusion. The temperature rise occurs during the 15 to 30 seconds needed to advance the ram prior to extrusion.

In the second method, which is used for hot extrusion of aluminum alloy powder billets, the powder is cold compacted and then hot pressed.[10] The hot pressed compact is extruded by techniques used for extruding cast aluminum alloy billets. Cold isostatically pressed compacts of molybdenum powder, preheated to the extrusion temperature, can be extruded without canning the compacts, according to the results of recent research.[11]

In most applications of hot extrusion of metal powders the third method is used. The powders are placed into a metallic capsule or "can," heated and extruded with the can. The "canning" step is generally also part of hot consolidation by hot isostatic pressing and hot forging of powders. It permits handling of powders which are toxic, radioactive, pyrophoric or easily contaminated by the atmosphere. In hot extrusion of powders, a green metal powder compact may be canned or the powder may be cold pressed into the metal can under moderate pressure as shown in Figure 13-3, with the can prevented from bulging by a packing die. For spherical powders which can be vibrated to a high green density, precompacting may not be necessary. An end plate with an evacuation tube is placed over the powder and welded to the can. The can is out-gassed by evacuation at room or elevated temperature and sealed off before can and powder are heated for extrusion. To prevent turbulent flow during extrusion, the end of the can is conical and fits into an extrusion die with a conical opening, as shown in Figure 13-4. To avoid folding, when the powder in the can is not packed very densely, a penetrator ram may be used as shown in Figure 13-5. It enters the inside diameter of the can and densifies the powder before actual

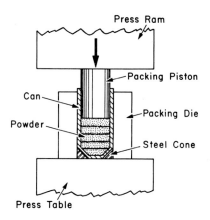

FIGURE 13-3 Packing of powder into a metal can.

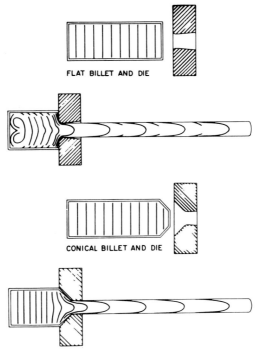

FIGURE 13-4 Extrusion with flat billet and die and with conical billet and die.

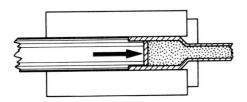

FIGURE 13-5 Penetrator technique in powder extrusion to avoid folding of can.

extrusion takes place. The material of the can must have as close as possible the same stiffness at the extrusion temperature as the powder to be extruded, it should not react with the powder and should be removable by etching or mechanical stripping. Copper and low carbon steel are most frequently used as material for cans.

When metal powders are canned and then hot extruded, the resulting extrusions will usually have simple circular, elliptical or rectangular cross

327

sections. However, a method of producing extruded structural shapes from superalloy powder, the "filled billet" technique,[12] was developed by Gorecki and Friedman. An enlarged replica of the desired shape is produced within a round extrusion billet as shown in Figure 13-6. The figure also shows the dimensions of the shape after extrusion. The ratio of the dimensions in the filled billet and in the final extrusion is determined by the extrusion ratio. The filler is produced from mild steel by milling or electric discharge machining. The cavity is filled with a powder of the superalloy composition to be produced. Most superalloy powders have spherical shape and can, therefore, be vibrated to densities above 60% of theoretical. The machined filler is placed in a carbon steel can. Superalloy powder is poured into the cavity, the billet is sealed, evacuated and heated for extrusion. The billet is extruded through a round orifice conical approach die. The filler is then dissolved, leaving the desired shape, 100% dense.

Hot extrusion of powders encapsulated in cans was first developed as a method for hot consolidating powders of beryllium and of dispersions of fissile material in a matrix of zirconium or stainless steel. The method has been almost universally used for copper base and nickel base dispersion strengthened alloys. This includes copper dispersion strengthened by

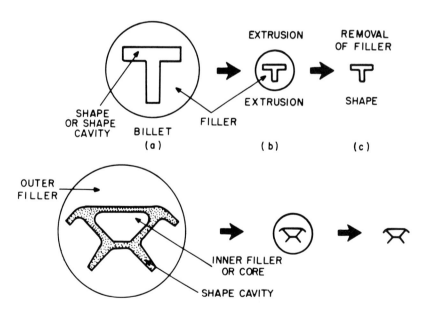

FIGURE 13-6   Outline of filled billet process for extrusion.

aluminum oxide,[13] in which an internally oxidized copper-aluminum alloy powder is extruded in a copper can at 930°C. The alloy TD nickel, a dispersion of thorium oxide in nickel, was consolidated from powder by hot extrusion.[14] Alloy powders of nickel with 15% molybdenum and varying amounts of thorium oxide were canned in mild steel and extruded into flat bars at 870°C at extrusion ratios ranging from 7.1 to 10.4.[15] For the first of the mechanically alloyed compositions of nickel base superalloys dispersion strengthened with yttrium oxide, Benjamin[16] reported extrusion at 2150° F or below with extrusion ratios of 12:1 or greater.

A hot extrusion process for producing seamless tubing from stainless steel powder was developed in Sweden.[17] In this process an argon or nitrogen atomized spherical stainless steel powder is filled into preformed molds of mild steel. By careful control of filling, densities of 70% of theoretical can be obtained. The molds weighing 35 to 120 kg are cold isostatically pressed in groups of 20-40 molds at pressures of 400 to 500 MPa (58 to 72,000 psi) to a density of 85 to 90% of theoretical and are then hot extruded at temperatures near 1200°C using a glass lubricant as in the Sejournet process. With extrusion ratios of 4:1 or higher, completely dense seamless tubing is produced. The process has been applied to austenitic stainless steels and particularly to a ferritic steel with low interstitial content.

IV. Hot Isostatic Pressing

Hot isostatic pressing was first intensively studied in the late 1950's and the 1960's at Battelle Memorial Institute as a technique for hot consolidating refractory metal powders, ceramic powders and cermets.[18,19] In recent years the technique has been commerically exploited for superalloys and tool steels.[20] High purity powders with spherical shapes produced by several methods had become available for these materials. They are ideally suited for hot isostatic pressing. Development on hot isostatic pressing of spherical titanium alloy powder[21] is underway at present.

Experiments on hot isostatic pressing from powder have been performed on alloys other than superalloys, tool steels and titanium alloys. In the case of maraging steels,[22] mechanical properties equivalent to those of wrought products have been reported.

Apparatus for hot isostatic pressing large components, generally a cold wall autoclave, is quite expensive. It consists of a water cooled large pressure vessel within which a resistance heated furnace, thermally insulated from the pressure vessel, is arranged. The furnace heats the material to be hot pressed. A gas, commonly argon, applies pressure up to 100 MPa (15,000 psi) for isostatic pressing. The principal problems in the design for the presses are: providing means of opening the vessel quickly for loading and unloading and ensuring uniform temperature distribution within the hot zone. Two methods of opening and closing the pressure vessel are in use, one with a threaded closure as illustrated in Figure 13-7, the other with a yoke closure as seen in Figure 13-8. With a threaded closure the load must be introduced from the top. With a yoke closure the load may be introduced into the pressure vessel from the bottom while it is separated from the top and bottom yoke closures and the yoke closures then moved sideways to close the pressure vessel, as seen in Figure 13-9. Both the top and bottom yoke closure and the pressure vessel may be provided with a wire wound frame to take up the high stresses exerted by the gas under pressure. The difficulty with non-uniform temperature distribution is due to heat convection in the dense pressurized gas between the cold vessel wall and the furnace. It can be overcome by compartmentalizing the space near the heater and by providing suitable insulation between heater and wall and heater and work. For hot isostatic pressing at temperatures below 1250°C,

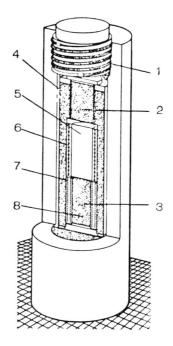

FIGURE 13-7   HIP installation with threaded closure; 1-Resilient thread, 2 and 3=ceramic plugs, 4-ceramic lining, 5-HIP zone, 6-heater, 7-current and thermocouple leads, 8-heater support.

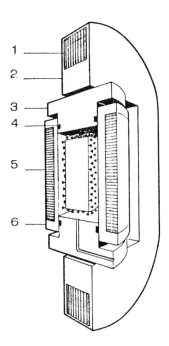

FIGURE 13-8 HIP installation with yoke closure; 1-wire wound furnace, 2-yoke segment, 3-end closure, 4-seal, 5-wire winding on HIP chamber, 6-steel core of pressure vessel.

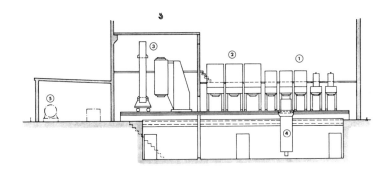

FIGURE 13-9 HIP production line. 1, 2-first and second stage preheating furnaces, 3-HIP chamber and yoke frame, 4-underfloor manipulator for loading and unloading, 5-compressor.

which includes the hot consolidation of high speed steel and of superalloy powders, the temperature in the furnace in the pressure vessel is maintained more or less constant near the hot isostatic pressing temperature not only during the actual pressing but also between cycles. The load to be pressed is preheated in a separate furnace, as seen in Figure 13-9, introduced into the pressure vessel, hot pressed and unloaded while still hot. In this way hot pressing cycles on the order of 2 to 4 hours can be maintained. For hot isostatic pressing at higher temperatures, as for refractory metals or cemented carbides, the load must be introduced into the pressure vessel while the interior of the vessel is cold; furnace and load are heated together. After hot isostatic pressing, load and furnace must be cooled before the pressure vessel can be opened.

One method of encapsulating the metal powder for hot isostatic pressing is similar to canning for hot extrusion. Welded sheet metal cans are used, which must be carefully tested for leaks, because thermally induced porosity ("TIP") in the hot pressed compacts is caused by even small leaks. In filling large cans, segregation of particle sizes must be carefully avoided. As pointed out in the introduction to this chapter, cans need not have simple cylindrical shapes near that of the final component. Figure 13-10 illustrates schematically the shape of the container for a turbine disk.[23]

Another type of encapsulation is in glass containers. The glass is chosen so that it deforms uniformly by viscous flow at the pressing temperature. For superalloy powders Vycor glass has the necessary viscosity at the hot-pressing temperature. Containers are produced by slip-casting glass powder and firing the casting. Again, after being filled with powder the container is evacuated and sealed. Glass containers, just like metal cans, may be given a shape similar to that of the final product. They also strip themselves automatically from the hot-pressed compact by shattering when they cool.

A variation of hot isostatic pressing of tool steel, called "CAP," consolidation at atmospheric pressure, has been developed by Universal Cyclops Corporation.[24] Spherical tool steel powder, produced by nitrogen atomization is treated with an aqueous solution of a chemical, which coats the powder particles and promotes hot consolidation. The treated loose powder is filled into cylindrical molds of a glass composition having the desired viscosity at the hot isostatic pressing temperature, such as a borosilicate glass. The powder in the mold is thoroughly degassed and evacuated and the mold sealed with a torch. The evacuated molds are introduced into the hot-isostatic pressing furnaces at temperatures of 1090 to 1200°C (2000 to 2200°F). No external pressure application is needed, since the difference between the atmospheric pressure outside and the vacuum inside the molds is sufficient to hot consolidate the powder. 100 kg

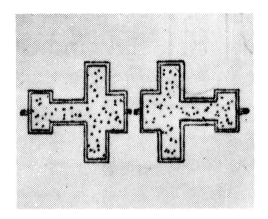

FIGURE 13-10 Schematic of shape of mild steel can for hot isostatic pressing of jet engine compressor disk.

(220 lb) tool steel preforms for hot rolling into flat and round bar and 250 kg (600 lb) preforms for forging have been produced by this method.

In addition to its commercial use for superalloys and tool steel powder, hot isostatic pressing is an important method of fabrication for beryllium, particularly the newer high purity beryllium powder produced from electrolytic flake by vacuum melting, casting, chipping, impact attritioning and sieving through a -44 μm screen.[25] This powder is consolidated by first cold isostatic pressing in evacuated elastomeric bags at 410 MPa (60,000 psi) and then hot isostatic pressing at 105 MPa (15,000 psi) in evacuated and sealed welded steel cans with peak temperatures of 915 to 1160°C. By producing compacts of high density by cold isostatic pressing, their heat conductance during hot isostatic pressing is greatly increased and the hot pressing process facilitated. For the same reason, cold isostatic processing before canning is sometimes used even for the spherical high speed steel powder.[26]

V. Hot Forging

Metal powders, especially spherical powders which have a high tap density, can be forged directly when encapsulated in a steel can. Gessinger and Cooper[27] have reported on direct closed die forging of a chrome-nickel steel powder produced by the rotating electrode method and canned in evacuated containers. This method is, however, not widely used.

References

1. W. W. Beaver and F. H. Larson, The powder metallurgy of beryllium in "Powder Metallurgy," ed. by W. Leszynski, New York (1961), p. 743-773.
2. H. D. Hanes, Beryllium processing - the foundation of structural powder metallurgy, paper presented at 4th International Conference on beryllium, London, England, October (1977).
3. F. V. Lenel, Resistance sintering under pressure, Trans. AIME, vol. 203, p. 158-167, (1955).
4. R. W. Boesel, M. I. Jackson, I. S. Yoshika, Spark sintering tames exotic P/M materials, Materials Engineering, Vol. 70, No. 4, p. 32-35 (1969).
5. C. G. Boetzel, V. S. deMarchi, Electrically activated pressure sintering (spark sintering) of titanium-aluminum-vanadium alloy powders, Mod. Dev. in Powder Metallurgy, Vol. 4, ed. by H. H. Hausner, p. 127-150 (1971).
6. K. Akechi and Z. Hara, Preparation of boron fiber reinforced aluminum by a resistance sintering process, Trans. Japan Inst. Metals, vol. 20, p. 51-56, (1979).
7. P. Loewenstein, L. R. Aronin and A. L. Geary, Hot extrusion of metal powders in "Powder Metallurgy," ed. by W. Leszynski, N.Y., (1961), p. 563-583.
8. R. S. Busk and T. E. Leontis, The extrusion of powdered magnesium alloys, Trans. AIME, vol. 188, p. 297-306, (1950).
9. R. S. Busk, The pellet metallurgy of magnesium, Light Metals, Vol. 23, p. 197-200, (1960).
10. J. P. Lyle, Jr., and W. C. Cebulak, Fabrication of high strength aluminum products from powder, in "Powder metallurgy for high-performance applications," ed. by J. J. Burke and V. Weiss, Syracuse, (1972), p. 231-254.
11. S. M. Tuominen and J. M. Dahl, Properties of unsintered molybdenum powder extrusions, presented at 108th annual meeting of AIME, New Orleans, Feb. (1979).
12. A. S. Bufferd, Complex superalloy shapes, in "Powder metallurgy for high-performance applications," ed. by J. J. Burke and V. Weiss, Syracuse, (1972), p. 303-316.
13. A. V. Nadkarni, E. Klar and W. M. Shafer, A new dispersion strengthened copper, Metals Eng. Quart., vol. 16, No. 3, p. 10-15, (1976).
14. F. J. Anders, G. B. Alexander and W. S. Wartel, A dispersion strengthened nickel alloy, Metal Progress, vol. 82, No. 6, p. 88-91 and 118, 122, (1962).
15. R. F. Cheney and W. Scheithauer, Jr., The high-temperature strength of oxide strengthened nickel alloys, Mod. Dev. in Powder Metallurgy, vol. 5, ed. by H. H. Hausner, p. 137-148, (1971).
16. John S. Benjamin, Dispersion strengthened superalloy by mechanical alloying, Met. Trans., vol. 1, p. 2943-2951 (1970).
17. C. Aslund, A new method for producing stainless steel seamless tubes from powder, Preprints, vol. 1, p. 278-283, 5th European Symposium on Powder Metallurgy, Stockholm, (1978).
18. H. D. Hanes, Hot isostatic pressing of high performance materials, in

"Powder metallurgy for high performance applications," ed. by J. J. Burke and V. Weiss, Syracuse, (1972), p. 211-230.

19. E. S. Hodge, Elevated-temperature compaction of metals and ceramics by gas pressure, Powder Metallurgy, vol. 7, No. 14, p. 168-201, (1964). See also references 24 and 25 in Fischmeister (following reference).

20. H. Fischmeister, Isostatic hot compaction - a review, Powder Metallurgy Int., vol. 10, p. 119-123, (1978).

21. R. H. Witt, O. Paul, M. T. Ziobro and F. D. Barberio, Current development in hot isostatic pressing (HIP) of titanium, Mod. Dev. in Powder Metallurgy, vol. 8, ed. by H. H. Hausner and W. E. Smith, p. 316-328, (1974).

22. R. M. German and J. E. Smugeresky, Effects of hot isostatic pressing temperature on the properties of inert gas atomized maraging steel, Mat. Sci. Eng., vol. 36, No. 2 p. 223-230 (1978).

23. S. H. Reichman, Low cost P/M superalloy applications in turbines, Int. J. of Powder Metallurgy, vol. 11, No. 4, p. 277-284, (1975).

24. U. S. Patent 3,704,508, V. N. di Giambattista, Process for compacting metallic powders. Dec. (1972).

25. G. J. London, G. H. Keith and N. P. Pinto, Grain size and oxide content affect beryllium properties, Metals Eng. Quart., vol. 16, No. 4, p. 45-57, (1976).

26. K. Zander and I. Strömblad, The ASEA-Stora Process for the production of high alloy steel, ASEA Journal, vol. 45, No. 2, p. 53-60, (1972).

27. G. H. Gessinger and P. D. Cooper, Direct forging of high temperature steel powder, Powder Metallurgy Int., vol. 6, No. 2, p. 87-89, (1974).

14

Powder Metallurgy of the Refractory and Reactive Metals

Metals which have melting points above 2000°C are called refractory metals. They include W, Re, Ta, Mo, Nb, Hf, Os, Ru and Ir. A related group of metals includes the reactive metals, titanium and zirconium, which have somewhat lower melting points. For many years these metals could not be produced in massive form by fusion metallurgy, but only by powder metallurgy. Indeed the first industrial powder metallurgy process was the manufacture of tungsten wire from tungsten powder, which was developed by William Coolidge early in this century. Beginning in the 1940's, methods for producing ingots were developed for all the refractory and reactive metals, first by vacuum arc melting and later by electron beam melting. Compacts pressed from powder or sponge are the starting material for these melting methods, but they are not considered part of the powder metallurgy of the metals. Even though refractory and reactive metals can now be melted, powder metallurgy remains an important method of fabrication for some of them, in particular, tungsten, molybdenum and their alloys. Cast ingots of tungsten and molybdenum are coarse-grained. Upon freezing, any oxygen in the molten metal precipitates as oxide at the grain boundaries. These oxides make it impossible to hot work the ingots. Even with careful control of interstitial elements, ingots of tungsten and molybdenum and their alloys can be hot worked only with great caution. On the other hand, tantalum, the only other refractory metal used industrially as commercially pure metal, and also the reactive metals, titanium and zirconium and their alloys, are now fabricated primarily by vacuum arc and electron beam melting. These metals dissolve considerable amounts of oxygen in their lattices in solid solutions and their ingots can be worked more readily than tungsten, molybdenum and their alloys.

The first section of this chapter will be concerned with tungsten wire products. Coolidge's classical method of fabricating tungsten wire from powder, which is still of great commercial importance, will be reviewed. Also included will be the use of dopants and their effect upon the structure and properties of tungsten wire and the use of tungsten wire for applications other than incandescent lamp filaments.

The second section will be concerned with tungsten products other than wire including also tungsten-rhenium alloys and heavy alloy. Only contact materials and similar applications, in which a tungsten skeleton is infiltrated, and porous tungsten products will be omitted, since their discussion is reserved for chapter 24 on electrical applications and chapter 15 on porous metals.

The third section will deal with molybdenum and its alloys. In the fourth section, those applications of tantalum for which powder metallurgy is the method of fabrication and in the last section the powder metallurgy of the reactive metals, titanium and zirconium, will be discussed.

I. Tungsten Wire Products

Lamp filaments from carbon which Edison had developed in the late 1890's were very fragile. There was, therefore, great interest in developing filaments from a high melting point metal which would be less fragile. Since tungsten is the metal with the highest melting point, determined efforts were made in the first decade of the 20th century to develop methods for tungsten lamp filaments. A number of methods were developed, but the Coolidge method, described and patented in 1910 and 1913,[1] was so superior to the other methods that it was the only one which survived and is used even today with only minor modifications for producing lamp filaments.

The tungsten powder, from which tungsten wire is fabricated, must be carefully controlled with regard to its purity and its particle size distribution. For many tungsten wire products it is also necessary to make small additions to the material in order to control the grain-structure and grain-size of the wire produced. The additions are called dopants. The two most important ones are the so-called KAS dopant, which is a combination of K_2O, Al_2O and SiO_2, and thorium oxide. The effect of the dopants upon the microstructure and properties of tungsten wire is discussed at the end of this section.

The principal method of controlling the purity of tungsten powder is the purification process for the raw material from which the powder is reduced. This material is ammonium para-tungstate (ATP), a hydrated product with formulas in the range from $5(NH_4)_2O . 12WO_3 . 11H_2O$ to $5(NH_4)_2O . 12WO_3 . 5H_2O$. Pure APT is prepared from crude tungstic acid by digesting it in aqueous ammonia, filtration of the ammonium tungstate solution and crystallization of APT.[2] The particle size of tungsten powder is determined by the particle size of the APT or the tungsten oxide from which it is produced and by the conditions under which they are reduced to tungsten powder, in particular the temperature range of the reduction process and the moisture content of the reducing hydrogen.

The details of the reduction process of ammonium paratungstate or of tungsten oxide WO_3 to tungsten powder depend upon the purpose for which the powder is to be used.[3] For tungsten products other than doped wire, in particular for large tungsten compacts and for tungsten carbide, relatively coarse powders are produced by reduction from WO_3 at temperatures up to $1000°C$ in stationary or rotary furnaces. When doped tungsten wire is to be produced, the APT is reduced with hydrogen in a first reduction step at approximately $600°C$ to the so-called blue oxide. The "blue oxide" is blue in color and has an oxygen content approximated by the formula $WO_{2.96}$. It still contains small amounts of ammonia and water. The dopant is introduced into the blue oxide as an aqueous solution of salts of potassium, aluminum and silicic acid and the oxide dried.[4] A typical dopant composition in weight % of the tungsten powder is 0.275 weight % K_2O, 0.38 weight % SiO_2 and 0.05 weight % Al_2O_3.[5] The reduction of the blue oxide to stable α-W may take place by several different paths depending on the dopant added and the temperature of reduction, which may be as high as $750°C$.[6] Much of the potassium in the dopant is lost during reduction. The reduced doped tungsten powder is washed with hydrofluoric acid, which removes some of the silica and alumina in the dopant. After washing, the powder is dried and may be given a further reduction treatment.

The next step in the production of tungsten rod to be drawn into wire is to press the powder into compacts with a square cross section up to $1'' \times 1''$ (25 x 25 mm) and up to 36" (914 mm) long, weighing as much as 6 kg. In a split die, hydraulic pressure is applied not only to the top and bottom punches, but also to hold the side plates of the die in position during compacting. When the pressure is released, the side plates can be removed and the green compacts lifted out of the die. This is necessary because of the very poor strength of green tungsten powder compacts. The green compacts are presintered in dry hydrogen at temperatures near $1200°C$. This increases their strength sufficiently so that they can be given the final sintering

FIGURE 14-1 "Treating bottle" for sintering tungsten bars by direct current passage (Courtesy: GTE-Sylvania).

treatment by passing a low voltage, high amperage current through them. For this purpose the upper and lower ends of the compact are clamped into contacts; the upper water cooled contact is stationary, while the lower contact is flexible to allow for the shrinkage of the compacts during sintering. Compacts and contacts are enclosed in a so-called "treating bottle," shown in Figure 14-1. They are sintered in a hydrogen atmosphere. The treating bottle is water cooled. The schedule of the current passing through the compact is carefully controlled. The maximum temperature reached exceeds 3000° C, the total sintering time is 15 to 30 minutes. The linear shrinkage of the compacts is about 15 percent, the density increases from 11 to 17g/cm³.

Of the dopant added to the blue oxide little survives the subsequent treatments, in particular the final sintering treatment, except for a critical quantity of potassium on the order of 60-100 ppm.

The ends of the rods which were clamped into contacts and are therefore not sintered are knocked off. The sintered rods are completely brittle at room temperature and must be worked at elevated temperatures near 1600°C. The first step is generally rolling in a rolling mill down to 1 cm diameter rods. Rolling is less labor intensive than the hand swaging process originally used as the first step in hot working. Rolling is followed by continuous hot swaging, in which feed rolls pull the bars through a heating furnace and a swaging machine. The machine consists of shaped dies made of hardened steel which are rotated around the bars and forced together by cams subjecting the bars to about 10,000 hammer blows per minute.

The swaged bars are drawn into wire. Cemented carbide dies are used down to 0.25 mm diameter and diamond dies for smaller diameters. Colloidal graphite is used as a lubricant, the drawing temperature is gradually lowered from 800 to 400°C. After drawing, the wire is degreased. Tungsten wire for incandescent lamp filaments is coiled on a straight iron core which is dissolved in acid. Quite commonly "coiled coil" filaments are used, in which the coiled wire is wound on molybdenum core wire, heated in a tube furnace to about 1600°C in order to recrystallize it. The molybdenum is later dissolved in a mixture of hot sulfuric and nitric acid.

During swaging and subsequent drawing of tungsten a fibrous microstructure, shown in Figure 14-2, is produced. At the service temperature of incandescent lamps this highly cold worked structure recrystallizes. In undoped tungsten wire the equiaxed tungsten grain structure seen in Figure 14-3 is produced. In KAS doped tungsten, on the other hand, the recrystallized microstructure consists of grains highly elongated in the direction of the wire axis, as seen in Figure 14-4. This

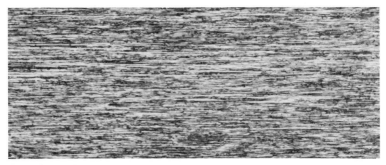

FIGURE 14-2   Microstructure of .007″ (178 μm) diameter tungsten wire for incandescent lamp as drawn, x250 (Courtesy: General Electric Company, Refractory Metal Products Dept.).

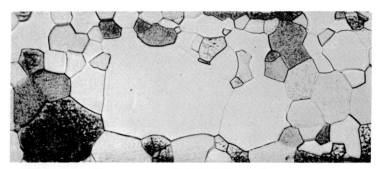

FIGURE 14-3   Microstructure of .007″ (178μm) diameter undoped pure tungsten wire, recrystallized, x250 (Courtesy: General Electric Co., Rrfractory Metal Products Dept.)

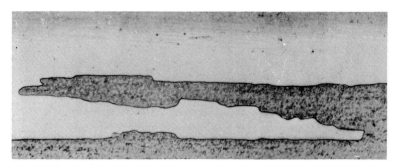

FIGURE 14-4   Microstructure of .007″ (178 μm) diameter doped tungsten wire, recrystallized, x250 (Courtesy: General Electric Company, Refractory Metal Products Dept.).

structure prevents the offsetting and sagging observed in undoped tungsten wire, which drastically reduces the life of coiled lamp filaments. As a quantitative measure of the resistance to failure at high temperature, the stress dependence of the rupture life in hours of 178 μm diameter doped and undoped tungsten filaments at a temperature of 3000° K may be taken. As shown in Figure 14-5,[7] the life to rupture of doped wire is considerably longer and much less dependent upon stress than that of undoped wire.

Studies of the last ten years have considerably enhanced the understanding of the effect of dopants upon the structure and properties of tungsten wire. Electron micrographs of the structure of recrystallized tungsten wire show rows or stringers of inclusions 5-30 nm in diameter in doped but not in undoped wire. Moon and Ku[8] were able to show that these inclusions are actually bubbles, since, as seen in Figure 14-6, the shadows of the inclusions in a replica of the fracture of recrystallized wire are all cast in

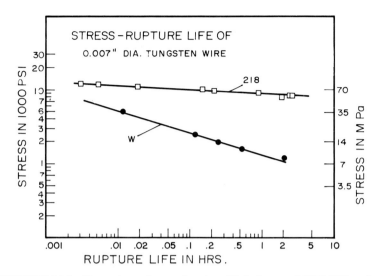

FIGURE 14-5 Stress dependence of rupture life in hours of .007″ (178 μm) tungsten wire at 3000 K. "W" is pure tungsten wire, "218" is doped tungsten wire.

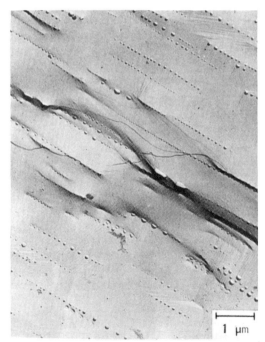

FIGURE 14-6 Fracture replica of 0.88 mm diameter wire of doped tungsten annealed 5 minutes at 2500°C.

the same direction. According to Moon and Ku, when wires containing needle-like pores are annealed, the pores break up into rows of bubbles. Studies by energy dispersive X-ray spectrometry, by secondary ion mass spectrometry, by Auger electron spectrometry[9] and finally by selected area diffraction on transmission electron micrographs[10,11] made it very probable that metallic potassium is contained in these bubbles. They inhibit the normal recrystallization of the cold worked tungsten fiber structure observed in undoped tungsten at temperatures above 1100°C. Instead the fiber structure is maintained up to temperatures of 2150°C. At these temperatures a sudden exaggerated growth of a few grains in the wire cross section occurs. Due to the presence of the bubble rows the resultant structure consists of the elongated grains of the non-sag structure.

Doping tungsten powder with 1 to 2% of thorium oxide in the form of solid oxide, of a hydroxide sol or of thorium nitrate inhibits grain-growth in recrystallized tungsten wire. Thoriated tungsten wire[12] is used mainly in applications, such as atomic hydrogen welding electrodes, electrodes for plasma spraying and xenon lamp, where the increased thermal electron emission of the material as compared to undoped tungsten is utilized.

Other uses for tungsten wire include coils used in vacuum metallizing, particularly for aluminum[5] and metal to glass seals, where the low coefficient of expansion of tungsten is important.

II.   Other Tungsten Products and Tungsten Alloys

The method for sintering tungsten powder compacts by passing an electric current through them is applicable only to rod shaped compacts with a uniform and not too large cross section and is therefore primarily used for tungsten compacts to be fabricated into wire. In Europe even these compacts are sintered by indirect heating.[4] As can be seen from Figure 8-6 in chapter 8, compacts from sufficiently fine tungsten powder will shrink to high densities above 90% of theoretical when they are sintered by indirect heating for extended periods at temperatures as low as 1800°C. In producing tungsten products, other than wire, particularly those with large dimensions, such as rocket nozzles, compacting by cold isostatic pressing is combined with indirect sintering in furnaces with tungsten heating elements. A press for isostatic pressing of tungsten and molybdenum compacts is shown in Figure 14-7. Figure 14-8 is a photograph of a furnace for indirect heating of tungsten compacts. The tungsten heating elements in such a furnace may be woven from tungsten wire or may be made from tungsten sheet as in the schematic drawing, Figure 14-9, for indirect sintering of tungsten compacts. The radiation shields surrounding the

FIGURE 14-7 Installation for isostatic pressing of tungsten and molybdenum powder compacts (Courtesy: GTE-Sylvania).

tungsten heating element are not shown in the figure. The tungsten sheet used for such furnaces is rolled from tungsten compacts in mills specially designed for the necessary high rolling pressure and the high temperature needed in rolling tungsten.

In chapter 2, two methods of producing tungsten powder, other than by hydrogen reduction of tungsten oxide or ammonium para-tungstate, were

FIGURE 14-8 Furnace for indirect sintering of tungsten and molybdenum powder compacts (Courtesy: GTE-Sylvania).

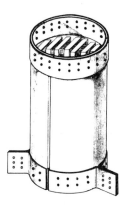

FIGURE 14-9 Schematic of arrangement of heating elements and tungsten powder compacts for sintering by indirect heating.

mentioned. The coarse spherical tungsten powder developed by Allied Chemical Company cannot be consolidated by cold pressing, but has been successfully hot isostatically pressed in molybdenum cans at temperatures of 1480 to 1590° C (2700-2900° F) and pressures of 70 to 140 MPa (10 to 20,000 psi).[13] Rocket nozzles, nose cones and forging billets were produced by this method. The very fine tungsten powder, produced by hydrogen reduction of tungsten chloride,[14] has been of primary interest as a raw material for cemented tungsten carbides.

Before tungsten alloys are discussed, the chemical and mechanical properties of pure tungsten should be briefly treated. Tungsten oxidizes in air at temperatures above 400°C. When heated in air above 800°C, the metal is rapidly destroyed by the formation of the volatile trioxide WO_3. Tungsten is stable at elevated temperatures in hydrogen, nitrogen and vacuum. In the recrystallized condition tungsten is brittle at room temperature. There is a sharp ductile-brittle transition temperature, which strongly depends upon impurities. For pure tungsten the transition temperature in a tensile test is near 300°C. The transition temperature decreases with increasing amounts of deformation, so that highly worked tungsten wire and strip is ductile at room temperature. Because of the fiber structure of cold worked tungsten, grain boundary fracture perpendicular to the fiber direction is avoided. For the room temperature tensile strength of highly worked tungsten wire, values as high as 5500 MPa (800,000 psi) have been reported. In the recrystallized condition the room temperature tensile strength is 1000 MPa (150,000 psi). For applications such as lamp filaments or heating elements the strength of tungsten at elevated temperatures is of greatest importance; it was illustrated by the data on stress rupture life of tungsten wire in Figure 14-5.

Among the alloys of tungsten in which a second element is in solid solution in the tungsten lattice, tungsten-molybdenum alloys with 30% tungsten are used for apparatus in contact with liquid zinc and zinc alloys, which do not attack this alloy.[15] When the properties of alloys of tungsten with rhenium were first studied, it was found that an alloy with 30% Re could be worked much more readily than pure tungsten;[16] it could be rolled at 1000°C as much as 86% before cracking occurred. Alloys with 20 and 10% Re are intermediate in ductility between pure tungsten and the 30% alloy. The reason why the temperature of transition from ductile to brittle behavior is lowered drastically by the addition of rhenium to both tungsten and molybdenum appears to be the ability of these alloys to deform at low stresses by mechanical twinning instead of only by slip. Unfortunately, the very high cost of rhenium makes its use as an alloying element for tungsten feasible only in commercial applications where specific properties of the products are greatly improved by the rhenium additions. Tungsten-rhenium alloys are used for that part of the rotating targets of X-ray tubes upon which the X-rays impinge. An addition of 5% Re to W decreases drastically the roughening of the target and increases the life of the targets considerably. W-Re alloys with 3, 5, 25 and 26% Re are also in common use for thermocouple wires for temperatures up to 3000 K. Finally alloys with 3% Re have found use in lamps where vibration resistance is of importance.

Alloys of 90 or more weight % tungsten with nickel and copper or nickel and iron are called "heavy alloy." As discussed in chapter 11, compacts from mixtures of fine tungsten powder and nickel and copper or nickel and iron powder can be readily sintered to densities within 1% or less of theoretical density at temperatures near the liquidus temperatures of the binder alloys nickel-copper or nickel-iron. This is in contrast to compacts from straight tungsten powder, which need much higher sintering temperatures. Although the compacts exhibit considerable linear shrinkage, they maintain their shape. Heavy alloy is widely used for applications where the high density of the alloys, in the range from 16.8 to 18.5 g/cm³, is needed. They include radioactive shielding applications, such as isotope source holders, parts for remote control handling equipment and beam collimators for linear accelerators. Inertial applications for heavy alloy are vibration dampeners, air-frame counter weights, gyroscope rotors, weights for self-winding watches and control instruments for guided missiles. An important military use is in artillery shell cores. In spite of their lower electrical conductivity compared with infiltrated tungsten-copper alloys they also find uses in electrical contacts and spark gap electrodes. While the first "heavy alloy" investigated was an alloy of tungsten, nickel and copper,[17] binder metal compositions containing iron and nickel,[18] sometimes with the addition of cobalt,[19] are now more widely used. The latter compositions are generally more ductile with elongations from 3 to 16% depending on the amount of binder metal ranging from 3 to 10% by weight. The tensile strength of heavy alloy is in the range from 760 to 970 MPa (110 to 140,000 psi), its Young's modulus from 270 to 340 GPa (40 to 50 x 10⁶ psi).[21] The good room temperature ductility of heavy metal with an iron-nickel binder is attributed to the fact that the hardness of the iron-nickel binder increases rapidly during deformation and matches the hardness of the tungsten phase.[20] Under these conditions the tungsten grains in the matrix can be deformed as much as 10%, which is more than in tungsten single crystals. In alloy metal with a nickel-copper binder, the development during sintering of the brittle imtermetallic compound Ni_4W is possible which detracts from the ductility of the alloy.[21] Because of the importance of heavy alloy for military applications efforts to optimize its ductility and yield strength through variation in composition, e.g. additions of rhenium and platinum[22] or of chromium, cobalt, manganese and copper[23] and by strain-hardening and aging treatments[24] have been undertaken.

III. Molybdenum

The methods of purifying molybdenum powder are similar to those for tungsten powder.[25] Technical molybdenum oxide, MoO_3, is obtained by roasting molybdenite MoS_2. This technical molybdenum oxide can be purified by sublimation, but the highest purity molybdenum is produced by dissolving the technical oxide in ammonia and crystallizing ammonium-dimolybdate (ADM), $(NH_4)_2O$. $2MoO_3$ from the solution. Either the ADM, the hydrated oxide or the oxide may be reduced in hydrogen to molybdenum powder, generally in two steps. At 580°C the red molybdenum oxide, MoO_2, is obtained, which is further reduced at approximately 900°C to the metallic powder.

Compacts are pressed from molybdenum powder either in rigid dies or isostatically. The compacts have a considerably higher green strength than tungsten powder compacts and no presintering is necessary. Because of its lower melting point, molybdenum powder compacts can be sintered indirectly more easily and cheaply than tungsten compacts. Direct resistance sintering is no longer used for molybdenum. Strip, sheet, foil, bar and wire are fabricated from sintered compacts by the methods used for steel, including forging, drawing, rolling and extruding. In the commercial fabrication of molybdenum products, powder metallurgy competes with the arc casting process. The mechanical properties of products by the two processes are equivalent, as seen in the ASTM specifications for molybdenum strip, sheet and foil[26] and for bar and wire,[27] in which the requirements for the room temperature mechanical properties are identical for cast and for powder metallurgy molybdenum. In strip, sheet and foil they depend upon the temper of the material and vary from 410 to 585 MPa (60-80,000 psi) ultimate tensile strength, 310 MPa (45,000 psi) yield strength and 5-10% elongation for the softest temper to 930 to 1140 MPa (135 to 165,000 psi) tensile strength, 760 MPa (110,000 psi) yield strength and 2-5% elongation for the hardest temper. In stress relieved round rods the properties depend upon the diameter and may vary from 585 MPa (85,000 psi) minimum tensile strength, 450 MPa (65,000 psi) minimum yield strength and 15% minimum elongation for rod 0.5 mm to 3.2 mm (0.020″ to 0.125″) diameter to 450 MPa (65,000 psi) minimum tensile strength, 380 MPa (55,000 psi) minimum yield strength and 10% minimum elongation for rod between 73 and 82 mm (2 7/8″ and 3 1/4″) diameter. Pure recrystallized molybdenum has a ductile-brittle transition temperature near room temperature. Molybdenum oxidizes in air at temperatures above 400°C even more readily than tungsten. The principal applications for unalloyed molybdenum include wire, rod and sheet in incandescent lamps and electronic tubes, heating elements in electric furnaces, electrodes in electrically heated glass melting furnaces and wire for spray coating steel. They are produced more commonly by powder metallurgy than by arc casting.

The molybdenum alloy of greatest technological importance is the so-called "TZM", which contains 0.5% Ti, 0.07% Zr and about 0.02% carbon. It has a higher recrystallization temperature and higher strength and hardness, both at room and at elevated temperatures, than unalloyed molybdenum and has also adequate ductility. Its improved properties are attributed to the dispersion of complex carbides in the molybdemun matrix. TZM was first developed as a cast alloy, but sintered TZM is produced both in Europe[28] and in this country.[29] It is compacted at pressures of 12 to 25 tsi from a mixture of molybdenum powder, titanium and zirconium hydride powders and carbon black and sintered at a temperature of 2200°C in dry hydrogen. It is forged at temperatures below 1200°C. The reactions occurring during sintering between Mo, Ti, Zr and carbon, the residual oxygen in the molybdenum and the sintering atmosphere are not completely understood. The question has been raised whether the composition and structure of the final product might be more easily controlled in the cast than in the P/M alloy. On the other hand, for one of the principal applications for TZM, i.e. forging dies for isothermal forging of superalloys at elevated temperatures, quite large blocks of TZM are needed. Cylinders with a diameter of 30″ from this alloy have been produced by powder metallurgy. It seem unlikely that such large blocks can be readily produced by arc casting since the very coarse grained as-cast structure will have to be refined by working before suitable stock for isothermal forging dies can be obtained. Efforts to improve the high temperature strength of powder metallurgy TZM alloys has led to alloys in which the titanium and zirconium carbide is replaced by hafnium carbide.[30] Such an alloy may contain 1.0% of hafnium and 0.06% of carbon. Its tensile strength at 1316°C is 85,000 psi compared with 70,000 psi for TZM and its creep rate at 1204°C and a stress of 48,000 psi is 0.038%/hr compared with 0.05%/hr for TZM. Methods for doping molybdenum powder with potassium and silicon have been developed in order to obtain a similar non-sag microstructure as in doped tungsten powder. This so-called HT-molybdenum is, however, much less widely used than the KAS doped tungsten. Alloys of molybdenum and rhenium are much more ductile than pure molybdenum. An alloy with 35% Re can be rolled at room temperature to more than 95% reduction in thickness before cracking. For economic reasons, molybdenum-rhenium alloys are not widely used commercially. Alloys of molybdenum with 5 and 41% Re are used for thermocouple wires.

IV. Tantalum[31,32]

The most important application for tantalum is in electrolytic capacitors. Tantalum is also used in the form of strip, sheet, tubing, wire and fabricated shapes in the chemical industry because of its corrosion resistance, in the electronic industry as emitters in electronic tubes and as high temperature heating elements and radiation shields. For reasons outlined in the introduction to this chapter, powder metallurgy as a fabrication method is not important for large tantalum products, which are fabricated from cast ingots. Only smaller products, such as foil, may be fabricated from vacuum sintered tantalum compacts. The production of tantalum powder by alkali metal reduction or electrolysis in a molten bath is described in chapter 2. The powder is pressed into compacts with a rectangular cross section 1" x 4" and 50" long and sintered by direct current passage according to the Coolidge method at low voltages and high current densities in a high vacuum at temperatures in the range of 2500 to 2700°C. A sketch of the sintering furnace is shown in Figure 14-10. During the sintering process, such impurities as residual salts and hydrogen are removed at low temperatures, at higher temperatures a lower oxide of tantalum distills off. The sintered compacts with a density of 90% of theoretical are cold worked to eliminate residual porosity and sintered a second time by current passage. They can then be readily fabricated into foil or wire primarily used in electrolytic capacitors. However, a second strictly powder metallurgical method for producing capacitors involves porous sintered anodes from tantalum powder. The anodes are compacted from a powder with a carefully controlled particle size distribution in the range from 5 to 40 μm.

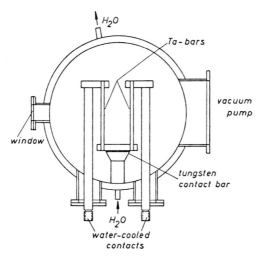

FIGURE 14-10 Schematic of sintering bell for sintering tantalum rods by direct current passage.

They have a low density of 8.5 to 11.5 g/cm³, 50-70% of theoretical and are sintered in a vacuum of 5×10^{-5} Torr at temperatures between 1800 and 2100°C. The dielectric in these capacitors is a Ta_2O_5 film produced by anodic oxidation. Either liquid or solid electrolytes are used. The capacitors from powder have the highest capacitance per unit weight of any capacitor material with values in the range of 7-11 $\mu F/g$ and have adequately low leakage rates.

V. Powder Metallurgy of Titanium, Zirconium and Their Alloys

Because of their relatively high cost, the use of titanium and zirconium and their alloys is restricted to applications where the outstanding properties of the metals and their alloys justify it. In the case of titanium and titanium alloys, their applications are, on the one hand, those based on the excellent corrosion resistance of titanium in many acids and in sea water and, on the other hand, those based on the high strength-density ratio of titanium alloys needed in aerospace applications. Zirconium and particularly the Zircalloy group of zirconium alloys find their primary use in nuclear applications where the low neutron capture cross section of the alloys combined with their mechanical and chemical properties justifies using them.

Since liquid titanium and zirconium react with most of the materials which might serve as crucibles for them, the usual methods of casting cannot be applied to these metals. It is therefore not surprising that powder metallurgy was the first method of consolidating titanium powder into a solid ductile material. The classical papers describing the process and the properties of the resulting product are those by Dean and associates.[33][34] Ti powder was produced by the Kroll process, as described in chapter 2. The coarse 30 mesh powder obtained was compacted at 690 MPa (50 tsi) into green compacts with 86% of theoretical density and sintered in a vacuum of 10^{-4} Torr at temperatures between 950 and 1000°C to a density of 93% of theoretical. During sintering the hydrogen and residual metallic magnesium in the powder were eliminated. The sintered compacts were densified by cold rolling and resintering. The powder metallurgy titanium could be cold and hot rolled, swaged into rods and drawn into wire. The work of Dean and coworkers never led to a commerical process for producing titanium in the form of sheet, strip, rod or wire since consolidation of titanium by powder metallurgy was soon superseded by vacuum arc casting of titanium produced by reacting $TiCl_4$ with either magnesium or sodium and purifying the resulting sponge by vacuum sublimation.

Powder metallurgy has been used for producing certain shapes from Zircalloy 2 powder for nuclear applications.[35] The powder with a composition 0.15% Fe, 0.05% Ni, 1.46% Sn, balance Zr was produced by hydriding the alloy, mechanically disintegrating the hydride and dehydriding as outlined in chapter 2. Completely dense shapes can be produced by hot pressing the powder in graphite molds. Properties matching nuclear wrought material with respect to impurity and gas content, mechanical properties and corrosion resistance were obtained. Tubing was produced by extruding the hot pressed compacts. This process is no longer used for commercial production.

The powder metallurgy of titanium and of titanium alloys[36,37,38] has found applications or potential applications primarily in three fields:

1. Titanium and titanium alloy shapes by cold compacting and sintering or cold compacting, sintering and hot forging which can be produced more economically from powder than from wrought titanium rod or sheet. For these applications low cost titanium powders are used, whose manufacture was described in chapter 2. One type of shapes produced by titanium powder metallurgy are fasteners, such as nuts and fixtures for use in highly corrosive environments, e.g. electroplating baths. They are compacted from elemental titanium powder and sintered in vacuum. For aerospace applications, where better mechanical properties are required, compacts may be pressed from a mixture of 90% titanium powder and 10% of a masteralloy powder with 60% aluminum and 40% vanadium. During vacuum sintering the alloy is homogenized. Typical properties of pressed and sintered C.P. titanium powder and of pressed and sintered mixtures of titanium and alloy powders are shown in Figures 14-11 and 14-12. To get completely dense parts, sintered titanium alloy preforms may be forged. Cold isostatic pressing is used more frequently for titanium and titanium alloy parts than for components from other powders. Examples are shown in chapter 6 in Figures 6-8, 6-9, 6-10 and 6-11.

2. Porous titanium products used primarily as filters for filtering highly corrosive fluids. These filters are produced by cold isostatic pressing from elemental titanium powder.[38] The method of fabrication has been discussed in chapter 6.

3. Airframe components from titanium alloy powders produced by hot isostatic pressing. The principal incentive for powder metallurgy processing is the possibility of requiring less material than when these components are produced by the conventional forging process. Because of the inherently high resistance to flow in forging dies, titanium forgings for use in manufacturing complex contoured components must be purchased greatly oversized with "buy to fly ratios" as high as 10:1. The principles

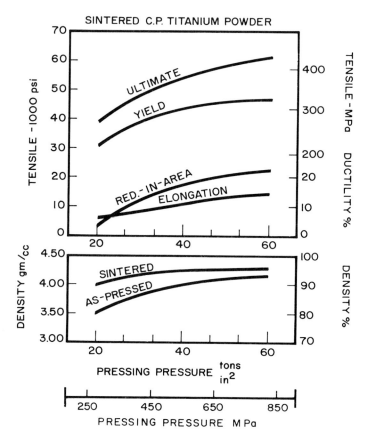

FIGURE 14-11 Density and tensile properties of titanium powder compacts as a function of compacting pressure.[37]

involved in producing components to near net shape by hot isostatic pressing have been discussed in chapter 13. Development work in hot isostatic pressing titanium alloys has been described by Witt et al.[39] and by Hillnhagen and Kramer.[40] In this work titanium alloy powders, primarily the 6 Al-4 V alloy, are used. The powders are produced by the hydriding-dehydriding as well as the rotating electrode process, both of which were described in chapter 2. The powders are hot isostatically pressed in molds from a variety of glasses, borosilicate, alumino-silicate and 96% silica glass. The mechanical properties are generally equivalent to those of conventional forgings. At the present time, work to produce specific military aircraft components such as fuselage braces and arrestor hook support fittings from titanium alloy powders by the HIP process are

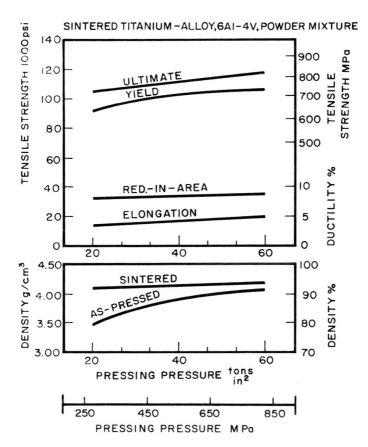

FIGURE 14-12 Density and tensile properties of compacts from a mixture of 90% titanium powder and 10% of a 60 Al-40 V master alloy.[37]

underway.[41] Also the possibility of assembling "HIP"ed components by electron beam welding is being explored.

References

1. W. D. Coolidge, Trans AIEE, vol. 29, Part II, p. 961-965 (1910), W. D. Coolidge, U. S. Patent 1,082,933, Dec. 30, 1913.
2. S. W. H. Yih and C. T. Wang, Tungsten, N.Y. (1979), p. 97ff.
3. Ref. (2) p. 131ff.
4. F. Heitzinger, The industrial significance of the most important high melting metals; Part II, Production and processing into pure metal alloys, Powder Metallurgy Int., vol. 10, No. 3, p. 136-138, (1978).

5. R. Eck, F. Kienzl, B. Tiles and H. Wagner, Fabrication, high temperature and corrosion properties of tungsten metallizing wire, Mod. Dev. in Powder Metallurgy, vol. 11, ed. by H. H. Hausner and P. W. Taubenblat, p. 91-108, (1977).

6. T. Wilker, C. Wert, J. Woodhouse and W. Morcom, Reduction of blue tungsten oxide, Mod. Dev. in Powder Metallurgy, vol. 5, ed. by H. H. Hausner, p. 161-169, (1971).

7. J. W. Pugh, On the short time creep rupture properties of lamp wire, Met. Trans., vol. 4, p. 533-538, (1973).

8. D. M. Moon and R. C. Koo, Mechanism and kinetics of bubble formation in doped tungsten, Met. Trans., vol. 2, p. 2115-2122, (1971).

9. H. G. Sell, D. F. Stein, R. Stickler, A. Joshi and E. Berkey, The identification of bubble-forming impurities in doped tungsten, J. Inst. of Metals, vol. 100, p. 275-288, (1972).

10. D. B. Snow, Dopant observations in thin foil of annealed tungsten wire, Met. Trans., vol. 3, p. 2553-2554, (1972).

11. D. B. Snow, The identification of second phases within bubbles in annealed doped tungsten wire, Met. Trans., vol. 5, p. 2375-2381, (1974).

12. H. Bildstein and R. Eck, Properties of highly deformed tungsten alloy for the vacuum technology, 9th Plansee seminar, May 1977, preprint No. 10.

13. E. S. Hodge, Elevated-temperature compaction of metals and ceramics by gas pressures, Powder Metallurgy, vol. 7, No. 14, p. 168-201, (1964).

14. L. Ramqvist, A new tungsten powder for producing carbides, Mod. Dev. in Powder Metallurgy, vol. 4, ed. by H. H. Hausner, p. 75-84, N.Y., (1971).

15. R. Eck, Corrosion of molybdenum-tungsten alloys produced by powder metallurgy in liquid zinc, Metall, vol. 32, p. 891-894 (1978).

16. R. I. Jaffee, C. T. Sims and J. J. Harwood, The effect of rhenium on the fabricability and ductility of molybdenum and tungsten, Plansee Proceedings 1958, ed. by F. Benesovsky, Vienna, (1959), p. 380-410. See also Yih and Wang, ref. 2, p. 338-346.

17. G. H. S. Price, C. J. Smithells and S. V. Williams, Sintered alloys, Part I, Copper-nickel-tungsten alloys sintered with a liquid phase present, Jnl. Inst. Metals, vol. 62, p. 239-264 (1938).

18. E. G. Green, D. J. Jones and W. R. Pitkin, Developments in high density alloys, "Symp. on Powder Metallurgy," Iron and Steel Inst., Special report No. 58, London (1956), 253-256.

19. G. Jangg, R. Kieffer, B. Childeric and E. Ertl, Investigations on tungsten heavy metals, Plannseeberichte für Pulvermetallurgie, vol. 22, No. 1, p. 15-28, (1974).

20. R. H. Crock and L. A. Shephard, Mechanical behavior of the two-phase composite, tungsten-nickel-iron, Trans, AIME, vol. 227, p. 1127-1134 (1963).

21. E. Ariel, J. Barta, D. Brandon, Preparation and properties of heavy metal, Powder Metallurgy Int., vol. 5, p. 126-129, (1973).

22. J. M. Dickinson, R. E. Riley, J. M. Taub, Strengthening of liquid sintered tungsten alloys, Mod. Dev. in Powder Metallurgy, vol. 8, ed. by H. H. Hausner and W. E. Smith, (1974), p. 329-352.

23. W. F. Chiao and F. R. Larson, Matrix chemistry and mechanical properties of high density tungsten alloys, Paper presented at 108th annual meeting of AIME, New Orleans, Feb. 1979.

24. W. G. Northcott, Jr., Strain hardening and ageing of liquid phase sintered W-Ni-Fe alloys, ibid.

25. W. Aschenbrenner, Occurrence and winning of tungsten, molybdenum, tantalum and niobium, ch. V, p. 65-69, in "Pulvermetallurgie und Sinterwerkstoffe," ed. by F. Benesovsky, Reutte, (1973).

26. ASTM standard specifications for molybdenum and molybdenum alloy strip sheet - foil and plate B 386-74.

27. ASTM standard specification for molybdenum and molybdenum alloy bar, rod and wire B 387-74.

28. R. Eck, Solid state reactions during the powder metallurgical production of the molybdenum alloy TZM, Planseeberichte für Pulvermetallurgie, vol. 20, No. 2, p. 95-109, (1972).

29. L. P. Clare and A. J. Vazquez, TZM molybdenum alloy via P/M, Mod. Dev. in Powder Metallurgy, vol. 5, ed. by H. H. Hausner, N.J., p. 259-271, (1971).

30. L. P. Clare and R. H. Rhodes, 9th Plansee Seminar, May 1977, Preprint No. D6.

31. C. A. Hampel, Tantalum, ch. 25, p. 469-481 in Rare Metals Handbook, ed. by C. A. Hampel, 2nd ed., N.Y. (1961).

32. G. Jangg and R. Eck, Niobium, tantalum, vanadium, how to work them and where to use them. HDT Veroffentlichungen, No. 136, p. 4-20, (1977).

33. R. S. Dean, J. R. Long, F. S. Wartman, E. L. Anderson, Preparation and properties of ductile titanium, Trans. AIME, vol. 166, p. 369-381, (1946).

34. R. S. Dean, J. R. Long, F. S. Wartman, E. T. Hayes, Ductile titanium - its fabrication and physical properties, ibid., p. 382-398.

35. Engineering Data, General Electric Zircalloy 2, Metallurgical Products Department, General Electric Co., Detroit, Mich.

36. G. I. Friedman, Titanium powder metallurgy, Int. J. of Powder Metallurgy, vol. 6, No. 2, p. 43-55, (1970).

37. S. Abkowitz, J. M. Siergiej, R. D. Regan, Titanium P/M preforms, parts and composites, Mod. Dev. in Powder Metallurgy, vol. 4, ed. by H. H. Hausner, N.Y., p. 501-511, (1971).

38. R. E. Garriott, E. L. Thellmann, Titanium powder metallurgy, a commercial reality, Mod. Dev. in Powder Metallurgy, vol. 11, ed. by H. H. Hausner and P. W. Taubenblat, p. 63-78, (1977).

39. R. H. Witt, O. Paul, M. T. Ziobro, F. D. Barberio, Current development in hot isostatic pressing (HIP) of titanium, Mod. Dev. in Powder Metallurgy, vol. 8, ed. by H. H. Hausner and W. E. Smith, (1974), p. 315-328.

40. E. Hillnhagen, K. H. Kramer, Properties of a powder metallurgical titanium alloy of high strength at elevated temperatures, 5th European Symposium on Powder Metallurgy, Stockholm, June 1978, Preprints, vol. 1, p. 222-226.

41. W. T. Highberger, Advances in manufacturing technology for titanium aircraft structures, Metal Progress, vol. 115, No. 3, p. 56-59, (1979).

15

Porous Metals

Powder metallurgy is by far the most common method of producing porous metallic products, in which the amount of porosity and the size and distribution of the pores is controlled. The three groups of porous metals produced commercially differ primarily in the amount of porosity contained. They are:

1. Metallic materials in which the porous structure serves as a reservoir for a lubricant, such as self-lubricating bearings

2. Metallic materials which have a controlled rate of permeation of fluids (liquids or gases) through the porous structure and which serve primarily as filters

3. Metallic materials which contain a very high internal surface area and serve as porous electrodes for batteries and fuel cells.

In addition, some potential applications of porous metallic materials which have been investigated in the laboratory will be discussed. They are tungsten ionizers with controlled porosities for cesium-ion engines and porous surgical implants. The production and properties of capacitors from tantalum powder, which should also be classified as porous metal products, have been discussed in chapter 14 on the powder metallurgy of refractory metals.

I. Self-lubricating Bearings

Self-lubricating bearings are the oldest industrial application of porous powder metallurgy materials going back to the middle 1920's when these bearings were developed independently in the Research Laboratories of General Motors Corporation in Dayton, Ohio,[1,2] and by the Bound Brook Oilless Bearing Company in New Jersey.[3]

The big advantage of porous bearings over solid bearings is the fact that the porosity in the bearings acts as its own oil reservoir. The principles involved in impregnating the pores of a bearing with lubricant are the same as those discussed in chapter 12 for the infiltration of a porous skeleton with a liquid metal. However, little basic work has been done on the process of impregnating and of forming a film between bearing and journal in self-lubricating bearings. When the journal in an impregnated self-lubricating bearing starts to turn, friction develops, the temperature rises and oil is forced out of the pores because of the greater coefficient of expansion of the oil compared with the metal and because of the hydrodynamic pressure differential in the oil film between journal and bearing. When rotation stops and the bearing cools, the oil is reabsorbed by capillary action. For many self-lubricating bearings[4] the lubricant contained in the pores of the bearing suffices for the life-time of the bearing. Other bearings are equipped with oil filled porous wicks located on the outside diameter of the bearings which feed additional oil through the bearing wall. The common shapes of self-lubricating bearing, as seen in the collection of bearings in Figure 15-1, are hollow cylinders which are force fitted into a housing, flanged bearings, thrust washers and self aligning bearings which have a part spherical external shape to allow aligning movement after assembly in a housing.

FIGURE 15-1 Assembly of bronze self-lubricating bearings (Courtesy: Keystone Carbon Company).

The original and still the most widely used composition for these bearings is a 90 Cu-10 Sn bronze, with or without an addition of graphite. Bearings with an iron powder basis, either straight iron or with an addition of graphite or of copper powder, are generally stronger than copper base bearings, but are considered inferior regarding their bearing properties, particularly under conditions of starved lubrication. A listing of compositions and of porosity ranges for the various grades of sef-lubricating bearings together with their areas of application are shown in Table 15-1, adapted from a paper by V. T. Morgan.[5]

The steps in producing porous bearings are:

> Powder selection
> Mixing
> Compacting
> Sintering
> Sizing
> Impregnation with lubricant

For the bronze bearings, mixtures of electrolytic, reduced or atomized copper powders, atomized tin powders and natural graphite powders are used. The particle size distribution of the copper powder, generally all through 150 mesh and between 30 and 65% through 325 mesh, and of the tin powder, generally finer than 325 mesh, has a strong influence upon the dimensional changes of the compacts during sintering, with the finer powder causing less growth of the compacts. Thorough mixing is very important to avoid defects due to melting of tin powder particle aggregates during sintering. Both reduced and atomized iron powders are used for the iron base bearings.

The powder mixtures are compacted on automatic compacting presses of the types described in chapter 5. The density of the green bearings is controlled by the compacting pressure.

In the early work on self-lubricating bearings it was assumed that pore formers, i.e. volatile substances which are added to the metal powders but are eliminated during sintering, were necessary to control the porosity of the materials. Later it was found that in most cases control of compacting and sintering conditions are sufficient for porosity control.

Bronze bearings are sintered at temperatures in the range of 815-895° C for relatively short times, with only 3-8 minutes at the sintering temperature. The microstructural changes taking place during sintering of bronze bearings are complex and have been discussed in chapter 11. The aim in producing bronze self-lubricating bearings is a microstructure of relatively large grains of homogeneous alpha copper-tin bronze interspersed with

Table 15-1

No.	Composition	Notes	Porosities
1	89 Cu, 10 Sn, 1 Graphite	General purpose bronze. Tolerant to unhardened shafts. Wide range of porosities available.	18 - 30%
2	85 Cu, 10 Sn, 5 Graphite	High graphite bronze. Lower load capacity. More tolerance toward oil starvation	12 - 24%
3	86 Cu, 10 Sn 3 Pb, 1 Graphite	Leaded bronze. More tolerance towards misalignment	18%
4	Pure soft iron	Cheaper than bronze. Unsuitable for corrosive conditions. Wide range of porosities available.	12 - 22%
5	98 Fe, 2 Cu	Increasing the Cu additions to Fe increases the hardness, strength and cost	17 - 20%
6	97 Fe, 3 Cu		
7	92 Fe, 8 Cu		
8	Fe, Cu, Sn; Graphite 80 Fe, 20 Bronze 50 Fe, 50 Bronze	Mixtures of iron, copper and tin to give performance properties between No. 1 and No. 4	
9	Soft iron with 2-3% graphite	Graphite improves marginal lubrication of the No. 4 composition and reduces load capacity	27%
10	88 Fe, 10 Cu, 2 Graphite	Graphite improves marginally lubrication of No. 5 to 7 and reduces load capacity	22%
11	Iron with 0.7% Graphite	Copper free, hardenable	18%
12	Iron-Copper with 0.7% Graphite	Hardenable, high strength porous steel	18 - 25%

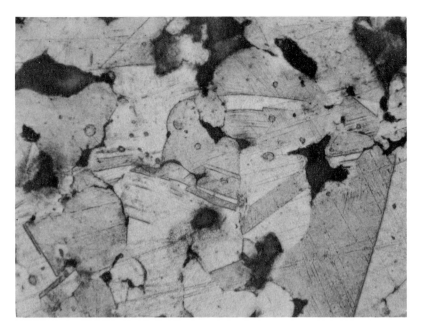

FIGURE 15-2 Microstructure of bronze self-lubricating bearing, 500x (Courtesy: Keystone Carbon Company).

pores and graphite particles as shown in Figure 15-2. Closely related to the microstructural changes during sintering are the dimensional changes, generally growth. These dimensional changes are critically dependent on size distribution of powder particles, green density of the compacts, rate of heating and time and temperature of sintering. The powder mixtures used in compacting porous bearings are blended by the bearing producer to obtain the desired strength and growth characteristics or are supplied by the powder producers as premixes. In order to get a similar amount of growth during sintering for small bearings for which the rate of heating is fast and for large bearings for which the rate is slow, the powder producers supply powder premixes with graded particle size distributions which produce uniform dimensional changes for different bearing sizes.

Iron base self-lubricating bearings are sintered at temperatures near 1120° C for 15-25 minutes. The dimensional changes during sintering of iron base bearings are generally small compared to those of bronze bearings.

For both bronze and iron base bearings the continuous furnaces, often mesh-belt furnaces, and protective atmospheres described in chapter 7 are used.

All self-lubricating bearings are sized after sintering in order to control their dimensions within the required close limits, generally ± .0005″ for outside and inside diameter for each inch of bearing diameter and ± .005″ for length. Sizing die and tools are shown in Figure 15-3.

After sizing, the bearings are impregnated with a lubricant, generally oil. The viscosity of the oil and the use of additives to inhibit corrosion, to increase film strength and to inhibit oxidation depend upon the particular application for which the bearings are used. To impregnate, the bearings are lowered in a basket into a vacuum chamber which is evacuated to approximately 1 mm Hg pressure. At this point, oil is permitted to enter the chamber. To complete impregnation, 60 psi of air or nitrogen is applied for 1 minute.

Since the material in porous bearings has lower compressive strength than that of solid material, the bearings may be inserted into the bearing housing by force fitting. The bearings are fitted on an insertion plug which is dimensioned to give the correct bore size which is on the order of .0015″ less than the original bore of the sintered bearing. During insertion the bearing

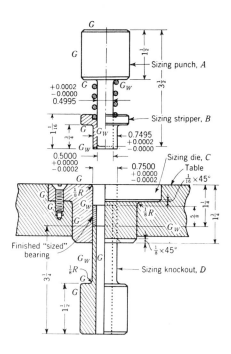

FIGURE 15-3 Sizing die tooling for plain cylindrical self-lubricating bearing.

wall is compressed, the bearing is aligned with the plug and an inside bearing diameter with a smooth mirror finish is produced. For extremely close tolerances a button burnishing tool may be used.

The American Society for Testing and Materials, the Society of Automotive Engineers and the Metal Powder Industries Federation have issued specifications[6] for self-lubricating bearings. The specifications show ranges of chemical composition, of impregnated density, of minimum values of interconnected porosity or oil content and of radial crushing strength of bearings. The oil content values assure that the bearings are satisfactory lubricant reservoirs. Although bearings are obviously not radially crushed in use, the strength values are a measure of how well the bearings are sintered.

The interconnected porosity or oil content is determined using Archimedes principle from the weight "B" of the bearing impregnated with oil weighed in air, the weight "A" of the dry bearing obtained by extracting the oil from the bearing and the weight "C" of the impregnated bearing weighed in water. With the known density of the oil "S", the oil content or interconnected porosity in volume % is given by

$$\text{Porosity} = \frac{B\text{-}A}{(B\text{-}C)S} \times 100$$

The values for porosity determined by this method may be compared with those calculated from the ratios of the dry densities of the porous bearings and the densities of the material in solid form. It is found that practically all of the porosity in porous bearings is interconnecting, i.e. the pores in the interior communicate with the outside surfaces of the bearings.

Since bearings are thin-walled compared with their diameter and fracture without significant plastic deformation, their radial crushing strength may be determined by compressing the test specimen between two flat surfaces with the direction of load perpendicular to the longitudinal axis of the specimen. The point where the pointer of the testing machine drops, because a crack is formed in the bearing, gives a measure of the radial crushing strength. From this load P, in lbs, the outside diameter of the bearing D, its wall thickness T and its length L, in inches, the so-called K factor is determined using the formula

$$K = \frac{P(D\text{-}T)}{LT^2}$$

Table 15-2 shows the data specified in ASTM specifications B 438, B 439 and B 612, which are similar to those in the corresponding SAE and MPIF specifications.

The load carrying capacity of self-lubricating bearings used to be considered inferior to that of solid bearings because the film of lubricant formed by hydrodynamic action was believed to be not as stable on a porous as on a

Table 15-2 Specifications for chemical composition, oil impregnated density, minimum oil content or interconnected porosity and minimum K-factor for self-lubricating bearings.

	Chemical composition	Impregnated density range	Minimum oil content or interconnected porosity	K-factor
	ASTM B 438			
Grade 1	87.5-90.5 Cu	5.8-6.2	27	15,000
	9.5-10.5 Sn	6.4-6.8	19	26,500
	up to 1.75%			
	Graphite			
Grade 2	82.6-88.5 Cu			
	9.5-10.5 Sn	6.6-6.9	19	26,500
	2.0- 4.0 Pb			
	ASTM B 439			
Grade 1	100 Fe			25,000
Grade 2	0.25-0.6 C	5.7-6.1	20	30,000
	bal Fe			
Grade 3	7.0-11.0 Cu			
	bal Fe			
		5.8-6.2	19	40,000
Grade 4	18.0-22.0 Cu			
	bal Fe			
	ASTM B 612			
	34-40 Cu			
	3.8-4.4 Sn	6.0-6.4	18	25,000-
	bal Fe			45,000

solid surface. This limitation in load carrying capacity is reflected in the recommended values shown in the ASTM standards, in which the capacity is given in lbs per projected unit area of the bering (length times inside diameter) in square inches as a function of shaft velocity. The values are shown in Table 15-3. However, recent investigations have shown that depending on proper bearing housing construction much higher load carrying capacities with PV values (pressure in psi, velocity in ft/min) as high as 150,000 are possible in self-lubricating bearings.

The use of these values of load as a function of velocity has definite limitations. Whether a bearing operates satisfactorily or not depends largely upon its running temperature. The temperature, in turn, is governed by

Table 15-3   Permissable loads for self-lubricating bearings

Bronze Bearings

Shaft velocity, ft/min, m/min	Permissable load, psi (MPa)	
	Low density	High density
Slow and intermittent	3200 (22)	4000 (28)
25 (7.6)	2000 (14)	2000 (14)
50-100 (15.2-30.4)	550 (3.9)	500 (3.4)
over 100 to 150 (30.4-45.7)	365 (2.5)	325 (2.2)
over 150 to 2000 (45.7-61)	280 (1.9)	250 (1.7)
over 200 (61)	P in psi = 50,000/V in ft/min	

Iron Base Bearings

	Copper free	Copper containing
Slow and intermittent	3600 (25)	8000 (55)
25 (7.6)	1800 (12)	3000 (20)
50-100 (15.2-30.4)	450 (3.1)	700 (4.8)
100-150 (30.4-45.7)	300 (2.1)	400 (2.8)
150-200 (45.7-61)	225 (1.6)	300 (2.1)
over 200	P in psi = 50,000 V in ft/min	

Iron-bronze Base Bearings

Slow and intermittent	4000 (2.8)
25 (7.6)	2000 (1.4)
50-100 (15.2-30.4)	400 (0.35)
100-150 (30.4-45.7)	300 (0.23)
150-200 (45.7-61)	200 (0.14)
over 200	P in psi = 40,000/V in ft/min
	but V should not exceed 400 ft/min

the balance between the heat generated and that conducted off through the shaft or otherwise. Both the heat generated, which is a function of the viscosity of the oil and the coefficient of friction between shaft and bearing, and the heat conducted off depend upon a number of bearing design factors, as was pointed out by V. T. Morgan.[4]

A principal field of use for self-lubricating bearings is fractional horsepower electric motors. A partial list of applications for these motors would include: electric fans and blowers, vacuum sweepers, washing machines, dish washers, clothes dryers, sewing machines, phonographs and record changers, hood and window raisers, food mixers, electric clocks, business machines, refrigerators, air-conditioning units, textile equipment and sundry instruments.

II. Porous Metallic Materials for Filters and Similar Applications[7,8]

The principal use of this type of porous materials is to remove solid particles from streams of liquids, such as oil, gasoline, refrigerants, polymer melts, aqueous suspensions or from streams of air or other gases. In these applications the metallic filters compete with porous filters from glass, ceramic, cellulosic materials or even screens. Metallic filters are resistant to heat; by proper selection of material, resistance to corrosion can be assured. Compared with screens, metallic filters offer a tortuous path which the fluid must follow through the filter; they provide depth filtration. Compared with glass, ceramic and cellulosic filters, the metallic filters have, on the one hand, good strength which prevents any particle of the filter material to break away in service and enter the stream of filtrate. On the other hand, they have ductility so that they do not fracture under mechanical or thermal shocks. They can be fitted into housings by appropriate metal forming operations such as rolling, machining, press fitting or welding. In applications where vibration is involved the limited fatigue resistance of porous filters must be taken into consideration. Metallic fitness can be fabricated into a variety of shapes, as will be seen from Figures 15-4 and 15-5 which show assemblies of bronze (Figure 15-4) and of stainless steel (Figure 15-5) filters.

There are many other applications for the type of porous metallic materials used for filters. Such applications are as separation media, e.g. for separating oil from water, as diffusion media, the so-called "spargers", for aerating liquids or for dispensing CO_2 in flowing liquids, for flow control media, such as controlled leaks or restrictors for snubbers or for silencers. A recent application is as high pressure support for reverse osmosis membranes. The materials have been suggested for transpiration cooling[9,10] for jet-engine afterburners and for deicing in aircraft wings.

Before the methods of fabrication and the properties of individual metallic filter materials are presented, those characteristics which are most important in filter materials are discussed. They are

    1. Adequate mechanical strength

    2. Retention of solid particles down to a specified size

    3. Fluid permeability which is the ability to pass a given volume of fluid of a specified viscosity through unit area and unit thickness of a filter under a given pressure gradient.

    4. Resistance to environmental attack.

Mechanical strength for a filter material is necessary for two reasons. It governs the possible ratio of area to thickness of large filters, especially

FIGURE 15-4 Assembly of filters from spherical bronze powder (Courtesy: Thermit, Inc.).

FIGURE 15-5 Assembly of filters from stainless steel powder (Courtesy: Newmet Products, Inc.).

369

where large pressures are involved, and it is a measure of the bonding between the metal particles. The interparticle bonds must be strong enough to prevent any particle from breaking away and entering the stream of filtrate.

Solid particle retention and permeability in filters are opposed to each other; the smaller the minimum size of particles retained in a filter, the lower is its permeability. In order to produce a filter which will pass a sufficiently large volume of fluid in a given time and yet retain particles to a small size, it is often necessary to design filters with large cross sectional areas. How this might be done is seen from an inspection of Figures 15-4 and 15-5.

Retention of solid particles may be determined by passing a suspension of particles of exactly known size, such as spheres of glass or a plastic through the filter and determine to which size these particles are retained by the filter. When suspensions of particles of a range of sizes are passed through a metallic filter of appreciable thickness, it is found that the filter will retain not only all particles above a given size, but also many particles smaller than this size.[11]

A widely used method for obtaining an estimate of particle retention is the determination of bubble test pore size.[11,12] A form of this test was standardized by ISO Committee 119 and has been published as ISO standard 4003. The sample to be tested is impregnated with and immersed in a test liquid. Air is introduced at the underside of the piece under gradually increasing pressure. The pressure at which bubbles are first emitted from the surface of the piece is a measure of the maximum pore size according to the equation

$$d = \frac{4\gamma}{p}$$

in which d is the maximim pore diameter in m, γ the surface tension of the test liquid in N/m and p the pressure difference across the test piece in pascal, equal to p_g - p_l where p_g is the gas pressure and p_l is equal to 9.81 ρh where ρ is the density of the liquid in kg/m^3 and h the distance between the upper surface of the test piece and the level of the test liquid. A comparison of the values of maximum pore diameter determined by this test and a test based on actual retention of particles shows that the bubble test gives values of maximum pore size considerably larger than the pore size determined by retention. For filters produced from uniform spherical particles the ratio of bubble test diameter and particle retention diameter is approximately 3.1, for porous metal made from irregular particles the ratio is approximately 5.

The fluid permeability of a permeable sintered metal material[11,13] may be determined by a test method which has also been standardized by ISO Technical Committee 119 as ISO standard 4022. The pressure drop and the volumetric flow rate are measured, when a test fluid of known viscosity and

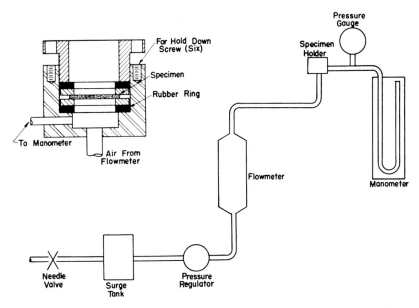

FIGURE 15-6 Schematic of apparatus for determining permeability of filter disks.

density is passed through a test sample. The principle is illustrated schematically in Figure 15-6 for the determination of the fluid permeability of a flat test piece. The ISO standard also shows how the fluid permeability of a hollow cylindrical test piece is determined.

When the flow of the fluid is strictly viscous, Darcy's law applies:

$$\frac{\Delta P}{e} = \frac{Q \cdot \eta}{A \, \psi_v}$$

in which ΔP is the pressure drop in Pa

e is the thickness of the test piece in m

A is its cross sectional area in m^2

η is the absolute dynamic viscosity of the test fluid in N.s/m^2 (kg/m.s)

Q is the volumetric flow rate of the fluid in m^3/sec equal to the mass flow rate of the fluid divided by its density ρ

ψ_v is the viscous permeability coefficient in m^2

The formula may be directly applied to measuring the permeability of incompressible fluids, i.e. liquids. However, air, whose density is a function of pressure, is much more commonly used for these tests. Darcy's law is then written:

$$\frac{P_1 - P_o}{e} = \frac{Q_m}{\rho} \cdot \frac{1}{A} \cdot \frac{\eta}{\psi_v}$$

in which Q_m is the mass flow rate of air, P_1 the pressure on the inlet, P_o the atmospheric pressure on the outlet side and ρ the density of the air at the average pressure $(P_1 + P_o)/2$ equal to

$$\rho = \rho_o \frac{P_1 + P_o}{2 P_o}$$

with ρ_o the density of the air at atmospheric pressure. For air, Darcy's law is then used in the form

$$\frac{P_1{}^2 - P_o{}^2}{2 P_o e} = \frac{Q_m}{\rho_o A} \cdot \frac{\eta}{\psi_v}$$

The flow of fluids through porous materials involves several mechanisms operating simultaneously; in addition to viscous flow, inertia and slip flow are important. Inertia flow is observed because of the loss of energy due to the changes in the direction of the fluid in passing through tortuous porosity and to the onset of local turbulance. Forchheimer[14] has shown that the presence of inertia flow can be taken care of by introducing an inertia permeability coefficient with units of length into the equation of fluid flow in the form

$$\frac{\Delta P}{e} = \frac{Q \cdot \eta}{A \cdot \psi_v} + \frac{Q^2 \cdot \rho}{A^2 \cdot \psi_i}$$

in which ψ_v and ψ_i are the viscous and the inertia coefficients of permeability respectively. By measuring the rate of flow of a given fluid at a series of pressure drops, both coefficients necessary to characterize the flow can be determined. In experiments with filters with three different pore sizes produced from 50 to 80 mesh, 100-200 mesh and 200-325 mesh stainless steel powder, using air as the fluid, downstream atmospheric pressure and pressure drops up to 60 psi (400 KPa) German[15] has shown that the Forchheimer[14] equation will give an excellent reproduction of the experimental data.

Slip flow applies to cases of flow where the mean free path of molecules between intermolecular collisions approaches the size of the pores in the filter. It is important for gases at low pressures and high temperatures and filters with very small pore sizes. Under these conditions inertia flow is absent. In the slip flow regime the permeability coefficient is pressure dependent and, as Klinkenberg[16] has shown, may be written in the form

$$\psi_s = \psi_v \left(1 + \frac{B}{P_{av}}\right)$$

where ψ_s and ψ_v are the slip and viscous permeability coefficients, P_{av} the average between inlet and outlet pressure $= \frac{P_1 + P_2}{2}$ and B the Klinkenberg factor, which is a constant for a given gas and porous material and has the dimensions of a pressure. By measuring over a range of different absolute pressures P_1 and P_2 and plotting ψ_s vs $\frac{2}{P_1 + P_2}$ a straight line is obtained, whose slope is $B.\psi_s$ and whose intercept is equal to the viscous permeability ψ_v. German[15] produced porous metal gas flow restrictors for mass spectrometer calibrations from -325 mesh water atomized 304L stainless steel powder by compacting, sintering, coining and annealing. He measured the flow rate of helium through these restrictors with inlet pressures from 10 to 600 psi absolute (70 to 4000 kPa) and vacuum on the outlet side. Using the slip flow correction of Klinkenberg for the viscous permeability coefficient, he was able to reproduce the flow pressure relationship for flows in the range from 10^{-4} to 10^{-1} cm^3/sec (10^{-10} to 10^{-7} m^3/sec).

The most widely used metallic filter materials are porous copper-tin bronze and porous stainless steel. There are many applications where one or the other of these materials is not attacked by the fluid to be filtered. For special cases, i.e. for filtration of highly corrosive fluids, titanium, monel, inconel and even precious metal filters are used. Different methods of fabrication from metal powders are used for bronze powders, on the one hand, and for powders of other metals and alloys, on the other. Filter materials in these two groups are, therefore, discussed individually.

A. Bronze Filters

Bronze filters are made by loose powder sintering of spherical bronze powder particles. The composition is 90-92% Cu, 8-10% Sn for atomized bronze powder. Filters from tin-coated cut wire have tin contents from 2.5 to 8%. In order to produce filters with the highest permeability for a given maximum pore size, powder particles of a reasonably uniform particle size are used.

Two methods of producing bronze powders with spherical particle shape are in use. A molten bronze alloy may be atomized under conditions which produce spherical particles or spherical copper powder particles may be

Table 15-4 Properties of four grades of filter materials produced by loose powder sintering spherical powders

Particle size of spherical powder particles		Tensile strength MPa	Recommended minimum thickness of filter m	Largest dimensions of particles retained μm	Viscous permeability coefficient ψ_v in m^2
mesh range	range in in μm				
20- 30	850-600	20-22	0.0032	50-250	2.5×10^{-4}
30- 40	600-425	25-28	0.0024	25-50	1×10^{-4}
40- 60	425-250	33-35	0.0016	12-25	2.7×10^{-5}
80-120	180-125	33-35	0.0016	2.5-12	9×10^{-6}

coated with a thin layer of very fine tin powder and presintered. The finer grades of spherical copper powder may again be produced by atomizing, but the coarser grades are generally obtained by chopping copper wire and tumbling the choppings. These methods have been discussed in more detail in chapter 2. The loose powders are sintered in graphite or stainless steel molds at temperatures near the solidus temperature of the bronze composition. The tin in the tin coated copper powder insures an excellent bond between the particles and also diffuses into the copper producing filters of a homogeneous α bronze composition. The sintering has been discussed in chapter 12 on loose powder sintering. In Table 15-4 the properties of four grades of bronze filter materials are presented.[17] By far the most common of these grades is the third. The two coarsest grades are no longer widely used. When the spherical powders are poured loose into the molds and vibrated, they reach the same density regardless of particle size; this density is approximately one half that of solid bronze. During sintering the filters shrink slightly. Tin-coated cut wire filters have sintered densities in the range from 4.6 to 5.0 g/cm^3, filters from atomized tin coated powders in the range from 4.6 to 4.9 g/cm^3 and filters from prealloyed bronze powder in the range from 5.0 to 5.2 g/cm^3. The shrinkage during sintering may be as high as 8%. To avoid excessive shrinkage filters from powders with fine particle size require lower sintering temperatures in the neighborhood of 815° C (1500° F). Because of the shrinkage during sintering, filters must be designed with a slight draft, so that they can be removed from the mold.

B. Filters from Powders of Stainless Steel[18] and Other Corrosion Resistant Metals and Alloys

While bronze filters are produced by loose powder sintering of spherical powders giving the best permeability for a given particle size, this production method is not practical for filters from other metals and alloys; therefore other methods of fabrication and non-spherical powders are used.

The most common composition of porous stainless steel filters is AISI 316L, i.e. 16-18% Cr, 10-14% Ni, 2-3% Mo and less than 0.1% C, balance iron, but other types of austenitic stainless steel powders are also fabricated into filters. The powders are produced by atomizing (see chapter 2). Particle size fractions of these powders are used for different grades of filters (see Table 15-5). The methods for fabricating stainless steel filter materials depend on the shapes of the filters. For making sheet, the loose powder mixed with a resin is spread in a mold, lightly pressed at a temperature where the resin is cured and the resulting sheet is sintered. During sintering the resin is decomposed. The porous sheet is densified by repressing and then resintered. It may be formed into hollow cylinders and seam-welded. A method of producing porous filters in the shape of hollow cylinders, which has been described in chapter 6, is cold isostatic pressing. It is used for filters from both stainless steel and titanium powder. Filters in the shape of hollow cylinders, particularly stainless steel filters with thin wall thicknesses, may be fabricated by cold extrusion with a plasticizer as also described in chapter 6.

Data on the fabrication and properties for a series of seven grades of filter material 1/16" thick from 316L stainless steel with decreasing pore size and permeability from one manufacturer (Mott Metallurgical Co., Farmington, CT) are given in Tables 15-5[15] and 15-6.[19] Table 15-5 shows the particle size, compacting pressure, sintering temperature and density of these grades together with the values of pore size as determined by the bubble test and the permeability as characterized by viscous (ψ_v) and inertia (ψ_i) coefficients of permeability. Table 15-6 gives the mechanical properties of these filter materials.

III. Porous Electrodes

The third group of porous metallic materials comprises those used as electrodes for alkaline batteries and for fuel cells.[20,21] The electrodes are in the shape of strip or sheet from a fraction of a mm to somewhat more than 1 mm thick. They are generally made of nickel. The chemical reactions occurring at the electrodes for batteries and for fuel cells are quite different and the requirements of materials for them, therefore, also differ.

Table 15-5 Data on fabrication of Mott Series A Filters. All samples from water atomized 316L stainless steel powder, lubricated with 0.3 wt.% zinc stearate for compacting and sintered in hydrogen for 3 hr.

Nominal Filter Rating μm	Powder Size Mesh	Compaction Pressure		Sintering Temperature °C	Density g/cm³	Bubble Test Maximum Pore Size μm	Permeability Coefficients	
		psi	MPa				ψ_v in m²	ψ_i in m
0.5	200/325	30,000	207	1290	5.93	10	7.7×10^{-14}	2.3×10^{-8}
2	200/270	20,000	138	1290	5.21	21	7.7×10^{-13}	2.0×10^{-7}
5	100/200	28,000	193	1340	4.82	26	2.0×10^{-12}	7.7×10^{-7}
10	50/100	15,000	103	1340	4.50	42	5.9×10^{-12}	2.1×10^{-6}
20	50/80	18,000	124	1370	4.19	58	1.8×10^{-11}	4.2×10^{-6}
40	30/50	15,000	103	1370	3.87	114	3.4×10^{-11}	1.3×10^{-5}
100	20/30	15,000	103	1370	3.48	347	1.4×10^{-10}	5.0×10^{-5}

Table 15-6 Mechanical properties of Mott Series A 316L stainless steel filters, 1/16" thick.

Nominal Filter Rating	Ultimate Tensile Strength		Yield Strength at 0.2% offset		Elongation %	Youngs Modulus	
	psi	MPa	psi	MPa		psi × 10⁻⁶	GPa
0.5	33,600	232	23,800	164	2.5	8.8	61
2	23,900	165	18,000	124	2.0	7.8	54
5	19,100	132	16,400	113	2.0	5.1	35
10	14,900	103	11,500	79	1.8	4.9	34
20	11,500	79	7,100	49	1.2	3.6	25
40	7,200	50	4,900	34	1.2	2.4	17
100	3,400	23	1,800	12	1.0	1.7	12

In electrodes for nickel/cadmium batteries, high porosities on the order of 70-90% are essential because the porous structure should accomodate as high a quantity as possible of the active mass, nickel hydroxide in the positive plate and cadmium hydroxide in the negative plate. Pore size distribution is not very important as long as the pores are not too large (because active material may fall out of them) or too small (because the active material cannot be incorporated in very small pores). Porous electrodes for alkaline batteries generally contain a central support which consists of a perforated strip, electroformed mesh or woven wire mesh which provides mechanical strength and the necessary high electrical conductivity to the electrodes.

Electrodes for fuel cells must form a three-phase boundary between the gas, the electrode and the electrolyte as shown schematically in Figure 15-7. The surface area of the interface should be as high as possible, because the higher the area the greater the activity of the cell. The electrolyte penetrates the pore space of the electrode but is pushed out by the gas pressure, and these two effects are controlled by the size, shape and distribution of the pores. In general, a narrow pore size distribution and a porosity of 50-60% are desirable. Ideally the pore shape should be conical, but since this is not possible to achieve, electrodes may be produced with a "macroporous" layer on one side and a "microporous" layer on the other side. Supports are also often incorporated in fuel cell electrodes. In order to operate fuel cells with porous nickel electrodes at lower temperatures ($< 100°$ C) and lower pressure, the electrodes may be activated by incorporating a powder of the nickel-aluminum Raney alloy into them.

Both the type of powder used and the method of fabrication are important in the production of porous electrodes. For alkaline battery electrodes which have to have very high porosities of 70-90%, the filamentary or fibrous type of carbonyl nickel powder (Inco grades 255 and 287) have the advantage of giving reasonably high strength to the sintered electrodes even

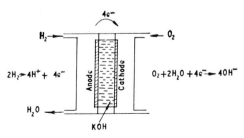

FIGURE 15-7 Principle of operation of the hydrogen/oxygen fuel cell.

at these high porosities. However, special grades of hydro-metallurgically produced nickel powder (Sherritt Gordon grade G) with very low apparent density, under 1 g/cm³, and fine particle size have also been used.[22] Of the types of fabrication techniques for porous electrodes which have been discussed in previous chapters, loose powder sintering and the slurry technique are particularly important for alkaline battery electrodes. Compacting in a rigid mold is of minor importance but powder rolling is widely used for both types of electrodes. In loose powder sintering the steps are:

> Fill bottom half of mold made of graphite or nickel
> Place support on powder surface
> Place second half of mold in position
> Fill mold completely and cover it
> Sinter at 900° C in reducing atmosphere

The slurry technique involves the stages of:

> Preparing a slurry of the right viscosity by suspending the powder
> in a solution containing methyl-cellulose
> Coating the support material by passing it through the slurry bath
> Drying and sintering the coated strip

Powder rolling for producing electrodes has been described in detail by Williams and Tracy.[23] Roll diameter, roll speed, roll gap, powder head and sintering conditions must be controlled in order to produce material of the desired thickness, porosity and pore size distribution. In order to achieve the high porosities, called for in alkaline batteries, a spacing agent, commonly methyl-cellulose, in quantities up to 40 volume percent is added to the filamentary type of carbonyl nickel powder and the strip is sintered at the low temperature of 700° C. A central support may be incorporated into the porous strip during rolling as shown in Figure 6-24.

In producing fuel cell electrodes by powder rolling, the equiaxed fine carbonyl powder rather than the filamentary type of powder may be used because the porosity needed for these electrodes is lower and the equiaxed powder gives finer pores of more uniform pore size. The fuel cell electrodes may be sintered at somewhat higher temperatures than alkaline battery electrodes. The maximum pore size in these electrodes as measured by the bubble test may be on the order of 3-8 μm.

The production of multilayer electrodes incorporating layers of differing porosity and pore size is a difficult problem. One method would be to powder roll the fine pore layer and then apply the coarse pore layer with a spray gun on the fine pore layer in the form of a slurry.[23] Another suggested method is to introduce two types of powder separated by the wire mesh support into the roll gap of the rolling mill and in this way roll directly a strip with a two layer structure and reinforcement.[24]

IV.   Tungsten Ionizers for Cesium-ion Engines

The operation of cesium-ion engines, which are to be used in the absolute vacuum of space environment, is based on the following principles: When vaporized cesium is passed through a high work function porous material, which is above a critical temperature, the cesium is ionized by contact with the material. By the use of focusing and accelerating electrodes, a beam of ions is ejected and a thrust of the order of milli-newtons is produced, which is suitable for the space environment, because it can be made continuous for long periods of time. Development work on such porous tungsten ionizers showed that their pore size must be very small, with several millions of pores per cm^2, so that the atoms of the cesium vapor are completely ionized. Neutral atoms leaving the ionizer are ionized outside the focusing field and these unfocused ions bombard the acceleration electrodes and destroy the engine. Also the structure of the ionizer must be stable at the operating temperature of 1150° C.

One method of producing porous ionizers developed by Turk of Hughes Research Laboratories[25] involves the use of spheroidized tungsten powder obtained by melting reduced tungsten powder in a plasma flame. The spheroidized powder is separated into narrow size fractions in the 2 to 7 μm range. Powder of these size fractions is consolidated into ionizers by warm pressing at pressures up to 75,000 psi in a tungsten carbide die at 400° C to densities of 75% of theoretical. The ionizers are sintered at temperatures between 1700 and 2300° C (the smaller the particle size, the lower the temperature) to densities 80% of theoretical.

Another approach to producing porous ionizers was reported by Hivert and Labbe of Onera.[26] They use a dispersion of thorium oxide in tungsten in order to produce a structure stable at the operating temperature. A pore former, methyl polymethacrylate, which also serves as a plasticizer, is incorporated into the tungsten in order to produce thin foils, 0.6 to 1.5 mm thick. After the pore-former-plasticizer is eliminated by heating in hydrogen, the foils are sintered in hydrogen at 1550° C. Ionizers with 30 million pores per cm^2 are produced.

V.   Porous Surgical Implants

Porous materials are of interest as metal prostheses, but the long-term corrosion and biocompatibility of these materials is still the subject of research. The metal prostheses include both totally porous implants and implants coated with a porous outer layer. The purpose of the coatings is to make it possible for bone tissue to grow into the pores of the porous coating. The mechanical properties (elastic modulus, yield strength and fatigue strength) of the porous material are intermediate between those of the material of the solid metal prostheses and those of bone. For these reasons the porous coating is expected to provide a good joint between solid prothesis and bone.

In the experiments of Dustoor and Hirschhorn[27,28] the prosthesis to be coated was produced by isostatic pressing 316L stainless steel powder in a rubber mold. After sintering in hydrogen at 1200° C this prosthesis was then coated by packing a viscous mixture of stainless steel powder and a thickener, an ammonium salt of alginic acid, around the prosthesis. The thickener is decomposed and the coating is sintered. In the experiments of Pilliar et al.[29] a solid cast substrate of a cobalt base alloy (30 Cr, 6 Mo, bal. Co) is used. It is coated with a porous layer by dipping or spraying the substrate with a mixture of metal powder of the same composition as the substrate and of an acqueous solution of a methylcellulose binder. After decomposing the binder, the coating is sintered for 3 hours in hydrogen. In order to make it possible for the bone tissue to grow into the coating, the coating must be highly porous and the pores must be sufficiently coarse in the range from 40 to 100 μm. Pilliar et al. have also experimented with porous surface coatings in cardiovascular applications where soft tissue grows into the porous coating.

References

1. U. S. Patent 1,556,658 H. M. Williams, General Motors Corp., Bearing of Cu, Sn and Graphite (1925).
2. U. S. Patent 1,642,347; 1,642, 348; 1,642,349 H. M. Williams and A. L. Boegehold (General Motors Corp.) Porous alloy structure, (1927).
3. U. S. Patent 1,607, 389 C. Claus (Bound-Brook Oilless Bearing Co.) Pressed metal articles (1926).
4. W. R. Toeplitz, Bearings from metal powders, Trans. AIME, vol. 161, p. 542-549, (1945).
5. V. T. Morgan, Bearing materials by powder metallurgy, Powder Metallurgy, vol. 21, No. 2, p. 80-85, (1978).
6. ASTM Standards: B 438 Copper-base sintered bearings (oil-impregnated), B 439 Iron-base sintered bearings (oil-impregnated), B 612 Iron-Bronze sintered bearings (oil-impregnated).
7. Curt Agte and Karel Ocetek, "Metallfilter," Berlin (1957).
8. G. R. Rothero, Porous media, Powder Metallurgy, vol. 21, No. 2, p. 85-89, (1978).
9. P. Duwez and H. Martens, The powder metallurgy of porous metals and alloys having a controlled porosity, Trans. AIME, vol. 175, p. 848-877, (1948).
10. F. V. Lenel and O. W. Reen, Porous stainless steel compacts for transpiration cooling, "Symp. on Testing Metal Powders and Metal Powder Products", ASTM Sp. Tech. Pub. 140, p. 24-42, (1953).

11. V. T. Morgan, Filter elements by powder metallurgy, "Symp. on Powder Metallurgy," The Iron and Steel Inst., London, p. 81-89, (1956).

12. ISO Standard 4003, Permeable sintered metal materials—Determination of bubble test pore size.

13. ISO Standard 4022 Permeable sintered metal materials—Determination of fluid permeability.

14. P. Forchheimer, Zeitschrift Verein deutscher Ingenieure, vol. 45, p. 1782, (1901).

15. R. M. German, Gas flow physics in porous metals, Int. J. of Powder Metallurgy & Powder Technology, vol. 15, p. 23-30, (1979).

16. L. J. Klinkenberg, American Petroleum Institute, Drilling production practice, p. 289, (1941).

17. Adapted from booklet on "Porex" filters, Moraine Products Division of General Motors Corp., (1943).

18. N. Nicholaus and R. Ray, Porous stainless steel—The unique filter material, Mod. Dev. in Powder Metallurgy, vol. 5, ed. by H. H. Hausner, New York, p. 187-199, (1971).

19. Mott Metallurgical Corp., Technical handbook for precision porous metal products.

20. V. A. Tracey, Production of porous nickel for alkaline-battery and fuel-cell electrodes: Practical and economic considerations, Powder Metallurgy, vol. 8, No. 16, p. 241-255, (1965).

21. V. A. Tracey, The properties and some applications of carbonyl nickel powder, Powder Metallurgy, vol. 9, No. 17, p. 54-71, (1966).

22. H. A. Hancock, D. J. I. Evans and V. N. Mackiw, Sintered plates from low density nickel powder, Int. J. of Powder Metallurgy, vol. 1, No. 2, p. 42-55, (1965).

23. N. J. Williams and V. A. Tracey, Porous nickel for alkaline battery and fuel cell electrodes, production by roll-compacting, Int. J. of Powder Metallurgy, vol. 4, No. 2, p. 47-62, (1968).

24. I. Amato, S. Corso, E. Sgambettera, Porous electrodes for Zn/air alkaline batteries, Mod. Dev. in Powder Metallurgy, vol. 11, ed. by H. H. Hausner and P. W. Taubenblat, Princeton, p. 331-339, (1977).

25. R. Turk, Tungsten ionizers with controlled porosity for cesium-ion engines, Mod. Dev. in Powder Metallurgy, vol. 2, ed. by H. H. Hausner, New York, p. 309-320, (1966).

26. A. Hivert and J. Labbé, Fabrication of ionizers for electrical propulsion, Preprints of the 9th Plansee Seminar, Reutte, Austria, vol. 1, No. 12, (1977).

27. M. R. Dustoor and J. S. Hirschhorn, Porous surgical implants, Powder Metallurgy Int., vol. 5, No. 4, p. 183-188, (1975).

28. M. R. Dustoor and J. S. Hirschhorn, Porous metal implants, Mod. Dev. in Powder Metallurgy, vol. 11, ed. by H. H. Hausner and P. W. Taubenblat, Princeton, p. 247-262 (1977).

29. R. M. Pilliar, D. C. MacGregor, I. Macnab and H. U. Cameron, P/M surface coatings on surgical implants, ibid. p. 263-278.

16

Cemented Carbides

Cemented carbides are a class of very hard, wear resisting materials produced by powder metallurgy. In Europe they are generally called "Hard metals" or "Hardmetals". A number of monographs[1,2,3,4] and reviews[5,6,7] on cemented carbides have been published. Cemented carbides were developed in Germany in the early 1920's. Karl Schröter[8] was the principal inventor. They were first produced commerically by Fried. Krupp in 1927. The original cemented carbides consisted of tungsten carbide and cobalt with a structure of tungsten carbide grains embedded in a cobalt matrix. This is still the most important composition from the point of view of quantities commercially produced. Cemented carbides combine the very high hardness and wear resistance of their carbide phase with sufficient mechanical and thermal shock resistance due to the metallic binder phase to make them useful for many applications. The first cemented carbides were developed in response to the demands for a material sufficiently wear resistant for drawing dies for tungsten wire. It was soon discovered that the new material could be used for machining and that cemented carbides have higher hardness and wear resistance, including hardness at elevated temperatures, than high speed tool steels. Today, the use of cemented carbides for metal cutting purposes consumes about half of the material produced. One of the most important forms in which cemented carbides for use in metal cutting are used is as tool bits, which are brazed or clamped into a steel holder. Brazed tool bits are generally reground after they have become dull in use. Some of the clamped bits can be indexed so that several edges of the bit can be used successively. These indexed bits are often not reground after all edges have been used and have become dull, but are thrown away; they are designated "throw-away" bits. Other important uses of cemented carbides are in rock drilling and stone cutting, in metal-forming tools and structural components, in wear parts and abrasive grits. The discussion of cemented carbides is presented in the following sections:

> Classification of cemented carbides
> Production of cemented carbides
> Composition, properties and testing of cemented carbides
> Sintering mechanism in cemented carbides

In an appendix to the chapter, the use of cemented carbides and of other "cermets" (ceramic-metal combinations) for structural high temperature applications will be briefly discussed.

I. Classification of Cemented Carbides

A. Tungsten Carbide - Cobalt Grades

These grades consist of fine angular particles of tungsten carbide bonded with metallic cobalt. The amount of cobalt may vary from as little as 3% to 13% for grades used as cutting tools and up to 30% for wear parts. The average size of the carbide particles varies from less than 1 μm to 5 μm for machining grades and even 10 μm for wear parts. Increasing amounts of cobalt and increasing particle size of tungsten carbide lower the hardness but increase the mechanical shock resistance of the grades. Grades containing only tungsten carbide in the carbide phase are used for machining cast iron, non-ferrous metals and non-metallic materials, but generally not for machining steel. They are also the materials used for non-machining applications of cemented carbides.

B. Grades in Which Cobalt is the Binder Phase and tungsten Carbide the Major Constituent of the Carbide Phase, but Which also Contain Titanium Carbide, Tantalum Carbide or Combinations of these Carbides with Tungsten Carbide

These are grades which are primarily used for cutting steel. In this operation, the straight tungsten carbide grades are subject to cratering which is an erosion of the tool surface by the chip, caused primarily by a diffusion reaction between tool and chip at the high temperature reached in machining. How serious cratering is depends on the material being machined and the machining conditions. Grades of cemented carbides containing titanium carbide and grades containing tantalum carbide were developed in the early 1930's in order to combat cratering during the machining of steel. In recent years, grades containing both titanium and tantalum carbides, which are superior to those containing only one of the addition carbides, have been developed. The amount of titanium and tantalum carbide in these grades depends upon the severity of cratering and may be as high as 35% TiC and 7% TaC. One or two percent of TaC or sometimes VC or Cr_3C_2 may be incorporated in the WC during its production even in grades which are primarily straight WC-Co to inhibit growth of the carbide grains during sintering.

C. Titanium Carbide Base Cemented Carbides

Cemented carbides in which TiC forms the carbide phase were first developed in the 1950's as "cermets" for high temperature use. The intention was to produce materials which combine high strength at elevated temperature with reasonable mechanical and thermal shock resistance for use in such applications as turbine blades in jet engines. This development was essentially unsuccessful, although certain specialized applications for the materials were found.

Titanium carbide based grades bonded by nickel-molybdenum in which the molybdenum is divided between carbide and binder phase were developed for certain steel cutting applications, particularly for finishing.[9] These grades are somewhat less resistant to mechanical shock and thermal fatigue than cemented carbides with a tungsten carbide base. They are attractive because they do not require cobalt as a binder metal for which the supply situation is unstable. Research to improve the toughness of cemented titanium carbides by addition of nitrides or carbo-nitrides is being undertaken.[10]

Still another series of grades, based on titanium carbide as the carbide phase, are the so-called "Ferro-tic" alloys (Ferro-titanit in Germany). Their development also started out as an attempt to produce turbine blades by infiltrating a titanium carbide skeleton with a nickel base alloy infiltrant. The material was not successful for turbine blades and the direction of development shifted from infiltration with a nickel-base alloy to infiltration with iron-base alloys, in particular high-temperature steels, such as high-speed steel. Eventually, it was found that these alloys could be produced more simply by mixing the titanium carbide with the ingredients necessary for the steel base binder in powder form, compacting and liquid phase sintering. In the sintered condition the materials are machinable and can then be heat treated to give them hardnesses higher than those of heat treated tool steels, but lower than other cemented carbides. A large number of these "Ferro-tic" alloys, generally with binder contents in the 50-60% range, have been developed for use as stamping, blanking and drawing dies and machine components, where the ability to machine before hardening reduces the cost of production. Recently, a "ferro-tic" alloy with an age-hardenable maraging steel as a binder phase has been developed which appears promising for applications where combinations of high strength, corrosion and oxidation resistance are required, as in die-casting inserts and cores, extrusion dies, etc.

D. Cemented Chromium Carbides

One other grade of cemented carbide which has found industrial use is cemented chromium carbide. One of the grades of cemented chromium carbide contains 83% Cr_3C_2, 15% Ni and 2% W. Cemented chromium carbides combine excellent wear resistance with good corrosion and oxidation resistance and are used for applications where these properties are needed, e.g. in valve components and plungers. They are not tough enough for cutting tool applications.

II. Production of Cemented Carbides

Up to recently, cemented carbide producers customarily prepared their own carbide and cobalt powders, often starting with tungsten ore, titanium oxide, carbon black and cobalt oxide. The reason was that the quality of the final product is very dependant upon the properties, in particular particle size and particle size distribution and chemical purity of the carbide and cobalt powders. This situation has changed since carbide and cobalt powders with adequately controlled properties have become commercially available; therefore many carbide producers begin fabrication with purchased carbide and cobalt powders.

A. Production of the Raw Materials[7]

The methods of production of the individual carbide powders was discussed in chapter 2, but that of the mixed carbides, which are important for steel cutting cemented carbide grades, should be briefly mentioned. They may be produced from a mixture of tungsten powder, titanium dioxide, tantalum oxide, Ta_2O_5, and carbon black in induction heated vacuum furnaces operating at temperatures near 2000°C. The use of vacuum at the final stage of carburizing insures a sufficiently high carbon content. Alternatively, the mixed carbides may be produced as a step in purifying the individual carbides by heating a mixture of the carbides in a high vacuum to 2000-2500°C. This treatment also lowers the oxygen and nitrogen content of the mixture. The high temperature treatments are necessary to produce the desired solid solutions of tungsten carbide in titanium and tantalum carbide. In the structure of cemented carbides with 5 to 25 atomic % TiC, 3-15 atomic % TaC, 86-52 atomic % WC and 6-13 atomic % Co equal amounts of tungsten carbide should be in solid solution in the cubic (Ti,Ta)C phase and should exist as hexagonal WC, which in itself dissolves very little titanium and tantalum carbide. Another method of producing the WC-TiC solid solution is the so-called "menstruum method" in which the individual carbides are dissolved in liquid nickel and solid solution carbide crystals precipitated during cooling.[11]

The cobalt powder used for cemented carbides is reduced from cobalt oxide. Again, its purity and its particle size distribution critically affect the quality of the cemented carbides.

B. Production of Cemented Carbides

The first step in the production process is the blending of the powders and their mixing and milling. The powder mixtures are milled in ball mills. The balls are made of cemented carbide, the mills may be carbide lined, but are often also made of stainless steel, since slight contamination with iron and nickel does not seem to be injurious. In addition to conventional ball mills, high energy vibratory ball mills and attritor mills are used, with the choice of mill depending on the grade of carbide produced. During milling the carbide particles are thoroughly coated with cobalt. Milling is generally done in an organic liquid, e.g. hexane, and except for mixtures to be hot-pressed, a lubricant, most commonly paraffin, is added to the mixture. After milling the organic liquid must be removed by drying. The method of spray drying, in which hot nitrogen impinges upon a stream of the carbide suspension, has been found particularly useful because a composite of powder and lubricant is obtained in the form of powder aggregates which flow freely and can be fed directly into automatic compacting presses.

The milled powder is commonly consolidated by cold pressing. Blanks may be pressed in rigid dies in large hydraulic presses or occasionally pressed isostatically. When large numbers of compacts of a given size are needed, e.g. for tool tips, they are pressed in automatic compacting presses to closely controlled dimensions in carbide-lined dies. The compacting pressures are relatively low on the order of 100 MPa.

There are other methods of consolidating the powder. One is by hot pressing in induction or resistance heated graphite molds; it is generally used for large components, in which the fact that a new mold is needed for each compact is economically tolerable. Another method is by extrusion for which the cemented carbides are mixed with a plasticizer. This is used for shapes which are difficult to form by conventional pressing and where the inevitable distortion during presintering for removal of lubricant and during final sintering can be tolerated.

Before blanks, compacts or extrusions can be given a final high temperature sintering treatment, they must be presintered, which also serves as the treatment for removal of lubricant (dewaxing). Presintering is generally done in hydrogen; special equipment is available to precipitate the evaporated lubricant electrostatically.

After presintering cemented carbide blanks are strong enough so that they can be readily shaped into the desired form. This may be done with abrasive tools made of silicon carbide or preferably with diamond coated tools. In the early days of cemented carbide manufacture, most products were shaped from presintered blanks before being given a final sintering treatment. Today, compacting to shape is the preferred method of consolidation except where small numbers of a given shape or large compacts are required.

In determining the dimensions of a shape, formed from a presintered blank, or of compacts pressed to a given shape, the very large shrinkage during final sintering, which is on the order of 18 to 26% linear (45-60% by volume) must be taken into account.

Final sintering may be done in hydrogen in resistance heated, semi-continuous tube furnaces, for which the green compacts are packed in alumina or in graphite powder in graphite boats. The alternative is vacuum sintering, in which the compacts are laid out on graphite plates, which are stacked with spacers in the furnace and heated by radiation from induction or resistance heated graphite susceptors. Straight tungsten carbide-cobalt grades and those containing only small amounts of other carbides can be readily sintered in hydrogen. However, some cemented carbide producers use vacuum sintering for all of their products. Sintering furnaces for cemented carbides have been discussed in chapter 7.

Two important treatments of cemented carbide products subsequent to final sintering are hot isostatic compaction and coating. In hot isostatic pressing, pressure on the order or 100-150 MPa is applied by an inert gas, generally argon, while the product is heated to a temperature near the sintering temperature. In this process any remaining small amounts of porosity in the cemented carbide are eliminated, provided the pores are not connected with the outside of the compact. The treatment is particularly important for applications where a perfect surface finish is required, as in drawing dies or in rolls for rolling mills.

The coating process by chemical vapor deposition of titanium carbide on cemented carbides grew out of processes originally developed for coating steel with titanium carbides and nitrides.[12] Cemented carbides, particularly tool bits, are coated with a layer of titanium carbide, on the order of 5 μm thickness. The most widely used coating process is chemical vapor deposition using titanium chloride and a hydrocarbon vapor with hydrogen as the carrier gas. The thin TiC coating provides the material with excellent cratering resistance and much improved wear resistance. It is of particular importance for throw-away tool bits which are not reground, since in regrinding the TiC layer would be removed. In addition to titanium carbide coatings, titanium nitride and combination titanium carbide-nitride coatings have been developed.

III. Composition, Properties and Testing of Cemented Carbides

Table 16-1 from J. A. Brookes' World Directory and Handbook of Hardmetals[4] shows the composition of typical carbide grades for machining. Also shown are the designations of the grades according to the ISO Application Code and the U. S. Industry Code. These codes are not specifications, but merely application guides. In the ISO code, e.g., the K grades are straight tungsten carbide grades, the P grades highly alloyed grades for machining ferrous metals with long chips and the M grades multiple purpose grades. Within each group designated by letter, increasing numbers indicate increasing toughness and decreasing wear-resistance. In testing cemented carbides, the most important property is their hardness, which is usually taken as a measure of their wear resistance, although the two properties do not always go parallel. In the United States hardness of cemented carbides is measured on the Rockwell A scale (diamond indenter, 60 kg load) often to the nearest 0.1 point hardness number, while Vickers hardness numbers are widely used in other countries. Vickers indentation hardness measurements have also been taken on single crystals of WC which have shown that the hardness of WC is quite anisotropic, varying from Vickers 2000 on the base plane to a range of 1300 to 1800 on the prism plane. The indentation hardness values obtained on tungsten carbide-cobalt grades of cemented carbides are not affected by the hardness anisotropy of tungsten carbide single crystals, since the orientation of the tungsten carbide crystals in the material is random. However, it depends upon cobalt content and WC grain size. Figure 16-1[13] shows how the Rockwell A hardness of grades with different cobalt contents depends on the WC grain size. A material with fine grain size may have a hardness 3 to 4 points higher on the R_A scale than one with coarse grain size. The dependence of hardness and particularly of wear resistance of cemented carbides upon the size of the carbide grains in the structure has led to several efforts to develop cemented carbides with carbide grain sizes smaller than 1 μm, generally called micro-grain cemented carbides. Special methods have been developed to produce ultra-fine grain material starting with a very fine tungsten powder by reduction of tungsten chloride (see chapter 2). However, quite fine grain carbides can also be produced

Table 16-1 Typical* carbide grades for machining.

| Designations | | Compositions | | | | |
ISO Application Code	U.S. Industry Code	WC	TiC	Ta(Nb)C	Co	NiMo
PO1.2	C8	-	80	-	-	20
PO1.3	C8	50	35	7	6	-
PO5	C7	78	16	-	6	-
P10	C7	69	15	8	8	-
P15	C6	78	12	3	7	-
P20	C6	79	8	5	8	-
P25	C6	82	6	4	8	-
P30	C5	84	5	2	9	-
P40	C5	85	5	-	10	-
P50	-	78	3	3	16	-
M10	-	85	5	4	6	-
M20	-	82	5	5	8	-
M30	-	86	4	-	10	-
M40	-	84	4	2	10	-
K01	C4	97	-	-	3	-
K05	C4	95	-	1	4	-
K10	C3	92	-	2	6	-
K20	C2	94	-	-	6	-
K30	C1	91	-	-	9	-
K40	C1	89	-	-	11	-

*Very considerable variation between manufacturers is possible.

conventionally. After milling with cobalt powder and during sintering, there is a tendency for grain growth of the carbide grains which can be inhibited by additions of grain growth inhibiting tantalum, vanadium or chromium carbide. The American Society for Testing and Materials has issued a recommended practice for evaluating the apparent grain size and distribution of cemented carbides (ASTM-B 390) according to which the metallographic structure of the carbides at 1500x is compared with standard micrographs for cemented carbides with 6, 10 and 18% cobalt.

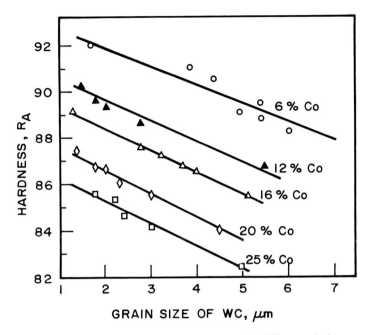

FIGURE 16-1 Hardness R_A of WC-Co alloys with different cobalt contents as a function of WC grain size.

Typical micrographs of straight tungsten carbide-cobalt cemented carbides are shown in Figures 16-2 to 16-10. Structures with increasing cobalt content, 3, 6, 10, 16 and 25%, and for the 6 and 10% cobalt grades, those with fine (micro-grain), medium and coarse carbide particles are shown. Particularly, the coarse structures clearly show the prismatic shape of the carbide grains. In general, the size of the tungsten carbide grains in cemented carbides is controlled by the size of the carbide particles in the powder mixture from which they are compacted. This is in contrast to the size of the tungsten grain in heavy alloys which are much larger than the tungsten particles in the powder mixture. Figures 16-11 to 16-13 show microstructures of cemented carbides which contain TiC and TaC in addition to WC. Going from Figures 16-11 to 16-13 the percentage of the cubic carbides increases and that of WC decreases with corresponding changes in the morphology of the carbide particles. The lower the WC content, the greater is the proportion of cubic rounded grains of (Ti,Ta)C with WC in solid solution and the smaller the proportion of the hexagonal angular grains of WC, in which very little TiC and TaC is soluble in solid solution. Figure 16-14 shows the microstructure of a cemented carbide

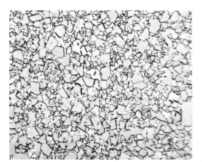

FIGURE 16-2 Microstructure of WC-Co alloy with 3% Co, x1500 (Courtesy: General Electric Co., Carboloy Systems Dept.).

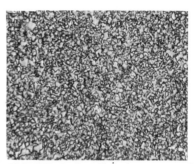

FIGURE 16-3 Microstructure of WC-Co alloy with 6% Co, fine WC grain size x1500 (Courtesy: General Electric Co., Carboloy Systems Dept.).

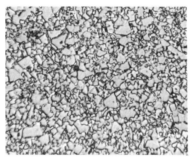

FIGURE 16-4 Microstructure of WC-Co alloy with 6% Co, medium WC grain size, x1500 (Courtesy: General Electric Co., Carboloy Systems Dept.).

FIGURE 16-5 Microstructure of WC-Co alloy with 6% Co, coarse WC grain size, x1500 (Courtesy: General Electric Co., Carboloy Systems Dept.).

containing only TiC as the carbide phase with 40% of a 62.5% Ni - 37.5% Mo binder phase.

A very important property of cemented carbides is their porosity, which should be as small as possible; in other words, the cemented carbides should be near theoretical density. Again, the American Society for Testing and Materials has issued a recommended practice (ASTM B 276) for porosity in which the metallographic structure is compared with standard micrographs at 200x. Three types of porosity, small, large and rosette type and three grades of porosity, small, medium and large, are designated in this recommended practice.

A second important mechanical property of cemented carbides is their transverse rupture strength. A test for this property for cemented carbides

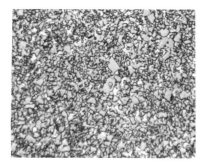

FIGURE 16-6 Microstructure of WC-Co alloy with 10% Co, fine WC grain size, x1500 (Courtesy: General Electric Co., Carboloy Systems Dept.).

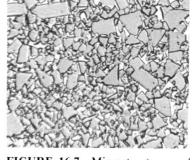

FIGURE 16-7 Microstructure of WC-Co alloy with 10% Co, medium WC grain size, x1500 (Courtesy: General Electric Co., Carboloy Systems Dept.).

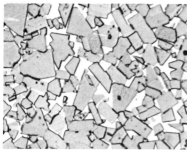

FIGURE 16-8 Microstructure of WC-Co alloy with 10% Co, coarse WC grain size, x1500 (Courtesy: General Electric Co., Carboloy Systems Dept.).

FIGURE 16-9 Microstructure of WC-Co alloy with 16% Co, x1500 (Courtesy: General Electric Co., Carboloy Dept.).

FIGURE 16-10 Microstructure of WC-Co alloy with 25% Co, x1500 (Courtesy: General Electric Co., Carboloy Systems Dept.).

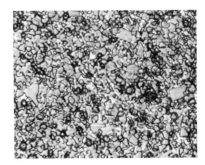

FIGURE 16-11 Microstructure of cemented carbide alloy with 76.4% WC, 4% TiC, 8.7% TaC, 10.9% Co, x1500 (Courtesy: General Electric Co., Carboloy Systems Dept.).

FIGURE 16-12 Microstructure of cemented carbide alloy with 72% WC, 8% TiC, 11.5% TaC, 8.5% Co, x1500 (Courtesy: General Electric Co., Carboloy Systems Dept.).

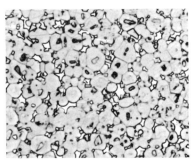

FIGURE 16-13 Microstructure of cemented carbide alloy with 64% WC, 25.5% TiC, 4.5% TaC, 6% Co, x1500 (Courtesy: General Electric Co., Carboloy Systems Dept.).

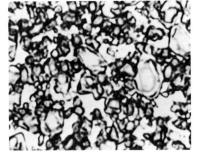

FIGURE 16-14 Microstructure of cemented carbide alloy with 60% TiC, 25% Ni, 15% Mo, sintered 1 hour at 2500°F (1370°C), Ferricyanide etch, x1000 (Courtesy: Ford Motor Co.). Co.).

has been standardized by the American Society for Testing and Materials (ASTM B 406). It depends upon both cobalt content and grain size. Gurland and Bardzil[13] have rationalized this dependence. They plotted transverse rupture strength vs. mean free path in the cobalt binder phase. This mean free path depends on both the cobalt content and the WC particle size. As shown in Figure 16-15 for cemented carbides with cobalt contents between 6 and 25% they found a maximum in the transverse rupture strength vs. mean free path plot. They explain this maximum by attributing the ascending portion of the curve where strength increases with

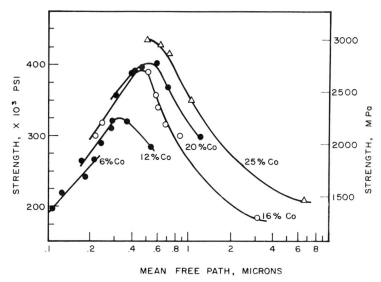

FIGURE 16-15 Transverse rupture strength of WC-Co alloys as a function of the mean fee path through the Co-phase.

increasing binder layer thickness to the effect of decreasing plastic constraint in the layer of the ductile phase. As plastic flow becomes easier, local stress concentrations are relieved and both crack initiation and crack propagation are impeded. The descending portion of the curve is explained in accordance with theories of dispersion strengthening: the strength decreases with increasing path length between the strengthening second phase particles. Improving the performance of cemented carbides as cutting tools by increasing the transverse rupture strength has been an important goal in the development of new grades in the laboratories of cemented carbide producers.

The transverse rupture strength depends, on the one hand, on the size and distribution of flaws within the material, which in the case of cemented carbides closely relates to the size and distribution of pores, and, on the other hand, on the resistance of the material to crack propagation from the flaws. Because, in the performance of cemented carbides both factors are important, transverse rupture strength is widely used as a commerical test. The types of fracture toughness tests, generally used for steels, in which a defined crack is introduced into the material have also been applied with suitable modifications to cemented carbides, but are not widely used in routine testing. Instead a simpler test for toughness[14] has been developed in which the length of the cracks is measured when a hardness indentation is made in cemented carbides.

IV. Sintering Mechanism in Cemented Carbides

The pseudobinary equilibrium diagram of WC and Co shows a eutectic between Co and WC at 1320°C and 35 wght. percent WC as shown in Figure 16-16. When a compact from a mixture of tungsten carbide and cobalt powder is heated, a liquid phase if formed by reaction between the tungsten carbide and cobalt. Sintering of cemented carbides is therefore an example of liquid phase sintering. As pointed out above, the growth of WC particles during liquid phase sintering is generally much more restricted than in other liquid phase sintering systems. This has been attributed to the fact that the kinetics of growth by solution and reprecipitation are controlled not by diffusion of tungsten and carbon through the liquid phase, but by the phase boundary reaction at the tungsten carbide-cobalt phase boundary. The question whether the WC particles in cemented carbides form a rigid skeleton or not has been hotly debated. Since the carbide particles in the structure of cemented carbides appear as rectangles and triangles rather than as circles, it is evident that the interfacial tension between single crystals of WC and liquid cobalt is anisotropic. Furthermore, during cooling from the sintering temperature to room temperature, much of the W and C, dissolved in the liquid cobalt, will precipitate on the existing WC grains. The carbide grains of irregular shape recrystallize by a solution-reprecipitation mechanism and thereby develop

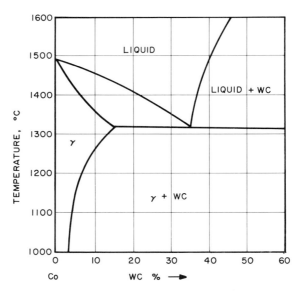

FIGURE 16-16   Pseudobinary equilibrium diagram of WC and Co.

low energy, crystallographically oriented habit faces, and eventually coalesce. This is a gradual process, while the cemented carbide structure is at the sintering temperature. Coalescence and grain growth is an undesirable phenomenon since both strength and hardness deteriorate as the structure coarsens.

Figure 16-17[15] shows a vertical section drawn at a given cobalt content through the ternary W-Co-C diagram. It becomes clear that the range in carbon content of the two-phase composition WC + beta (cobalt solid solution) is quite narrow. For an alloy with 16% cobalt, at 1300°C, this range is between 5.05 and 5.17% C, corresponding to a carbon content of the WC between 6.01 and 6.14%. At higher carbon contents free graphite is present, at lower carbon contents the so-called eta phase of the approximate composition W_3Co_3C is present. Both high and low carbon contents are undesirable. Free graphite decreases strength and hardness.

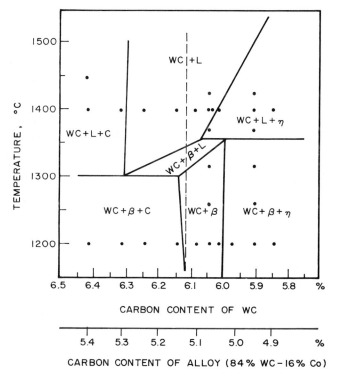

FIGURE 16-17 Vertical section through the ternary W-C-Co ternary diagram at 16% Co.

When the eta phase is present, excessive grain growth of carbide particles is often observed; the eta phase also decreases the toughness of cemented carbides markedly. It was long believed that during cooling from the sintering temperature, all of the tungsten carbide in solution in the cobalt is precipitated. Newer investigations[16,17,18] have shown that both tungsten and carbon remain in solution in the cobalt at room temperature. The amount of tungsten remaining in solid solution in the cobalt depends upon the carbon content, with higher amounts of tungsten in solution in alloys low in carbon and vice versa. For this reason, the control of carbon, even within the narrow range of the two-phase region of the equilibrium diagram, is important since the amount of tungsten in solid solution in the binder phase affects the mechanical properties of alloys.

Appendix

"Cermets" for high temperature applications

In the early 1950's, research efforts were directed towards developing materials for jet-engine turbine blades and similar applications with better creep and stress-rupture properties than the nickel and cobalt base superalloys then available.[19,20] One approach was through combinations of a ceramic phase having high strength and oxidation resistance at elevated temperatures and a metallic phase imparting thermal and mechanical shock resistance to the material. The combinations most thoroughly studied were cemented titanium carbides and chromium-aluminum oxide materials. In the case of the cemented titanium carbides, materials with 40 to 65 wght. % binder metal were investigated, in which the binders were nickel-molybdenum, nickel-molybdenum-aluminum and nickel-cobalt chromium alloys. The lower the binder content the higher were the stresses for a 100 hour stress rupture life at 1600 and 1800°F (870 and 980°C) but lower binder content materials did not have sufficient mechanical shock resistance for the applications for which they were intended. The chromium-aluminum oxide materials, generally with 30 wght. % chromium, were superior to the cemented tungsten carbides with regard to stress-rupture properties, but their mechanical and thermal shock resistance was quite insufficient for such applications as jet engine turbine blades. A study of the creep properties of chromium-alumina combinations with up to 50% chromium[21] has been concerned primarily with an understanding of the mechanism of creep in these materials.

Although the hopes for developing ceramic-metal combinations for jet engine blades were not fulfilled, other specialized high temperature applications for this group of materials, in particular for thermo-couple protection tubes, have been developed. Combinations of molybdenum metal and zirconia with up to 50 volume % ZrO_2[22] are used as protecting sheaths for thermo-couples used in measuring temperatures in liquid iron and steel baths. Molybdenum powder and zirconium oxide powder are mixed and the necessary thin-walled tubes fabricated by cold extrusion of the powders mixed with a plasticizer. For other applications pressing in rigid dies or cold hydrostatic pressing is used. The compacts which have porosities of 30 to 40% are sintered at approximately 2000°C. Practically theoretical density is obtained during sintering. The material has high resistance to corrosion by the liquid metal and by slags, good thermal conductivity and adequate thermal and mechanical shock resistance for the application.

References

1. P. Schwarzkopf and R. Kieffer, Cemented Carbides, N.Y. (1960).
2. R. Kieffer and F. Benesovsky, Hartstoffe, Vienna, (1963).
3. R. Kieffer and F. Benesovsky, Hartmetalle, Vienna, (1965).
4. K. J. H. Brookes, World Directory and Handbook of Hardmetals, 2nd edition, London, (1979).
5. E. M. Trent, Cutting tool materials, Metall. Reviews, No. 127, vol. 13, p. 129-144, (1948).
6. H. E. Exner and J. Gurland, A review of parameters influencing some mechanical properties of tungsten-carbide cobalt alloys, Powder Metallurgy, vol. 13, No. 25, p. 13-30, (1970).
7. F. Benesovsky, Carbide, in "Ullmanns Enzyklopadie der technischen Chemie," 4th ed., Vol. 9, p. 122-136, (1975).
8. K. Schröter, Hard metal alloys, U. S. Patent 1,549,615 (1925).
9. M. Humenik and N. M. Parikh, Cermets: I Fundamental concepts related to microstructure and physical properties of cermet systems, J. Am. Cer. Soc., vol. 39, No. 2, p. 60-63, (1956).
10. A. Hara, T. Yamamoto, Y. Doi, H. Sakanoue, N. Takahashi and T. Nomura, The performance of new solid TiC cutting tool, "T 12 A", Sumitomo Electric Technical Review, no. 17, Oct. 1977, p. 50-58.
11. P. M. McKenna, Tool materials (Cemented Carbides), ch. 20 in "Powder Metallurgy", ed. by J. Wulff, Cleveland, p. 454-469, (1942).
12. A. Münster and W. Ruppert, Z. Elektrochemie, vol. 57, p. 558-564, 564-571, 571-579, (1963).
13. J. Gurland and P. Bardzil, Relation of strength, composition and grain-size of sintered WC-Co alloys, Trans, AIME, vol. 203, p. 311-315 (1955).
14. S. Palmquist, Crack formation energies in Vickers hardness impressions as a measure for the toughness of cemented carbides, Arch. f. d. Eisenhüttenwesen, vol. 33, No. 9, p. 629-633 (1962).
15. J. Gurland, A study of the effect of carbon content on the structure and

properties of sintered WC-Co alloys, Trans. AIME, vol. 200, p. 285-290, (1954).

16. O. Rüdiger, D. Hirschfeld, A. Hoffmann, J. Kolaska, G. Ostermann and J. Willbrand, Compositon and properties of the binder metal in cobalt bonded tungsten carbide, Mod. Dev. in Powder Metallurgy, vol. 5, ed. by H. H. Hausner, N.Y., (1971), p. 215-224.

17. O. Rüdiger, G. Ostermann, and J. Kolaska, Some new results and problems in cemented carbide research, Techn. Mitt. Krupp Forschungs Ber., vol. 28, No. 2, p. 33-59 (1970).

18. O. Rüdiger, D. Hirschfeld, A. Hoffmann, J. Kolaska, G. Ostermann, and J. Willbrand, Compositon and properties of the binder metal in WC-Co alloys, ibid., vol. 29, p. 1-14, (1971).

19. M. G. Ault and G. C. Deutsch, Applicability of powder metallurgy to problems of high temperature materials, Trans. AIME, vol. 200, p. 1214-1226, (1954).

20. F. V. Lenel, Powder metallurgy - now, Proceedings ASTM, vol. 55, p. 655-688, (1955).

21. G. Engelhardt and F. Thümmler, Creep deformation of Al_2O_3-Cr- cermets with Cr-content up to 50 vol. %, Mod. Dev. in Powder Metallurgy, vol. 8, ed. by H. H. Hausner and W. E. Smith, Princeton, (1974), p. 605-626.

22. F. Heitzinger, Molybdenum and zirconia, a new metal-ceramic material for new applications, ibid., p. 371-390.

CHAPTER

17

Structural Parts

The term "P/M structural part" is used in this chapter for components which are produced from metal powder by compacting and sintering to closely controlled dimensions. They are made by powder metallurgy because this technique is more economical than competing techniques, such as casting, machining or cold forming from bar stock, while at the same time, the mechanical properties achieved are adequate for the applications for which they are used. Certain magnetic parts, both soft magnetic parts from iron, iron-silicon, iron-phosphorus and iron-nickel and hard magnets, such as the "Alnico" magnets, are produced from powder because closely controlled dimensions and adequate magnetic properties can be achieved at a cost saving by this technique compared with casting, machining or cold forming. The reason why these parts are produced from powder is, therefore, similar to that for "P/M structural parts" except that adequate magnetic rather than mechanical properties are required. These parts will be discussed in chapter 23 on magnetic applications.

The term "P/M structural parts" does not include parts which are made by powder metallurgy because this technique imparts properties which cannot be achieved by fusion metallurgy, such as the high wear resistance of cemented carbides or of metal-diamond combinations, the frictional properties of metallic friction materials, the combination of electrical and frictional properties of metallic brushes for motors and generators or the porosity of self lubricating bearings or filters. Many structural parts have some porosity, but, in most applications, it is present as a consequence of fabrication from metal powders and not because it is particularly desirable.

Also excluded from "P/M structural parts" are semi finished products such as sheet or strip which is powder rolled (roll compacted) from metal powders or products hot consolidated from metal powders by hot pressing, hot extrusion, hot isostatic pressing or hot forging. Some recent developments have led to components which are hot isostatically pressed to near net shape; however, they will not be considered in this chapter on "P/M structural parts", but in chapter 21. Certain components from refractory and reactive metal powders, in particular molybdenum and titanium powder, are produced by compacting and sintering to closely controlled dimensions and should probably be considered "P/M structural parts," but their discussion is included in the chapter on the powder metallurgy of refractory and reactive metals, chapter 14.

Although compacting and sintering are the basic operations by which "P/M structural parts" are produced, intermediate or subsequent fabrication steps are often involved. They include presintering, sizing, coining, repressing, resintering, infiltration, machining, heat treating including steam treating, joining, plating etc. These steps are included in this chapter. Techniques in which a preform is produced by compacting and sintering and is subsequently precision forged to closely held dimensions and near theoretical density are reserved for their own chapter, "Hot forging and cold forming of preforms," chapter 18. References to typical applications of structural parts and of hot forged preforms, a comparison of the economics of production of parts by P/M and by competing methods and of the energy requirements of these processes are the subjects of chapter 19.

The technology of P/M structural parts was developed in the U.S.A. in the late 1930's and grew out of that for self-lubricating bearings which are also produced by compacting, sintering (and sizing) to closely controlled dimensions. The first "P/M structural parts" were oil pump gears for the Oldsmobile automobile.[1] The principal differences in fabricating self-lubricating bearings and oil pump gears from metal powders were in:

1. the composition of the powder mixture
2. the tooling design
3. the type of furnace used.

1. Instead of a mixture of copper, tin and graphite powders used for self-lubricating bearings, mixtures of sponge iron and graphite powders were used for oil pump gears.

2. The tooling consisted of a die-barrel, upper and lower punches and a core rod. Unlike tooling for selflubricating bearings, the diebarrel had the negative of the involute shape of the gear which was produced by broaching and the punches were ground to the involute gear shape. This made it possible to compact the gears to a true involute, a shape which is difficult to

machine. The same type of compacting presses as for large self-lubricating bearings were used. The gears had a uniform cross section in the direction of pressing and were single-level compacts. Tooling and presses for two-level parts, such as flanged self-lubricating bearings, were already available. 3. The iron-base compacts had to be sintered in atmosphere controlled furnaces operating near 1100°C, while the bronze self-lubricating bearings are sintered near 800°C. Furnaces operating at these higher temperatures had become available for use in producing assemblies of steel parts by "copper brazing."

Compacting of P/M structural parts larger and more complex in shape than the oil pump gears requires larger and more complex presses and tooling which were discussed in chapter 5. G. Zapf[2] developed a numerical system by which the complexity of the shape of structural parts can be recorded. It includes categories for parts which are rotation symmetric, quasi symmetric and non symmetric; have one, two or more levels; are circular; or have profiles on the outside or the inside; or have profiled flanges. The basic shape limitations in pressing parts directly on automatic presses without subsequent machining were outlined in chapter 4 and apply, of course, to P/M structural parts.

The first P/M structural parts had the chemical composition of plain carbon steel. Gradually other compositions were introduced. Probably 90% of all metal powder going into structural parts is iron powder.

The following four sections of this chapter will be concerned with the mechanical properties and specifications for structural parts.

Because of the importance of parts based on iron powder the first section will be devoted to mechanical properties of these structural parts. The section will take the place of a more general discussion of the mechanical properties of ferrous powder metallurgy products and will be confined to those compositions and treatments which have found applications for structural parts. The effect of density, of additions of alloying elements and of heat treatment upon ultimate tensile strength or transverse rupture strength and on tensile elongation will be treated. A discussion of materials from prealloyed and partially prealloyed powders will be included. The role of tests for tensile yield strength, impact energy and fatigue strength of materials for structural parts from iron powder in assessing their mechanical behavior will be reviewed.

The second section will be on the requirements for specifications of P/M structural parts in general.

The specifications developed in MPIF standard 35 and the corresponding ASTM specifications for

P/M iron

> P/M steel
> P/M copper-iron, copper-steel and iron-copper
> P/M iron-nickel and
> P/M nickel-steel

will be presented in the third section.

The fourth section will be devoted to the properties and the specifications for P/M structural parts from copper and copper alloys, stainless steel and aluminum alloys.

In the fifth and last section of this chapter secondary operations will be discussed.

I. Mechanical Properties of Parts Based on Iron Powder

In chapter 8, graphs were presented in Figures 8-22 to 8-25 for the tensile strength, hardness, elongation and impact strength measured on unnotched specimens as a function of density for compacts of iron powder sintered one hour at 1100°C. The graphs indicate that the most important requirement to get better properties is high density or low porosity of the sintered compacts. For most P/M structural parts from iron powder this high sintered density is achieved by using powders with high compressibility and pressing at high compacting pressures, thereby achieving a high green density. Dimensional changes during sintering are generally kept to a minimum. For densities above about 7.0 g/cm³, repressing and resintering may be necessary. Similar values of mechanical properties are obtained at a given level of sintered density regardless of whether the density was achieved by simple compacting and sintering or by repressing and resintering after sintering. How high a density can be achieved when pressing a structural part from a given powder and at a given pressure also depends upon the shape of the part as was explained in chapter 4.

The strength of compacts from straight iron powder is limited. Higher strength P/M structural steel parts are produced from mixtures of iron powder and powders of alloying elements, of which the most important are carbon in the form of graphite powder, copper powder and nickel powder. The effect of increasing amounts of combined carbon upon the transverse rupture strength of compacts sintered to a density of 6.3 g/cm³ is shown in Figure 17-1.[3] The transverse rupture strength is approximately double the tensile strength for materials which in the transverse rupture strength test deform little before breaking. The strengthening effect of carbon additions is due to the increasing amounts of pearlite in the microstructure with

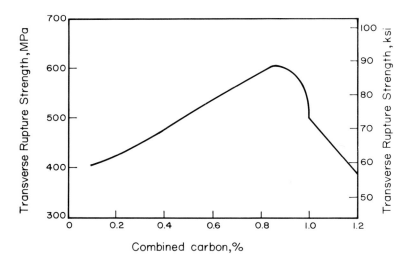

FIGURE 17-1 Transverse strength of pressed and sintered iron-carbon alloys with a density of 6.3 g/cm³ as a function of carbon content.

increasing carbon contents. Hypereutectoid P/M steels have lower strength as-sintered because proeutectoid cementite forms at austenite grain boundaries, which frequently connect adjacent pores.

As in steel parts from mixtures of iron and graphite powder, properties of alloy steel structural parts again depend strongly on the sintered density of compacts. In Figure 17-2³ the tensile strength, yield strength, elongation and unnotched Charpy impact energy of compacts of a mixture of iron powder with 4% nickel powder and sufficient graphite powder to produce combined carbon contents of 0, 0.4 and 0.8% carbon are plotted vs. sintered density. The very strong effect of combined carbon content is again evident. In compacts sintered to a density of 7.0 g/cm³ the tensile strength increases from 300 MPa with no combined carbon to 500 MPa with 0.8% combined carbon and the yield strength from 170 to 280 MPa. On the other hand, the ductility, which is relatively low because of the porosity of the compacts even without combined carbon, is further lowered when the carbon content is raised. At a density level of 7.0 g/cm³ the elongation decreases from 4% with no carbon to 2% with 0.8% combined carbon. To obtain density levels above 7.0 g/cm³ carbon-containing compacts must be presintered after compacting at a temperature where carbon does not go into solution, repressed and then given a final sintering treatment. The effect of varying the nickel content in the range up to 7% nickel, again at combined carbon contents of 0, 0.4% and 0.8% and at a density of 7.4 g/cm³, is shown in

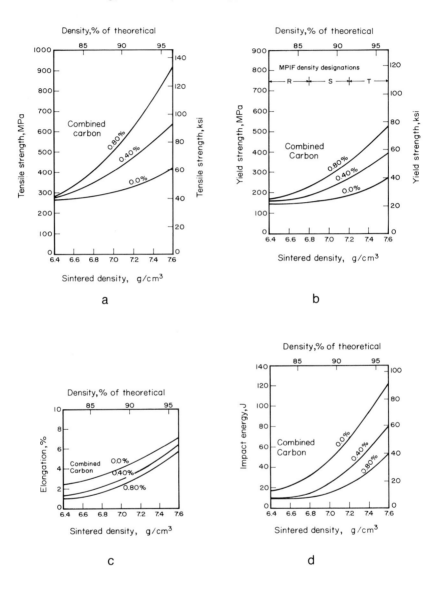

FIGURE 17-2 Mechanical properties of 4% nickel steels from a mixture of iron, nickel and graphite powders with 0, 0.4 and 0.8% combined carbon, as a function of density. (a) tensile strength, (b) yield strength, (c) elongation, (d) unnotched Charpy impact energy.

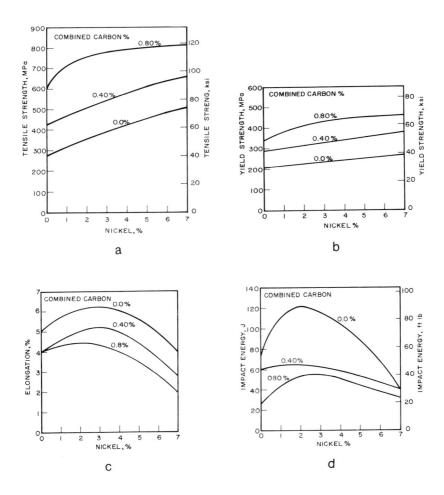

FIGURE 17-3 Mechanical properties of nickel steels from a mixture of iron, nickel and graphite powder, sintered to densities between 7.2 and 7.6 g/cm³ as a function of nickel content (a) tensile strength, (b) yield strength, (c) elongation, (d) unnotched Charpy impact energy.

Figure 17-3.[3] At this high density level, 2 to 3% nickel added to iron or iron-graphite mixtures raises the ductility and toughness of the repressed and sintered compacts, an effect familiar from wrought nickel steels. The rapid increase in tensile strength for eutectoid steels (0.8% combined carbon) from 610 to 760 MPa when 2% nickel is added to the mixture is due to the higher hardenability of nickel steels. Harder and stronger transformation

407

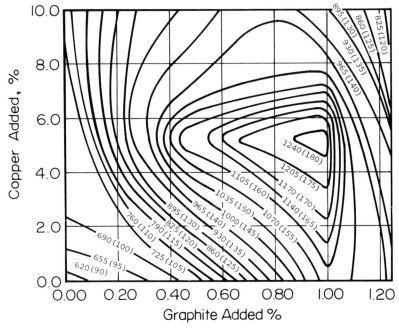

FIGURE 17-4 Transverse rupture strength of compacts from a mixture of iron, copper and graphite powder, sintered to a density of 6.8 g/cm³ in endothermic gas; lines represent compositions having the same transverse rupture strength, given in MPa with ksi equivalent values in parentheses; combined carbon in alloys about 80% of amount of graphite included in mixture.

products are produced during cooling in the cooling zone of the furnace in nickel containing steels compared with nickel-free steels.

The combined effects of graphite and copper powder additions on compacts sintered in an endothermic atmosphere to a density of 6.8 g/cm³ is shown in Figure 17-4.[3] The transverse rupture strength is plotted as a function of the amounts of graphite and copper powder added to the iron powder. The amount of combined carbon in these compacts is about 80% of the graphite added. The highest value of transverse rupture strength of 1240 MPa is obtained with additions of 0.9% graphite and 5% copper. A steel of this composition is quite brittle and is used in applications where toughness is not required. The high strength of iron-copper-carbon compacts in the as-sintered condition is probably due to the combined effect of copper causing precipitation strengthening and increasing the hardenability of the steel.

Structural parts with copper powder additions have a tendency to grow rather than shrink during sintering. This phenomenon has been discussed in chapter 11. The tendency to grow can be counteracted either by additions of graphite powder or by the use of an iron powder in which the particles are highly porous, such as the powder produced by hydrogen reduction of mill scale.[4] In compacts from a mixture of this iron powder and of copper powder the copper is absorbed upon melting into the porous structure of the particles rather than penetrating between particles, thereby wedging them apart and causing growth.

The data in Figures 17-1 to 17-4 refer to the mechanical properties of structural parts in the as-sintered condition, i.e. cooled in the cooling zone of the sintering furnace from the sintering temperature, which is always in the austenitic region, to room temperature. The rate of cooling roughly corresponds to a rate between annealing and normalizing in wrought steels. Carbon containing parts can, of course, be heattreated by quenching from the austenitic range producing martensite and tempering the martensitic structure. Low density P/M structural steel parts are quite brittle in the as-sintered conditions and are made even more brittle by heat treating. For this reason the increase in tensile strength observed in medium carbon wrought steel in the quenched and tempered, compared with the annealed or normalized conditions, is not observed in P/M structural steel parts of low density below 6.8 g/cm.[3] The martensite formed in heat treating increases the hardness of the parts but not their strength. The porosity in P/M structural steel parts also has a strong effect upon their hardenability since it lowers the thermal conductivity of the material. This is illustrated in Figure 17-5.[3] Slugs were pressed from an iron-graphite powder mixture which after sintering gave a eutectoid structure similar to a 1080 wrought steel except for the lower manganese content. The samples had densities of 6.0, 6.4, 6.8 and 7.1 g/cm.[3] Modified Jominy bars were end-quenched in a water column on one end of the bars. Apparent hardness readings on the Rockwell A scale are plotted as a function of the distance from the quenched end. All the samples were martensitic at the quenched end, but the apparent hardness decreases with decreasing density because of the effect of porosity upon the apparent hardness, as discussed in chapter 8. The horizontal bars, which represent the approximate distance over which the average amount of martensite in the microstructure exceeded 50%, indicate the rapid decrease in hardenability with decreasing density. For comparison purposes, the hardenability band and the depth to which 50% martensite is found in end-quenched wrought 1080 steel is also plotted.

For P/M structural steel parts fabricated to higher sintered densities by pressing, sintering, repressing and resintering, excellent mechanical

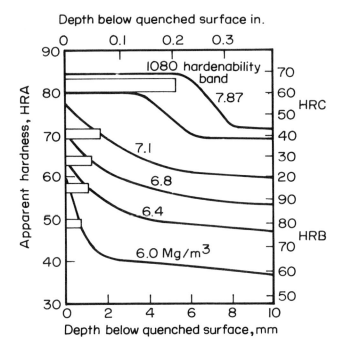

FIGURE 17-5 Effect of porosity of hardenability; compacts of a mixture of iron powder with 0.8% graphite were pressed and sintered to various densities, austenitized and end quenched; apparent hardness reflects both depth of hardening and porosity; horizontal bars represent approximate distance over which microstructure contained more than 50% martensite. Hardenability band for 1080 steel (with higher manganese content) included for comparison.

properties may be achieved by heat treatment. Figure 17-6[3] shows curves for the tensile strength and elongation of heat treated specimens produced from a mixture of iron, nickel, molybdenum and graphite powder to the composition of 4640 steel (2% Ni, 0.3% Mo, 0.4% C) as a function of their sintered density. The effects of tempering temperature upon the properties of specimens from two steels fabricated to a density of 7.4 g/cm³ and quenched from 870°C are shown in Figures 17-7[3] and 17-8.[3] The steel in Figure 17-7 contained 0.45% combined carbon, its properties are compared with those of wrought 1045 steel. The steel in Figure 17-8 was produced from a mixture of iron, nickel and graphite powder to a 0.45% combined

Density, % of theoretical

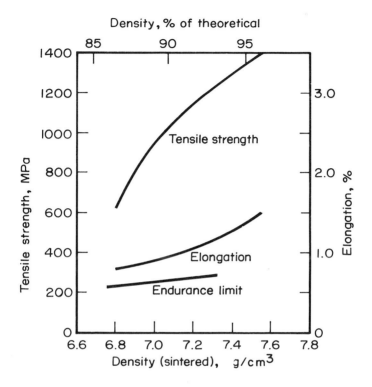

FIGURE 17-6 Tensile strength and elongation of specimens from a mixture of iron, nickel, molybdenum and graphite powder with the composition of 4640 steel (2% Ni, 0.3% Mo, 0.4% C) after pressing, sintering, repressing, resintering and heat treating as a function of sintered density. Specimens which were not repressed and resintered before heat treatment had similar properties.

carbon and 2% nickel composition. For comparison purposes, the tensile strength of wrought 4640 steel as a function of tempering temperature is shown.

As discussed in chapter 7, controlling the combined carbon content in P/M structural parts produced from mixtures of iron and graphite powder with or without the addition of copper or nickel offers certain difficulties because the sintering atmospheres must be fine tuned to avoid either decarburization or carburization during sintering. For this reason, carbon-free ferrous structural parts were widely used in Europe, in which combinations of copper powder and nickel powder additions and a sintering temperature above 1200°C provided the required high

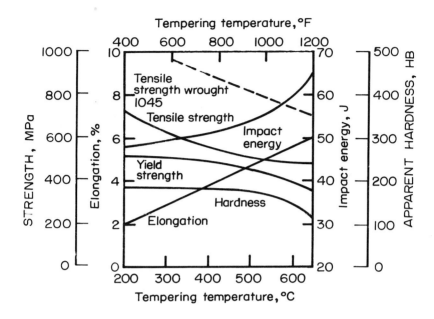

FIGURE 17-7 Effect of tempering temperature on mechanical properties of hardened and tempered sintered steels. 0.37 to 0.47% C, 7.2 to 7.6 g/cm³ density; quenched from 870°C (1600°F). Values for tensile strength of wrought 1045 steel for comparison.

mechanical properties. Figure 17-9[5] is a graph of the tensile strength and elongation of compacts of iron powder with additions of copper powder, nickel powder or both as a function of sintered density. A tensile strength of 730 MPa with an elongation of 6% can be obtained in structural parts containing 5% Ni, 4.5% Cu with a density of 7.4 g/cc. This material was widely used in P/M structural parts for German automobiles.

Another carbon-free group of compositions for P/M structural parts used in Europe are those containing additions of copper and tin powders to iron powder.[6] To get good ductility the ratio of copper to tin in these compositons should be in the range of 3:1 to 9:1. Because a liquid phase is formed at relatively low temperatures, the compositions can be sintered at lower temperatures than other iron powder mixtures. A mixture of iron powder with 4% copper and 1% tin pressed and sintered for 20 minutes at 950°C to a density of 7.0 g/cm³ gives specimens with a tensile strength of 350 MPa and 8% elongation.

Carbon, nickel, copper and in some cases molybdenum are the alloying

412

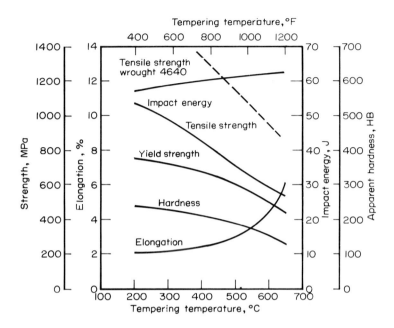

FIGURE 17-8 Effects of tempering temperature on mechanical properties of hardened and tempered sintered nickel steels; 0.39 to 0.47% C, 1.75-2.25% Ni, 7.2 to 7.6 g/cm³ density; quenched from 870°C (1600°F). Values for tensile strength of wrought 4640 steel for comparison.

elements which have traditionally been used in P/M structural parts from iron powders. In recent years phosphorus, on the one hand, and manganese, chromium and vanadium, on the other, have been introduced as alloying elements in iron base structural parts, primarily in Europe. Phosphorus is added to iron powder in the form of fine ferrophosphorus powder.[7] Powders with 20-25% P and those with 15% P have been used; the latter have the advantage of being less abrasive. When compacts from a mixture of iron and ferrophosphorus powder are heated to 1050°C, a eutectic liquid with 10% phosphorus is formed which is distributed by capillary action and thereby assists the diffusion of phosphorus into the iron particles. Much of the diffusion takes place in the α phase, since only alloys with less than 0.6% P contain γ phase. The diffusion in the α phase is much more rapid and leads to a high rate of sintering and spheroidization of the pores. The most widely used iron-phosphorus alloys contain 0.3 to 0.6% P. The tensile strength, elongation and dimensional changes of

413

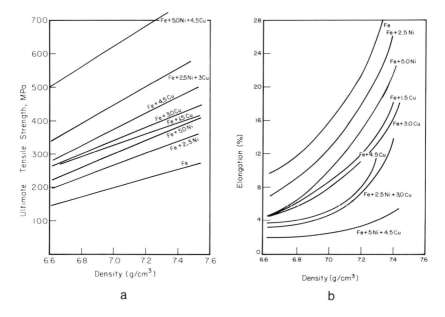

FIGURE 17-9   Tensile strength and elongation of binary Fe-Cu and Fe-Ni and of ternary Fe-Cu-Ni sintered alloys as a function of density.

compacts of reduced iron powder with increasing amounts of phosphorus, compacted at 589 MPa (42.7 tsi) to a density of 6.9 g/cm³ and sintered in endogas and in dissociated ammonia for 24 minutes at 1120°C are shown in Figure 17-10. The larger amount of shrinkage in compacts with 0.6% or more P may be counterbalanced by adding either copper powder or small amounts of graphite. A single-pressed and sintered compact of atomized iron powder with 0.35% carbon and 0.80% P developed a tensile strength of 600 MPa (87,000 psi) and an elongation of 7% at a sintered density of 7.05 g/cm³. It grew only 0.2% during sintering one hour at 1120°C. In addition to their good tensile strength and ductility P/M structural parts from iron-ferrophosphorus mixtures have the advantage of lower cost compared with those from iron-copper or iron-nickel compositons.

The use of the alloying elements manganese, chromium and vanadium in iron-base P/M structural parts has been avoided until recently because of the high affinity of these elements to oxygen, which makes it difficult to sinter, without oxidation, parts containing these elements in the usual sintering atmospheres. Introducing these alloying elements in the form of master alloys containing carbon has been suggested in Germany.[5] Very fine

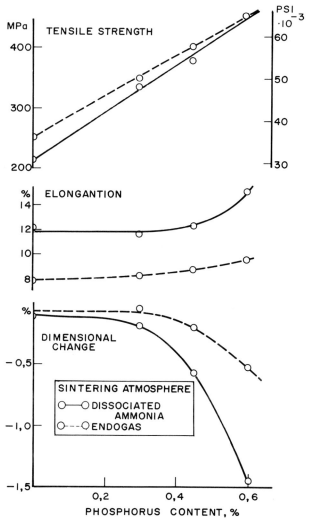

FIGURE 17-10 Mechanical properties and dimensional changes during sintering of iron-phosphorus alloys from a mixture of sponge iron and ferrophosphorus powder compacted at 589 MPa and sintered at 1120°C in dissociated ammonia and endogas respectively. Approximate density 6.9 g/cm³.

powder with particle sizes in the range of 5-10 μm of two of these master alloys has become available. They are "MCM" containing 20% Mn, 20% Cr, 20% Mo and 7% C and "MVM" containing 20% Mn, 20% V, 20% Mo and 5% C. Because in these master alloys the alloying elements Mn, Cr, V and Mo are in the form of carbides, they do not oxidize during heating of compacts containing them to the sintering temperature. Mixtures of atomized iron powder with up to 7% of the master alloy MCM or with 1 to 3% of MCM and an additional 0.6% graphite have been compacted at 600 MPa and sintered in dissociated ammonia in a walking beam furnace at 1280°C using boats with an iron-aluminum getter powder. Tensile strength values up to 750 MPa with an elongation of 2.5% were obtained by adding 7% MCM to iron powder. Alloys of iron powder with 2% MCM and 0.6% graphite gave tensile strength of 500-600 MPa and 4% elongation, and similar values were obtained with additions of 2% MVM and 0.6% graphite. These alloys can also be readily heat-treated and have superior hardenability compared to alloys with 2 to 4% nickel and 0.6% carbon.

The iron-base structural parts discussed so far were produced either from mixtures of elemental powders (Fe, Ni, Cu, Graphite, Sn, Mo) or from mixtures of pure iron powder with master alloy powders (ferrophosphorus, Mn-Cr-Mo-C or Mn-V-Mo-C masteralloys). The method of producing iron powder by water atomizing liquid steel made it possible to add alloying elements to the liquid steel and thereby produce powders in which the alloying elements are homogeneously distributed. A number of prealloyed steel powder compositions produced by water atomizing are commercially available, the best known of which has a composition with 2% Ni and 0.5% Mo roughly corresponding to the 4600 steel composition. These prealloyed steel powders are produced with very low carbon contents. The carbon is added to the powder in the form of graphite before compacting, since incorporating the carbon in the composition of the prealloyed powder would drastically increase its hardness and therefore decrease its compressibility. Most of the prealloyed steel powders use primarily nickel and molybdenum as alloying elements. The other common alloying elements in low alloy steel, i.e. Mn, Cr and V, make it more difficult to anneal the alloy powders after atomizing or to sinter compacts from the alloy powders in the usual sintering atmospheres without the danger of producing oxides of Mn, Cr or V.

In comparing the properties of compacts from mixed elemental powders and from prealloyed powders it must be kept in mind that prealloyed powders have a lower compressibility than elemental powder mixtures because the alloying elements in the lattice of iron increase its hardness by solid solution strengthening even in the complete absence of carbon in the alloy. This means that higher pressures are necessary to press, to a given density, compacts from prealloyed steel powders than those from mixtures of pure atomized iron powder and of the alloying elements. During sintering of compacts from a mixture of elemental powders partial homogenization takes place. The factors determining the rate of homogenization have been discussed in detail in chapter 10. Although the carbon will be distributed homogeneously in the austenite, heterogeneities with respect to the distribution of the other alloying elements will be found. Because of this heterogeneous distribution, the metallographic structure of the sintered compacts will also be heterogeneous, which in turn will have a strong effect upon their mechanical properties. These effects have been studied in detail.[8,9,10]

When wrought medium carbon alloy steels are heat-treated, it is well known that completely homogeneous alloys have a better hardenability than those which are heterogeneous. Better hardenability has also been found in the heat treatment of parts from prealloyed powder compared to those from mixed elemental powders. Nevertheless, under many sintering conditions and with many subsequent heat treatments, the mechanical properties of the partially inhomogeneous compacts from mixtures of elemental powders have been found to be comparable and sometimes even superior to those of compacts from prealloyed powders even when compared on the basis of equal sintered densities. For this reason and because of the higher cost of prealloyed powder compared with mixtures of elemental powders, the use of prealloyed steel powders for the conventional fabrication of structural parts has been quite limited. The prealloyed powders have been used primarily for fabrication of structural parts by hot-forging preforms.

A middle position between compacts from prealloyed powders and those from mixtures of elemental powders both in cost and in properties is taken by compacts from partially prealloyed powders. The powders are produced by bonding the alloying elements to the iron powder by heating the mixture in a reducing atmosphere. Finer powders of the alloying elements, because they promote homogenization of the alloy during sintering, are used than those in the usual mixtures. The alloying ingredients may in certain cases be added as oxides, which are reduced during the prealloying heat treatment. Partially prealloyed powders are available with both reduced and atomized iron powder as a base, in which 1.75% Ni, 1.50% Cu and 0.5% Mo or 4.0% Ni, 1.5% Cu and 0.5% Mo are bonded to the iron powder particles. As in the case of prealloyed powder, carbon is introduced by mixing graphite with the partially prealloyed powder. The advantages of partially prealloyed powders are that the powders have as high a compressibility as the mixtures of elemental powders and that the segregation due to faulty mixing which if often encountered in mixtures of elemental powders is avoided. Compacts from prealloyed powder show more uniform properties than those from powder mixtures.[11] The more intimate mixture of the alloying ingredients also makes it possible to obtain higher strength and hardness in as-sintered compacts if the rate of cooling in the cooling zone is slightly increased.[12]

When the mechanical properties of materials used for P/M structural parts are determined, it has been customary to measure tensile strength or transverse rupture strength and elongation. As Fischmeister et al.[13] pointed out: "Ultimate tensile strength is of interest when a component is meant to undergo more than negligible deformation in service. Sintered steel - at least the more sophisticated grades - are normally used for high-precision components, and even a very slight plastic deformation will render them useless. Elongation is important in connection with plastic-forming operations, but sintered steel components are made to finished shape..... The modern designer of a precision component must look for data on yield strength or proof stress, toughness, surface hardness, and fatigue strength."

Tensile yield strength at 0.2% offset is included in the tables for mechanical properties in the specifications for materials for structural parts issued by the Metal Powder Industries Federation (MPIF Standard 35). The data indicate typical values that may be expected from special subsize test specimens, as will be discussed in the next section. Similar tables for mechanical properties are appended to the ASTM specifications for standard parts, but many of these tables do not include values for tensile yield strength. There are good reasons for omitting them particularly for materials with high porosity. As has been pointed out repeatedly, plastic yielding is observed in tensile testing porous materials, particularly pure iron, at stresses well below the macroscopically determined yield strength.[14,15] Because of local high stresses in the vicinity of pores which act as stress raisers, crack formation may be observed at stresses barely above one half of the value of the 0.2% yield point. Crack formation also contributes to the elongation observed in tensile testing of porous materials. Nevertheless, the stress-strain curves for specimens, which are loaded repeatedly, with load release before fracture has occurred, are identical with the curve for a single load application. This is of considerable consequence for the application of P/M structural parts.

The customary tests for toughness of wrought materials, particularly for heat treated steels, have been the Charpy and the Izod impact tests on specimens with a notch. Charpy and Izod notch tests on porous materials result in extremely low and widely scattered values for impact energy. For this reason tests on unnotched Charpy or Izod specimens have been set up and the results of these tests are included in the tables of mechanical properties in MPIF specifications (standard 35) of materials for structural parts. This data may give some indication of the resistance to suddenly applied loads, but what is really needed is an investigation in porous materials of fracture toughness of the type widely studied in wrought materials.

As for yield strength and impact energy, data for fatigue strength are also included in the tables for mechanical properties in MPIF standard 35. Many of these data represent results of rotating beam tests in which the endurance limit at 10^7 cycles was determined. For most porous iron-base materials a constant ratio of approximately 0.38 between the endurance limit and the ultimate tensile strength was found. Constant ratios, which may be higher or lower than 0.38, have also been found for certain classes of wrought steels. However, iron-base sintered materials have peculiarities in their fatigue behavior, as was observed by C. Razim.[16] He points out that the constant ratio of fatigue strength to tensile strength applies to tests on smooth specimens. The fatigue strength of notched specimens from wrought steel increases much less with increasing tensile strength than for smooth specimens, the less so with increasing sharpness of the notch. For specimens having cracks corresponding to very sharp notches, the same fatigue strength is found for high strength as for lower strength materials with a limiting value near 90 MPa, a stress level where fatigue cracks do not grow regardless of the sharpness of the crack. The pores in porous materials have a tendency to act as cracks and may contribute to crack growth in fatigue tests. Razim determined the tensile strength and fatigue strength of materials supplied by a number of manufacturers who produce structural parts to the German specifications

Sint B 11   (0.86-1.28% C, 0.11-0.19% Ni, 1.56-4.98% Cu, density 5.27-6.45% g/cm³)

Sint B 21   (0.8-1.19% C, 0.30-0.40% Ni, 5.32-10.90% Cu, density 6.25-6.36 g/cm³)

Sint C 11   (0.85-1.36% C, 0.13-0.21% Ni, 1.60-4.36% Cu, density 6.65-6.96 g/cm³)

Sint C 21   (0.87-1.19% C, 0.26-0.37% Ni, 5.38-10.62% Cu, density 6.60-6.96 g/cm³)

Sint D 11   (0.86-1.36% C, 0.07-0.18% Ni, 1.45-4,89% Cu, density 7.0-7.30 g/cm³)

He observed that the range of tensile strength of specimens from various suppliers increases with increasing alloy content and particularly with increasing density. The average fatigue strength also increases with increasing alloying content and density, but so does the range of values for fatigue strength. He comes to the conclusion that for structural parts from iron powder:

The lower limit of fatigue strength is about 160 MPa regardless of the type of material (range of alloy content and of density)

The upper limit of fatigue strength increases with increasing density and therefore increasing tensile strength

The range over which the values of fatigue strength vary increases with increasing quality of the material (alloy content and density).

II. Requirements for Specifications of P/M Structural Parts

Efforts to write specifications for sintered metal powder structural parts by Committee B-9 of the American Society for Testing and Materials go back more than 30 years. In developing the specifications it was soon decided that they should contain requirements for the chemical composition and the density of the parts, although it was found difficult to agree upon the range of variation in composition and in density permissible for any one particular class of material. The specifications should, however, also contain requirements for the mechanical properties of the material. It is quite often difficult or impossible to machine standard specimens, be it for tensile, transverse rupture or radial crushing strength, from P/M structural parts of complex shapes. An alternative would be to require the parts manufacturer to compact standard specimens from the same powder mixture to the same green density as the structural parts in question, sinter these specimens under the same conditions as the parts and submit them to the customer for verification of mechanical properties. Apart from the fact that such a practice would lead to abuses, it would also not be a fair test of the mechanical properties of the parts, which may have very different section thicknesses than the standard specimens. It was finally decided to insert clauses into the specifications[17] regarding the mechanical properties reading:

"1. The manufacturer and the purchaser shall agree on qualification tests for the determination of mechanical properties.

2. These tests shall be performed on production parts.

3. These tests shall be determined after consideration of the function of the part.

4. The limits shall be agreed upon by the manufacturer and the purchaser.

5. All shipments of parts subsequent to the establishment of testing conditions shall conform to the limits agreed upon."

Examples of such qualification tests would be compression or bending tests on parts usually held in fixtures designed for the purpose and observing whether fracture occurs before the part has exhibited strength or ductility adequate for the purpose intended. Impact properties may be determined by dropping a weight upon a part to be tested. The weight, height of drop, point of impact and the method of supporting the part should be prescribed.

One drawback of qualification tests is the expense of developing them, which is warranted only when large quantities of parts with critical requirements for mechanical behavior are involved. Another drawback is that they do not give the designer any indication of the mechanical properties which a certain class of sintered metal powder structural parts should develop. To overcome this second drawback it was decided to add to the specifications appendices in which data for mechanical properties of specimens having densities and chemical compositions in the range of the specification are listed. These properties are tensile strength and elongation determined on flat specimens $\frac{1}{4}''$ x $\frac{1}{4}''$ in cross section with a $1''$ gauge length and compressive yield strength determined on round specimens $\frac{1}{2}''$ in diameter and $\frac{3}{4}''$ long. The specimens are molded to size and not cut from commercial parts or machined from sample blanks. Hardness values are apparent hardness usually measured on one of the Rockwell scales. The properties listed in the ASTM and in the more complete MPIF specifications are "typical properties." The wording of the appendix makes it very clear that the data do not constitute a part of the specification, but merely indicate to the purchaser the mechanical properties that may be expected. The question whether the tables of properties should show typical or minimum properties is still under debate.

III. Specifications for Structural Parts from Iron Powder

Tables 17-1, 17-2 and 17-3[3] show the contents of MPIF specifications for structural parts from iron powder. In table 17-1 the compositions included in the specifications are shown, table 17-2 explains the designation of density ranges included in the specifications and table 17-3 gives the typical mechanical properties for each of the MPIF specifications. The MPIF specifications are considerably more complete than the corresponding ASTM specifications both as to compositions and to properties listed. ASTM and SAE specifications are included by reference. The MPIF specifications include the materials most widely used for structural parts produced in the United States. Besides the mechanical properties for parts in the as-sintered (AS) conditions, those for sintered and sized (SS), actually sintered and repressed parts and those for parts heat treated (HT)

Table 17-1   Compositions of ferrous P/M structural materials

Description	MPIF	Designation(a) ASTM	SAE	C	Ni	Cu	Fe
					MPIF composition limits and ranges, %(b)		
P/M iron	F-0000	B310, Cl A	853, Cl 1	0.3 max	· · ·	· · ·	97.7-100
P/M steel	F-0005	B310, Cl B	853, Cl 2	0.3-0.6	· · ·	· · ·	97.4-99.7
P/M steel	F-0008	B310, Cl C	853, Cl 3	0.6-1.0	· · ·	· · ·	97.0-99.1
P/M copper iron	FC-0200	· · ·	· · ·	0.3 max	· · ·	1.5-3.9	93.8-98.5
P/M copper steel	FC-0205	· · ·	· · ·	0.3-0.6	· · ·	1.5-3.9	93.5-98.2
P/M copper steel	FC-0208	B426, Gr 1	864, Gr 1, Cl 3	0.6-1.0	· · ·	1.5-3.9	93.1-97.9
P/M copper steel	FC-0505	· · ·	· · ·	0.3-0.6	· · ·	4.0-6.0	91.4-95.7
P/M copper steel	FC-0508	B426, Gr 2	864, Gr 2, Cl 3	0.6-1.0	· · ·	4.0-6.0	91.0-95.4
P/M copper steel	FC-0808	B426, Gr 3	864, Gr 3, Cl 3	0.6-1.0	· · ·	6.0-11.0	86.0-93.4
P/M copper steel	· · ·	B426, Gr 4	864, Gr 4, Cl 3	0.6-0.9	· · ·	18.0-22.0	75.1 min
P/M iron-copper	FC-1000	B222; B439, Gr 3	862	0.3 max	· · ·	9.5-10.5	87.2-90.5
P/M iron-nickel	FN-0200	B484, Gr 1, Cl A	· · ·	0.3 max	1.0-3.0	2.5 max	92.2-99.0
P/M nickel steel	FN-0205	B484, Gr 1, Cl B	· · ·	0.3-0.6	1.0-3.0	2.5 max	91.9-98.7
P/M nickel steel	FN-0208	B484, Gr 1, Cl C	· · ·	0.6-0.9	1.0-3.0	2.5 max	91.6-98.4
P/M iron-nickel	FN-0400	B484, Gr 2, Cl A	· · ·	0.3 max	3.0-5.5	2.0 max	90.2-97.0
P/M nickel steel	FN-0405	B484, Gr 2, Cl B	· · ·	0.3-0.6	3.0-5.5	2.0 max	89.9-96.7
P/M nickel steel	FN-0408	B484, Gr 2, Cl C	· · ·	0.6-0.9	3.0-5.5	2.0 max	89.6-96.4
P/M iron-nickel	FN-0700	B484, Gr 3, Cl A	· · ·	0.3 max	6.0-8.0	2.0 max	87.7-94.0
P/M nickel steel	FN-0705	B484, Gr 3, Cl B	· · ·	0.3-0.6	6.0-8.0	2.0 max	87.4-93.7
P/M nickel steel	FN-0708	B484, Gr 3, Cl C	· · ·	0.6-0.9	6.0-8.0	2.0 max	87.1-93.4

(a) Designations listed are nearest comparable designations; ranges and limits may vary slightly between comparable designations. (b) MPIF standards require that the total amount of all other elements be less than 2.0%.

Table 17-2   Density designations and ranges of ferrous P/M materials

MPIF suffix	ASTM type (a)	SAE type	Density MG/m³	Theoretical density, % (b)
	Designation			
N	I	1 (c)	less than 6.0	less than 76
P	II	2	6.0 to 6.4	76 to 81
R	III	3	6.4 to 6.8	81 to 86
S	IV	4	6.8 to 7.2	86 to 91
T	V (d)	5 (d)	7.2 to 7.6	91 to 97
U			7.6 to 8.0	over 97

(a) ASTM B426 only; different density ranges used in ASTM B310 and B484. (b) Density of pure iron is 7.87 Mg/m³. (c) Density range of 5.6 to 6.0 Mg/m³ is specified. (d) Minimum density of 7.2 Mg/m³ is specified.

by quenching in oil from 870° C and tempering one hour at 200° C are given. Compositions and specifications for parts from iron powder infiltrated with copper will be discussed in section 5.

IV. Properties and Specifications of P/M Structural Parts from Copper and Copper Alloys, Stainless Steel and Aluminum Alloys

A. Development of Copper and Copper Alloys

P/M structural parts from copper and copper alloys are next in importance to those from iron and steel. They include those from pure copper which have been developed fairly recently for applications where good electrical and thermal conductivity is important, and those from copper-tin bronzes, brass and nickel silver, which have been available for several decades. The following subsections will be concerned primarily with compositions which are widely used commercially for structural parts and for which specifications have been written or are in the process of being written. However, considerable development work has also been done on the powder metallurgy of those copper alloys which have not yet found applications in P/M structural parts, but which are commercially used in the cast or wrought condition because of their strength or their damping properties. This development work should be briefly mentioned.

Table 17-3 Typical mechancial properties of ferrous P/M materials (Ref. 3)

Designation	Suf-fix (a)	Condi-tion (b)	Tensile strength		Yield strength		Elongation in 25 mm or 1 in., %	Fatigue strength		Impact energy (c)		Apparent hardness	Elastic modulus	
			MPa	ksi	MPa	ksi		MPa	ksi	J	ft·lb		GPa	10⁶ psi
F-0000-	N	AS	110	16	75	11	2.0	40	6(d)	4.1	3.0	10 HRH	70	10.5
	P	AS	130	19	95	14	2.5	50	7(d)	6.1	4.5	70 HRH	90	13
	R	AS	165	24	110	16	5	60	9(d)	13	9.5	80 HRH	110	16
	S	AS	205	30	150	22	9	80	11(d)	20	15	15 HRB	130	19
	T	AS	275	40	180	26	15	105	15(d)	34	25	30 HRB	160	23
F-0005-	N	AS	125	18	105	15	1.0	45	7(d)	3.4	2.5	5 HRB	70	10.5
	P	AS	170	25	140	20	1.5	65	10(d)	4.7	3.5	20 HRB	90	13
	R	AS	220	32	160	23	2.5	85	12(d)	6.8	5.0	45 HRB	110	16
		HT	415	60	395	57	0.5	155	23(d)	...	...	100 HRB	110	16
	S	AS	295	43	195	28	3.5	110	16(d)	12	9.0	60 HRB	130	19
		HT	550	80	515	75	0.5	210	30(d)	...	...	25 HRC	130	19
F-0008-	N	AS	200	29	170	25	0.5	75	11(d)	2.7	2.0	35 HRB	70	10.5
		HT	290	42	...	...	<0.5	110	16(d)	...	...	90 HRB	70	10.5
	P	AS	240	35	205	30	1.0	90	13(d)	4.1	3.0	50 HRB	90	13
		HT	400	58	...	...	<0.5	150	22(d)	...	...	100 HRB	90	13
	R	AS	290	42	250	36	1.5	110	14(d)	4.7	3.5	65 HRB	110	16
		HT	510	74	...	...	<0.5	195	28(d)	...	...	25 HRC	110	16
	S	AS	395	57	275	40	2.5	150	22(d)	9.5	7.0	75 HRB	130	19
		HT	650	94	625	91	<0.5	245	36(d)	...	...	30 HRC	130	19
FC-0200-	P	AS	160	23	115	17	2.5	60	9(d)	7.5	5.5	80 HRB	90	13
	R	AS	205	30	145	21	4	80	11(d)	9.5	7.0	15 HRB	110	16
	S	AS	255	37	160	23	7	95	14(d)	23	17	30 HRB	130	19
FC-0205	P	AS	275	40	235	34	1.0	105	15(d)	4.7	3.5	35 HRB	90	13
	R	AS	345	50	260	38	1.5	130	19(d)	7.5	5.5	60 HRB	110	16
		HT	585	85	560	81	<0.5	220	31(d)	...	...	30 HRC	110	16
	S	AS	425	62	310	45	3.0	160	24(d)	13	9.5	75 HRB	130	19
		HT	690	100	655	95	<0.5	260	38(d)	...	...	35 HRC	130	19
FC-0208-	N	AS	225	33	205	30	<0.5	85	13(d)	3.4	2.5	45 HRB	70	10.5
		HT	295	43	...	...	<0.5	110	16(d)	...	...	95 HRB	70	10.5
	P	AS	310	45	280	41	<0.5	115	17(d)	4.1	3.0	50 HRB	90	13
		HT	380	55	...	...	<0.5	145	21(d)	...	...	25 HRC	90	13
	R	AS	415	60	330	48	1.0	155	23(d)	6.8	5.0	70 HRB	110	16
		HT	550	80	...	...	<0.5	210	30(d)	...	...	35 HRC	110	16
	S	AS	550	80	395	57	1.5	210	30(d)	11	8.0	80 HRB	130	19
		HT	690	100	655	95	<0.5	260	38(d)	...	...	40 HRC	130	19
FC-0505-	N	AS	240	35	205	30	0.5	90	13(d)	4.1	3.0	50 HRB	70	10.5
	P	AS	345	50	290	42	1.0	130	19(d)	6.1	4.5	60 HRB	90	13
	R	AS	455	66	380	55	1.5	170	25(d)	6.8	5.0	75 HRB	116	16

Table 17-3   (contd.) Typical mechanical properties of ferrous P/M material (Ref. 3).

Designation	Suffix (a)	Condition (b)	Tensile strength MPa	ksi	Yield strength MPa	ksi	Elongation in 25 mm or 1 in., %	Fatigue strength MPa	ksi	Impact energy (c) J	ft-lb	Apparent hardness	Elastic modulus GPa	10^6 psi
FC-0508-	N	AS	330	48	295	43	<0.5	125	18(d)	4.1	3.0	60 HRB	70	10.5
	P	AS	425	62	395	57	1.0	160	24(d)	4.7	3.5	65 HRB	90	13
		HT	480	70	480	70	<0.5	185	27(d)	...	...	30 HRC	90	13
	R	AS	515	75	480	70	1.0	195	29(d)	6.1	4.5	85 HRB	116	16
FC-0808-	N	AS	250	36	...	...	<0.5	...	...	...	...	55 HRB	...	...
FC-1000-	N	AS	205	30	...	...	0.5	...	...	...	...	70 HRF	...	...
FN-0200-	R	AS	195	28	125	18	4	75	11	19	14	38 HRB	115	17
	S	AS	260	38	170	25	7	105	15	43	32	42 HRB	145	21
	T	AS	310	45	205	30	11	125	18	68	50	51 HRB	160	23
FN-0205-	R	AS	255	37	160	23	3.0	105	15	14	10	50 HRB	115	17
		HT	565	82	450	65	0.5	225	33	8.1	6	32 HRC	115	17
	S	SS	345	50	215	31	3.5	140	20	24	18	70 HRB	145	21
		HT	760	110	605	88	1.0	305	44	22	16	42 HRC	145	21
	T	SS	420	61	255	37	4.5	165	24	43	32	85 HRB	160	23
		HT	925	134	725	105	2.0	370	54	38	28	46 HRC	160	23
FN-0208-	R	AS	330	48	205	30	2.0	130	19	11	8	62 HRB	115	17
		HT	690	100	650	94	0.5	275	40	8.1	6	34 HRC	115	17
	S	AS	450	65	280	41	3.0	180	26	19	14	79 HRB	145	21
		HT	930	135	880	128	0.5	370	54	16	12	45 HRC	145	21
	T	AS	545	79	345	50	3.5	220	32	30	22	87 HRB	160	23
		HT	1105	160	1070	155	0.5	415	60	24	18	47 HRC	160	23
FN-0400-	R	AS	250	36	150	22	5	95	14	22	16	40 HRB	115	17
	S	AS	340	49	205	30	6	140	20	47	35	60 HRB	145	21
	T	AS	400	58	250	36	6.5	160	23	68	50	67 HRB	160	23
FN-0405-	R	AS	310	45	180	26	3.0	125	18	14	10	63 HRB	115	17
		HT	770	112	650	94	0.5	310	45	8.1	6	27 HRC	115	17
	S	AS	425	62	240	35	4.5	165	24	20	15	72 HRB	145	21
		HT	1060	154	880	128	1.0	415	60	14	10	39 HRC	145	21
	T	AS	510	74	295	43	6.0	205	30	41	30	80 HRB	160	23
		HT	1240	180	1060	154	1.5	450	65	19	14	44 HRC	160	23
FN-0408-	R	AS	395	57	290	42	1.5	160	23	8.1	6	72 HRB	115	17
	S	AS	530	77	390	57	3.0	215	31	14	10	88 HRB	145	21
	T	AS	640	93	470	68	4.5	255	37	22	16	95 HRB	160	23
FN-0700-	R	AS	360	52	205	30	2.5	145	21	16	12	60 HRB	115	17
	S	AS	490	71	275	40	4	195	28	28	21	72 HRB	145	21
	T	AS	585	85	330	48	6	240	34	35	26	83 HRB	160	23
FN-0705-	R	AS	370	54	240	35	2.0	150	22	12	9	69 HRB	115	17
		HT	705	102	550	80	0.5	280	41	11	8	24 HRC	115	17
	S	AS	525	76	330	48	3.5	205	30	23	17	83 HRB	145	21
		HT	965	140	760	110	1.0	385	56	20	15	38 HRC	145	21
	T	AS	620	90	390	57	5.0	250	36	33	24	90 HRB	160	23
		HT	1160	168	895	130	1.5	500	65	27	20	40 HRC	160	23
FN-0708-(e)	R	AS	395	57	280	41	1.5	160	23	8	6	75 HRB	115	17
	S	AS	550	80	380	55	2.5	220	32	16	12	88 HRB	145	21
	T	AS	655	95	455	66	3.0	260	38	22	16	96 HRB	160	23
FX-1005-(e)	T	AS	570	83	440	64	4.0	...	...	19	14	75 HRB	135	20
		HT	830	120	740	107	1.0	...	...	9.5	7.0	35 HRC	135	20
FX-1008-(e)	T	AS	620	90	515	75	2.5	...	...	16	12	80 HRB	135	20
		HT	895	130	725	105	60.5	...	...	9.5	7.0	40 HRC	135	20
FX-2000-(e)	T	AS	450	65	...	...	1.0	...	...	20	15	60 HRB	...	...
FX-2005-(e)	T	AS	515	75	345	50	1.5	...	...	12.9	9.5	75 HRB	125	18
		HT	790	115	655	95	<0.5	...	...	8.1	6.0	30 HRC	125	18
FX-2008-(e)	T	AS	585	85	515	75	1.0	...	...	14	10	80 HRB	125	18
		HT	860	125	740	107	<0.5	...	...	6.8	5.0	42 HRC	125	18

(a) Indicates density range, (b) AS, as sintered; SS, sintered and sized; HT, heat treated, typically austenitized at 870°C (1600°F), oil quenched and tempered 1 hr at 200°C (400°F). (c) Unnotched Charpy test. (d) Estimated at 38% of tensile strength.

Matthews[18,19] described cupro-nickel alloys with 10 and 25% nickel and aluminum bronzes with 5 to 11% aluminum, balance copper. By including repressing and resintering steps in the P/M processing, high densities and strength properties approaching those of wrought material were obtained.

Studies on precipitation hardened copper alloys from powder go back to the work of Hensel et al.[20] in 1946. These authors produced nickel bronzes (Cu-Ni-Sn alloys), copper-chromium and copper-beryllium alloys from mixtures of the elemental powders, except in the case of beryllium in which nickel-beryllium and cobalt-beryllium master alloys were used. The powder mixtures were pressed, sintered, repressed to 95% minimum of theoretical density, quenched and aged. For the copper-chromium and the copper-beryllium alloys in the aged condition combinations of high strength and relatively high electrical conductivity were obtained. The possibility of using age hardened copper-nickel-tin alloys without and with quartenary alloying additions and age hardened copper-beryllium alloys for high strength copper base structural parts was studied in an extensive investigation by Fetz et al.[21] Elemental, partially and fully prealloyed powders were used. The results appear to indicate that high strength may be obtained, but that the processing steps necessary to obtain it make the commercial application of these alloys from powder uneconomical.

Alloys of manganese and copper have been proposed because of their noise damping properties. Powder metallurgy work on such alloys include an investigation on prealloyed powders with 40% manganese and 2% aluminum without and with small tin additions[22] and another investigation[23] on mixtures of manganese and copper powders with 60 to 75% manganese. In both investigations reasonably strong alloys with high damping capacity were obtained.

B. P/M Structural Parts from Pure Copper

The raw material for P/M structural parts from pure copper[24] must be a high purity copper powder. As discussed in chapters 8 and 9, green compacts from copper powder pressed to high densities tend to expand during sintering because of gas entrapment in the pores of the compacts. The usual practice in producing copper parts with high final densities is, therefore, to compact at moderate pressures of 15 to 18 tsi, to sinter at temperatures 50 to 150° C below the melting point of copper, to repress the sintered compacts to the desired high density and to resinter them to obtain an annealed structure, if this is desirable. A method of avoiding swelling during sintering of copper compacts even when they have been pressed at high pressure has been suggested by Hirsch.[25] The pressed green compacts are impregnated with an aqueous solution of a salt, e.g. lithium nitrate, which prevents swelling during sintering, but does not leave a residue deleterious to conductivity. The effect of density on ultimate tensile strength, elongation and electrical conductivity of copper powder is shown in Figure 17-11.

A proposed ASTM standard for sintered copper structural parts for electrical conductivity applications sets up the requirements for chemical

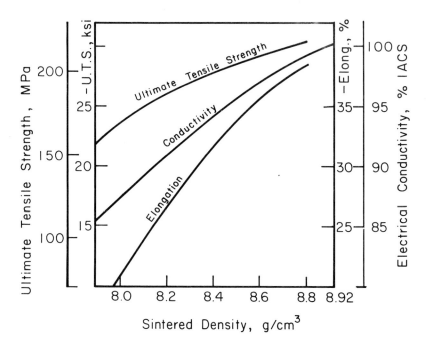

FIGURE 17-11 Effect of density upon ultimate tensile strength, elongation and electrical conductivity of compacts from electrolytic copper powder.

composition and density ranges shown in table 17-4. Typical tensile properties and electrical conductivity of copper powder metallurgy parts, shown in the appendix to the proposed specification, are reproduced in table 17-5.

In order to obtain the required high electrical conductivity in P/M copper parts, the percentage of impurity elements in solid solution in the copper should be very low. As little as 0.023% iron in solid solution in copper will lower its conductivity to 86% of that of pure copper. Small

Table 17-4   Chemical composition and density ranges of copper P/M parts

Cu, min. %	99.90
other, max. %	0.10
density, range, type I	7.8 - 8.3 g/cm³
type II	8.3 g/cm³ minimum

Table 17-5   Typical tensile properties and electrical conductivity of copper P/M parts

	Type I	Type II
Density	8.0	8.3
Ultimate tensile strength psi	23,000	28,000
MPa	159	193
Elongation %	20	30
Electrical conductivity % of ACS	85	90
Ohm^{-1}m^{-1}	0.493×10^8	0.522×10^8

Table 17-6   Composition and Properties of P/M Bronze
Composition of P/M Bronze, MPIF Designation CT-0010 86.3-90.5% Cu, 9.5-10.5% Sn, 0-1.7% graphite, 0-1.0% Fe Other elements, total by difference 0.5% max Properties of P/M Bronze CT-0010

Density range	Ultimate tensile strength		Compressive yield str. 0.2%		Elong. %	Comparable Specifications
	psi	MPa	psi	MPa		
N:5.6-6.0g/cm³ AS	8000	55	7000	48	1.0	SAE840
R:6.4-6.8g/cm³	14,000	97	11,000	76	1.0	SAE841
S:6.8-7.2g/cm³	18,000	124	17,500	121	2.5	SAE842
						ASTM B255 Type II

amounts of iron mechanically mixed with the copper powder lower the conductivity much less (2% loss in conductivity for 0.023% iron) unless the iron dissolves in the copper during sintering.

B.   P/M Bronze Structural Parts

These parts are generally produced from mixtures of copper and tin powder by methods very similar to those used for self lubricating bronze bearings. The compositions and properties listed in MPIF specification 35 are shown in table 17-6.

C.   P/M Brass and Nickel Silver
Structural Parts

In contrast to bronze structural parts, those from brass, leaded brass and nickel silver are produced from prealloyed atomized powder. The purpose of the lead in the lead bearing compositions is to make it easier to compact

Table 17-7   Composition and properties of sintered brass and sintered leaded brass

Density range	Ultimate tensile strength psi	MPa	Tensile yield strength 0.2% offset psi	MPa	Elong. %	Apparent hardness
Composition of P/M 90/10 Brass Designation CZ-0010						
88.0-91.0% Cu, 8.3-12.0% Zn, 0-0.3% Fe, Total other elements by difference 0.4% max.						
Properties of P/M 90/10 Brass CZ-0010						
T:7.2-7.6 g/cm³	20,000	138	9,000	62	13	Rockwell H57
U:7.6-8.0g/cm³	27,000	186	10,000	69	18	Rockwell H70
Composition of P/M 70/30 Brass, Designation CZ-0030						
68.5-71.5% Cu, 27.8-31.5% Zn, 0-0.3% Fe, Total other elements by difference 0.4% max						
Properties of P/M 70/30 Brass CZ-0030						
T:7.2-7.6g/cm³	31,000	214	13,000	90	20	Rockwell H76
U:7.6-8.0g/cm³	37,000	255	15,000	103	26	Rockwell H85
Composition of P/M 90/10 Leaded Brass, MPIF Designation CZP-0210						
86.0-90.0% Cu, 7.3-13.0% Zn, 1.0-2.0% Lead, 0-0.3% Fe Total other elements 0.4% max						
Properties of P/M 90/10 Leaded Brass, CZP-0210						
T:7.2-7.6g/cm³	18,000	124	7,000	48	14	Rockwell H46
U:7.6-8.0g/cm³	25,500	176	8,000	55	20	Rockwell H60
Composition of P/M 80/20 Leaded Brass, MPIF Designation CZP-0220						
77.0-80.0% Cu, 17.3-22.0% Zn, 1.0-2.0 Pb, 0-0.3% Fe, Total other elements 0.4% max						
Properties of P/M 80/20 Leaded Brass, CZP-0200						
T:7.2-7.6g/cm³	24,000	166	11,000	76	13	Rockwell H55
U:7.6-8.0g/cm³	28,000	193	13,000	90	19	Rockwell H68
W:8.0-8.4g/cm³	32,000	221	15,000	103	23	Rockwell H75
Composition of P/M 70/30 Leaded Brass, MPIF Designation CZP-0230						
66.5-70.5% Cu, 26.8-32.5% Zn, 1.0-2.0% Pb, 0-0.3% Fe, Total other elements 0.4% max						
Properties of P/M 70/30 Leaded Brass, CZP-0230						
T:7.2-7.6g/cm³	28,000	193	11,000	76	22	Rockwell H65
U:7.6-8.0g/cm³	34,000	234	13,000	90	27	Rockwell H76

the powder and to facilitate machining of the sintered parts in a secondary operation. Adding lithium stearate as a lubricant to the atomized powders instead of the zinc stearate used for parts from iron powder improves the mechanical properties of brass and nickel silver P/M parts as shown in the work of Zaleski and Powell.[26]

The composition and properties listed in MPIF specification 35 for the brasses and leaded brasses are shown in table 17-7 and those for nickel silver in table 17-8.

Table 17-8   Composition and Properties of P/M Nickel Silver.

Composition of P/M Nickel Silver, MPIF Designation CZN-1818
62.5-65.5% Cu, 16.5-19.5% Ni, Balance Zn, Other elements 1.0 max
Properties of P/M Nickel Silver CZN-1818

Density	Ultimate tensile strength		Compressive yield strength 0.1% offset		Elonga-tion %	Young's Modulus		Apparent hardness	Charpy unnotched impact strength ft. lbs.
	psi	MPa	psi	MPa		psi $\times 10^6$	GPa		
U:7.6-8.0 g/cm³	30,000	207	16,000	110	10	13	90	Rockwell H 75	9.0

Comparable specification - ASTM B 458, Grade 2, Type I

W:8.0-8.4 g/cm³	35,000	241	17,000	117	12	14	17	Rockwell H 85	12.0

Comparable specification - ASTM B 458, Grade 2, Type II

Table 17-9   Nominal Chemistry of P/M Grades of Stainless Steel Powders

Grade	Ni	Cr	Mo	Si	Mn	S	C	Fe
303L	12.5	17.5		0.7	0.35	0.2	0.02	Bal.
304L	10.5	18.5		0.7	0.25		0.02	Bal.
316L	13.0	17.0	2.1	0.7	0.20		0.02	Bal.
410L		13.0		0.7	0.40		0.02	Bal.

D.   P/M Stainless Steel Parts[27]

Both austenitic and martensitic stainless steel parts are fabricated from powders which are produced by water atomization of the molten alloys as described in chapter 2. The most common grades produced from powder are 316 L, 303 L, 304 L and 410 L whose compositions are shown in table 17-9.

Stainless steel P/M structural parts are used in preference to parts from other compositions because of their corrosion resistance, which makes it unnecessary to plate the parts for surface finish and gives them a pleasing

Table 17-10   Mechanical Properties of 316L Pressed to 6.85 g/cm³ and Sintered for 30 min at 2050° F (1120° C) in Various Atmospheres

Property	Sintered in dissociated ammonia	Sintered in hydrogen
Yield strength (0.2% offset)	39,8000 psi (274.4 MPa)	26,600 psi (183.4 MPa)
Ultimate tensile strength	53,000 psi (365.4 MPa)	41,800 psi (288.2 MPa)
Elongation in 1 in. (25.4 mm)	7.0%	10.9%
Apparent hardness	67 R_B	47 R_B

color and appearance, and because of their relatively high mechanical properties. Austenitic stainless steels are non-magnetic, the martensitic stainless steel 410 L may be heat-treated.

The most widely used grade of P/M stainless steel is 316 L which has the best corrosion resistance. Nevertheless, because of its porosity, the corrosion resistance is inferior to wrought material of the same composition depending upon the conditions under which it is processed and particularly its degree of porosity. For critical applications the corrosion resistance should be assessed by field testing.

With regard to processing, austenitic stainless steel parts should be sintered so as to avoid carbon pick-up particularly from the lubricant added to facilitate compacting, because carbon lowers the corrosion resistance. A separate de-waxing operation is often recommended. The mechanical properties of the austenitic stainless steels depend upon the atmosphere in which they are sintered. Sintering in dissociated ammonia introduces as much as 0.3% nitrogen in the composition, which increases the strength, but decreases the ductility of the material, as shown in table 17-10[27]. For material sintered in vacuum, the properties are similar to those of material sintered in hydrogen, unless the vacuum sintered material is allowed to pick up nitrogen during cooling. The properties of stainless steel types 303 and 316 are given in MPIF standard 35 for the density ranges P(6.0-6.4 g/cm³) and R(6.4-6.8 g/cm³) which are relatively low. They are shown in table 17-11. Parts can be readily fabricated to densities above 6.8 g/cm.³ The composition of the 303 grade contains 0.2% sulfur to improve the machinability of the steel. The values for elongation in table 17-11 are for material sintered in dissociated ammonia at the low temperature of 2050° F. By sintering at higher temperatures both the ductility and the corrosion resistance of austenitic stainless steels can be raised considerably. Specimens pressed to 6.85 g/cm³ and sintered 30 min at 2400° F will have elongations of 18% when sintered in dissociated ammonia and 26% when sintered in hydrogen. Sintering at 2400° F will, however, also cause the

Table 17-11   Properties of Austenitic Stainless Steels (MPIF Designations SS 303
and SS 316)

Density range	Type	Ultimate tensile strength		Tensile yield strength		Elongation %
		psi	MPa	psi	MPa	
P:6.-6.04 g/cm³	SS303	35,000	241	32,000	221	1
	SS316	52,000	359	47,000	324	2
R:6.4-6.8 g/cm³	SS303	38,000	262	32,000	221	2
	SS316	54,000	372	40,000	276	4

Table 17-12   Properties of Martensitic Stainless Steel MPIF Designation SS-410

Density range	Ultimate tensile strength		Tensile yield strength		Elongation %
	psi	MPa	psi	MPa	
N:5.6-6.0 g/cm³	42,000	290	41,000	283	0-1.0
P:6.0-6.4 g/cm³	55,000	379	54,000	372	0-1.0

parts to shrink from the die size by approximately 1.5%.

The martensitic grade of P/M stainless steel with the MPIF designation SS-410 can be heat treated by quenching and tempering if it contains sufficient carbon or nitrogen so that it transforms to austenite at elevated temperature. During normal sintering in dissociated ammonia enough nitrogen is introduced into the composition for this purpose. The as-sintered properties of grade SS-410 as given in MPIF standard 35 are shown in table 17-12.

E.   P/M Aluminum Alloy Structural Parts

The sintered properties of compacts from aluminum powder and from mixtures of aluminum powder with alloying ingredients were reported in several investigations in the 1940's.[28,29] The compacts were pressed at relatively high pressures, on the order of 35 tsi or higher, without the addition of a lubricant in a die with carefully lubricated diewalls. At these pressures compacts with relative densities of 95-99% are obtained. Metallic contact between the aluminum powder particles is established during compacting. Since the porosity in these high-density compacts is non-interconnecting, they can be sintered in air, in nitrogen or hydrogen. Pure aluminum powder compacts were sintered at 620°C, 40° below the melting point of the metal, mixtures of aluminum and elemental alloy powders,

Table 17-13   Compositions and Density Ranges for P/M Aluminum Alloy Structural Parts

Element	Composition % Grade 1	Grade 2	Grade 3
Copper	0.5 max	1.5-2.5	3.5-5.0
Magnesium	0.4-1.2	0.2-1.2	0.2-0.8
Silicon	0.2-0.8	0.5 max	1.2 max
Aluminum, min	96.0	94.3	91.5
Total of other elements determined by difference, max.	1.5	1.5	1.5

Density range
Type I — 2.30-2.45 g/cm³
Type II — 2.45-2.60 g/cm³
Type III — 2.6 g/cm³ min.

such as zinc, mangnesium and copper, were sintered at temperatures where a liquid eutectic phase was formed. Under these conditions mechanical properties were obtained closely approaching those of the corresponding wrought alloys. The principal reason why these laboratory investigations did not lead to commercial production of P/M aluminum structural parts was the fact that no satisfactory system of die wall lubrication became available which would permit compaction without galling and seizing.

P/M aluminum alloy structural parts have been successfully fabricated in commercial production for the last several years by mixing a lubricant with the powder mixture instead of using diewall lubrication, although attempts to reintroduce part production with diewall lubrication have also been renewed.

Because of the tendency to gall and seize when aluminum powder is pressed, fairly large amounts of lubricant are needed. In order to eliminate the lubricant during sintering, the parts must not have too high a density so that the lubricant can escape without disrupting the compacts. The compacts are pressed from a mixture of aluminum powder (a relative coarse grade with 20% particles larger than 150 microns to minimize galling), of elemental copper, magnesium and silicon powders and by 1½ by weight of lubricant at pressures of 30,000 psi (210 MPa). The green density should be low enough (not more than 93% equal to about 2.62 g/cc) so that the lubricant can escape during sintering. The compacts are to be sintered at a temperature at which reaction of the aluminum with the magnesium and/or copper and silicon forms a liquid phase. If the compacts are heated rapidly, the oxide layer present on the surface of the aluminum powder particles is thin enough so that the liquid phase which is formed spreads and thereby apparently mechanically removes much of the oxide present providing good contact. An atmosphere with a low dew point (max. -40°C) should be used; nitrogen or a nitrogen-hydrogen mixture is recommended. The low dew point is necessary so that no additional oxide is formed during sintering.

Three compositions and three density ranges for P/M aluminum alloy structural parts have been standardized in standard B 495 of the American Society for Testing and Material. They are shown in table 17-13.

Sintered P/M aluminum alloy structural parts are relatively soft and can readily be sized. This sizing also improves the grey finish of the as-sintered parts. Similar to wrought aluminum alloys, the P/M alloys may be strengthened by precipitation hardening. P/M aluminum alloy structural parts may also be surface treated by chemical polishing, by anodizing and by coloring.

The table of typical mechanical properties shown in the appendix of ASTM Standard B 595 is reproduced in table 17-14. It gives data not only for the as-sintered condition, but also for specimens solution treated, quenched and aged.

V. Secondary Operations

There are a number of subsequent operations to which P/M structural parts are subjected after sintering. They include:

>Sizing, coining and repressing, sometimes followed by resintering or annealing
>Impregnation
>Infiltration
>Heat treatment and steam treatment
>Machining
>Joining
>Plating and other surface finishing operations.

A. Sizing, Coining, Repressing and Resintering

In these operations the sintered structural part is inserted in a confined die and struck by a punch. The principal aim of "sizing" is to correct distortions which have occurred during sintering without large changes in the density of the part. This operation is sometimes called "coining," although coining actually refers to giving a profile to the surface of a part, like in coining a blank to produce a coin. In repressing sufficient pressure is applied to the

Table 17-14   Typical Properties for P/M Aluminum Alloy Specimens

Grade a	Type	Thermal Condition b	Ultimate tensile strength		Tensile yield strength (0.2% offset)		Elongation in 1 in. or 25 mm. %	Apparent Rockwell Hardness
			psi	MPa	psi	MPa		
1	I	T1	12,000	83	9,00	62	4.0	60-65 HRH
		T4	14,000	97	11,500	79	3.5	65-70 HRH
		T6	20,000	138	19,000	131	0.5	80-85 HRH
1	II	T1	18,500	128	10,000	69	6.0	80-85 HRH
		T4	23,000	159	15,000	103	5.0	50-55 HRE
		T6	30,000	207	28,000	193	2.0	65-70 HRE
2	I	T1	19,000	131	13,000	90	4.0	80-85 HRH
		T4	23,000	159	17,500	121	5.0	50-55 HRE
		T6	26,000	179	25,000	172	1.0	55-60 HRE
2	II	T1	20,000	138	15,500	107	4.0	80-85 HRH
		T4	24,500	169	18,500	127	3.5	55-60 HRE
		T6	28,000	193	27,000	186	2.0	60-65 HRE
3	I	T1	20,000	138	14,000	97	2.0	80-85 HRH
		T4	24,000	165	21,000	145	2.0	55-60 HRE
		T6	30,000	207	25,000	172	0.5	65-70 HRE
3	II	T1	22,000	152	17,000	117	3.0	85-90 HRH
		T4	26,000	179	22,000	152	2.5	55-60 HRE
		T6	35,000	241	33,000	228	1.0	70-75 HRE
3	III	T1	25,000	172	22,000	152	3.0	55-60 HRE
		T4	32,000	221	26,000	179	2.5	70-75 HRE
		T6	42,000	290	40,000	276	2.0	80-85 HRE

[a]Typical sintering atmosphere for these grades may be nitrogen, dissociated ammonia or vacuum.
[b]Description of Thermal Conditions:
  T1 as sintered
  T4 solution heat treated at 940 to 970° F (504 to 521°C), cold water quenched and aged minimum of 4 days at room temperature.
  T6 solution heat treated at 940 to 970° F (504 to 521°C), cold water quenched and aged 18 hr at 320 to 350° F (160 to 177°C).

presintered or sintered compact to increase its density. Resintering is a second sintering operation and also relieves the cold work introduced during repressing. The effect of repressing and resintering upon the mechanical properties has been presented in the previous sections.

B.   Impregnation

When structural parts also serve as bearings, they may be impregnated with a lubricant, just like self lubricating bearings. When a porous part is to be made impervious to liquids or gases, it is impregnated with a thermosetting polymer available as a viscous liquid. The polymer is changed to a solid by a low temperature treatment. This treatment is used to make P/M structural parts pressure tight and for parts to which a surface finishing operation, such as plating, is to be applied. The treatment prevents absorption or entrapment of the plating solution in the pores of the part.

C. Infiltration

The process of infiltration of iron base structural parts with copper alloys was discussed in chapter 12. The advantages of the process have been enumerated by Dundaller[30]:

1. Increased Mechanical Properties - Higher tensile strengths and hardnesses, greater impact energies and fatigue strengths are obtained through infiltration.

2. Uniform Density - Parts which contain non-uniform and/or heavy sections can be infiltrated to obtain more uniform density: infiltration tends to even out density variations.

3. Higher Density - Infiltration is a useful method to increase sintered part weight without increasing the size of the part. Given the press size limitations, restrictions in pressing technique and powder compressibility, many times it is easier to obtain high density through infiltration and, certainly when considering normal P/M operations, it would be difficult to obtain densities in excess of 7.2 g/cm³ without resorting to additional pressing and sintering operations. Infiltration makes densities in excess of 7.2 g/cm³ possible in a single pressing and sintering operation.

4. Removal of Porosity for Secondary Operations - Infiltration may be used in place of impregnation as a method to seal surface porosity so secondary operations such as pickling and plating may be performed without damaging the interior of the part and creating subsequent "bleeding" problems. It is also a method of sealing a part used for application in which no porosity is desired.

5. Selective Property Variation - It is possible by infiltrating only selected areas of a part to obtain, within limits, a controlled variation of properties in the part, e.g., variations in density, strength and hardness. This is known as localized infiltration. Infiltration to considerably less than the full density in the part (say to 7.1 g/cm³) is known as starve infiltration.

6. Assembly of Multiple Parts - Different sections of the final part, pressed separately, can be assembled by sintering the individual pieces together and bonding the pieces into one part through common infiltration.

As Durdaller points out, the porous iron or steel skeleton does not have to be completely filled with the infiltrant. The weight percentage of

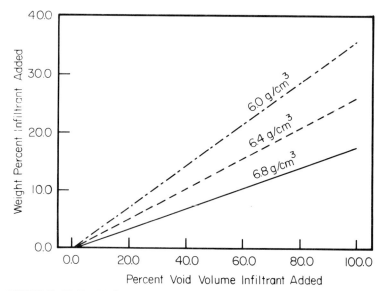

FIGURE 17-12   Weight percent of infiltrant which must be added to compacts of 6.0, 6.4 and 6.8 g/cc density in order to fill with infiltrant from 0 to 100% of the void volume in the compacts.

infiltrant needed for steel compacts of different densities in order to fill a given percentage of the void volume of the porous skeleton is shown in Figure 17-12.[30] However, when less than 40-50% of the void volume is infiltrated, the infiltrant may not penetrate through the entire cross section of the part to be infiltrated. On the other hand, achieving 100% infiltration of the void volume which corresponds to a density of approximately 7.8 g/cm³ often presents difficulties with control of the part dimensions.

The effect of infiltration upon transverse rupture strength of specimens of iron powder with varying amounts of graphite added pressed to 6.4 g/cm³ density, sintered and then infiltrated as a function of the percent void volume copper infiltrated is shown in Figure 17-13.[30] MPIF standard 35 includes specifications and typical properties for P/M infiltrated steels which are shown in table 17-15. All specifications are for the density range T:7.2-7.6 g/cm³. The lower copper content for grades FX 1005 and FX 1008 is obtained by using a skeleton of higher density then in grades FX 2000, FX 2005 and FX 2008, where a less dense skeleton is infiltrated with a greater amount of infiltrant. The advantage of starting with a denser skeleton are not only improved mechanical properties, but also lower cost, because of the lesser amount of expensive infiltrant needed. Properties for both the as-sintered and the heat-treated condition are shown.

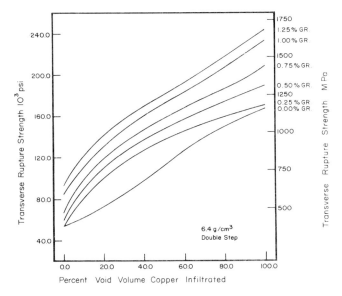

FIGURE 17-13 Transverse rupture strength of sintered steel from mixtures of iron powder with 0 to 1.25% graphite, compacted and sintered to 6.4 g/cm³ density as a function of the void volume of the compacts infiltrated with copper.

Table 17-15 Composition and Typical Properties of P/M Infiltrated Steels

		Designation(a)			MPIF composition limits and ranges, %(b)			
Description	MPIF	ASTM	SAE	C	Ni	Cu	Fe
P/M infiltrated steel FX-1005		· · ·	· · ·	0.3-0.6	· · ·	8.0-14.9	80.5-91.7
P/M infiltrated steel FX-1008		· · ·	· · ·	0.6-1.0	· · ·	8.0-14.9	80.1-91.4
P/M infiltrated steel FX-2000		B303, Cl A	870	0.3 max	· · ·	15.0-25.0	70.7-85.0
P/M infiltrated steel FX-2005		B303, Cl B	· · ·	0.3-0.6	· · ·	15.0-25.0	70.4-84.7
P/M infiltrated steel FX-2008		B303, Cl C	872	0.6-1.0	· · ·	15.0-25.0	70.0-84.4

(a) Amount of other elements less than 4%.

Designation	Sur-face fix (a)	Condi-tion (b)	Tensile strength MPa	ksi	Yield strength MPa	ksi	Elongation in 25 mm or 1 in., %	Fatigue strength MPa	ksi	Impact energy (c) J	ft·lb	Apparent hardness	Elastic modulus 10⁶ GPa	psi
FX-1005-(e)	T	AS	570	83	440	64	4.0	· · ·	· · ·	19	14	75 HRB	135	20
		HT	830	120	740	107	1.0	· · ·	· · ·	9.5	7.0	35 HRC	135	20
FX-1008-(e)	T	AS	620	90	515	75	2.5	· · ·	· · ·	16	12	80 HRB	135	20
		HT	895	130	725	105	60.5	· · ·	· · ·	9.5	7.0	40 HRC	135	20
FX-2000-(e)	T	AS	450	65	· · ·	· · ·	1.0	· · ·	· · ·	20	15	60 HRB	· · ·	· · ·
FX-2005-(e)	T	AS	515	75	345	50	1.5	· · ·	· · ·	12.9	9.5	75 HRB	125	18
		HT	790	115	655	95	<0.5	· · ·	· · ·	8.1	6.0	30 HRC	125	18
FX-2008-(e)	T	AS	585	85	515	75	1.0	· · ·	· · ·	14	10	80 HRB	125	18
		HT	860	125	740	107	<0.5	· · ·	· · ·	6.8	5.0	42 HRC	125	18

(b) AS: as sintered HT: Heat-treated (typically austenitized at 870°C, oil quenched and tempered 1 hr at 200°C. (c) Unnotched Charpy test.

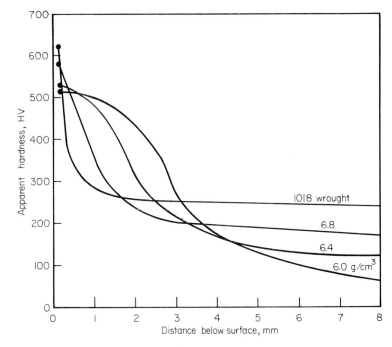

FIGURE 17-14 Effect of porosity upon the hardness and hardness distribution of compacts from plain iron powder, carbonitrided after sintering; hardness traverses reflect both depth of case and density of compacts. Hardness traverse for carbonitrided wrought 1018 steel for comparison.

D. Heat Treatment and Steam Treatment

P/M structural parts of suitable composition may be heat-treated in the same manner as wrought or cast parts. In heat-treating P/M steel parts austenitizing in salt baths and particularly in cyanide baths is generally avoided, because of the danger of trapping salt in the porous structure. The effect of porosity upon the hardenability of porous steels was illustrated in Figure 17-5. For higher density P/M steel parts the improvement in mechanical properties through heat treatment is shown in Figures 17-6, 17-7 and 17-8 and in table 17-3. P/M aluminum alloy parts may also be heat treated resulting in the typical mechanical properties shown in table 17-14.

The surface hardness of P/M steel parts may be increased by carburizing and carbonitriding. While in carburized and carbonitrided wrought steel parts the case depth is entirely a function of the diffusion of carbon and

Table 17-16   Effects of steam treating on density and hardness of ferrous P/M materials

Material	Density, Mg/m³ Sintered	Density, Mg/m³ Steam treated	Apparent hardness Sintered	Apparent hardness Steam treated
F-0000-N	5.8	6.2	7 HRF	75 HRB
-P	6.2	6.4	32 HRF	61 HRB
-R	6.5	6.6	45 HRF	51 HRB
F-0008-M	5.8	6.1	44 HRB	100 HRB
-P	6.2	6.4	58 HRB	98 HRB
-R	6.5	6.6	60 HRB	97 HRB
FC-0700-N	5.7	6.0	14 HRB	73 HRB
-P	6.35	6.5	49 HRB	78 HRB
-R	6.6	6.6	58 HRB	77 HRB
FC-0708-N	5.7	6.0	52 HRB	97 HRB
-P	6.3	6.4	72 HRB	94 HRB
-R	6.6	6.6	79 HRB	93 HRB

nitrogen in the solid state, in porous parts the carburizing or carbonitriding atmosphere penetrates into the pores and produces a case depth which is a function of the part density. This is illustrated in Figure 17-14[3] in which the effect of density on the case depth, measured as apparent Vickers hardness, in specimens carbonitrided at 870°C for 30 minutes and oil quenched is shown in comparison with the case depth of 1018 steel treated in the same manner. To produce a typical case and core structure during carburizing a density of 7.0 g/cm³ and above is needed.

In steam treating[31] the porosity of P/M structural parts is directly used. The parts are treated in dry steam at approximately 550°C. The steam reacts with the surfaces of the part, not only the outer surface, but also the inner surface along the pores connected with the outside. A layer of magnetic oxide (Fe_3O_4) is formed on the outside surface and as a skeleton throughout the interior of the part. Steam treating not only provides some corrosion resistance, but also changes the mechanical properties, increases the density, hardness, wear resistance and compressive strength of the parts. The effects of steam treatment are shown in table 17-16.[3] Steam treating is widely used for P/M structural parts, e.g. for shock absorber pistons.

E.   Machining[32,33,34,35]

Wherever possible P/M structural parts are compacted and sintered to final dimensions so that no subsequent machining is necessary. However, there are many cases where this is not possible and where the parts have to be machined, e.g. when holes at an angle to the direction of pressing or when tapped holes are required. For this reason, the machinability of sintered parts is important and has been repeatedly investigated. Reen[35] has described a drill life test which gives a reliable measure of the machinability of P/M parts. In general, neither the tool life is as long nor the surface finish as good in machining P/M structural parts as in machining wrought parts. Grades of both iron powder and stainless steel powder have been developed in which sulfur is incorporated in the structure of the powder particles to improve tool life and surface finish. In brass and nickel silver powders lead is added for the same purpose. The machining characteristics of sintered P/M steels can be improved by appropriate choice of cemented carbide grades and the grinding characteristics of the parts by the design and hardness of the grinding wheels.

F.   Joining

Many of the conventional joining operations for wrought materials may also be performed on P/M structural parts. Of the various welding methods electrical resistance welding (butt welding, projection welding) is probably better suited than oxyacetylene welding and arc welding, where there is danger of oxidation of the interior porous material. Argon arc welding has successfully been used for stainless steel structural parts. Copper brazing is best performed on copper infiltrated parts, in fact copper infiltration and copper brazing may be combined into one operation. Powder metallurgy parts may be joined by using somewhat different compositions for the components to be joined, one which slightly grows or expands during sintering, the other which slightly shrinks. The composition which grows is used for the inner portion of the assembly, the one which shrinks for the outer portion. The parts are assembled as compacted; during sintering an excellent joint is formed.

G.   Plating and Other Surface Finishing Operations

P/M structural parts may be plated by electroplating or other plating processes. In order to avoid penetration and entrapment of plating solutions in the pores of the part, they are generally impregnated before plating as described above.

References

1. F. V. Lenel, Oil pump gears, an example of iron powder metallurgy, in "Powder Metallurgy," ed. by J. Wulff, Cleveland, p. 502-511, (1942).
2. G. Zapf, Pulvermetallurgie, in "Handbuch der Fertigungstechnik," Carl Hanser Verlag, Munich, in print.
3. The chapter "Ferrous powder metallurgy materials" in Metals Handbook, Ninth ed., Vol. 1, p. 327-347, American Society for Metals, Metals Park, O. (1978) is the source of the following figures and tables:

   Fig. 17-1   from fig. 9, p. 339
   Fig. 17-2   from fig. 5, p. 336
   Fig. 17-3   from fig. 11, p. 341
   Fig. 17-4   from fig. 10, p. 340
   Fig. 17-5   from fig. 13, p. 342
   Fig. 17-6   from fig. 18, p. 345
   Fig. 17-7   from fig. 15, p. 344
   Fig. 17-8   from fig. 16, p. 344
   Fig. 17-14   from fig. 14, p. 343

   Table 17-1   from table 2, p. 333
   Table 17-2   from table 3, p. 333
   Tables 17-3 and 17-15   from table 4, p. 334-335
   Table 17-14   from table 8, p. 344
4. F. V. Lenel and T. Pecanha, Observations on the sintering of compacts from a mixture of iron and copper powders, Powder Metallurgy, Vol. 16, No. 32, p. 351-365, (1973).
5. G. Zapf and K. Dalal, Introduction of high oxygen affinity elements manganese, chromium and vanadium in the powder metallurgy of P/M parts, Mod. Dev. in Powder Metallurgy, Vol. 10, ed. by H. H. Hausner and P. W. Taubenblat, Princeton, p. 129-152, (1977).
6. F. J. Esper and R. Zeller, Elongation and tensile strength of Fe-Cu-Sn sintered bodies, Powder Metallurgy, Third European P/M Symposium, 1971, Conference supplement part 1, p. 311-322, (1971).
7. P. Lindksog, J. Tengzelius and S. A. Kvist, Phosphorus as an alloying element in ferrous P/M, Mod. Dev. in Powder Metallurgy, Vol. 10, ed. by H. H. Hausner and P. W. Taubenblat, Princeton, p. 98-128, (1977).
8. F. W. Heck, Sintered and heat treated 4640 steel, Mod. Dev. in Powder Metallurgy, Vol. 5, ed. by H. H. Hausner, N.Y., p. 453-469, (1971).
9. P. Lindskog and G. Skoglund, Alloying practice in the production of sintered steels, Powder Metallurgy, Third European P/M symposium, 1971, Conference supplement part 1, p. 371-396.
10. R. T. Holcomb and J. Lovenduski, The effect of density on the properties of prealloyed nickel-molybdenum low alloy steel powders, Mod. Dev. in Powder Metallurgy, Vol. 8, ed. by H. H. Hausner and W. E. Smith, Princeton, N.J., p. 85-107, (1974).

11. S. A. Kvist, P. Lindskog and L. E. Svensson, Improved tolerances of sintered ferrous components containing copper, Metal Powder Report, Vol. 32, No. 2, p. 38-45, Feb. (1977).

12. P. Lindskog and O. Thornblad, Ways to improve the strength of sintered components made from partially prealloyed steel powders, Powder Metallurgy Int., Vol. 11, No. 1, p. 10-11, (1979).

13. H. F. Fischmeister, K. E. Easterling, L. E. Larsson, L. Olsson, P. A., Nickolausson and K. Heurling, Comparison of the mechanical properties of sintered and sinter-forged Ni-Cu steels with cast and wrought alloys using yield-strength and toughness criteria, Powder Metallurgy, Third European P/M symposium 1971, Conference supplement part II, p. 530-560.

14. M. Eudier, The mechanical properties of sintered low-alloy steels, Powder Metallurgy, No. 9, p. 278-290, (1968).

15. H. E. Exner and D. Pohl, Fracture behavior of sintered iron, Powder Metallurgy Int., Vol. 10, No. 4, p. 193-196, (1978).

16. C. Razim, Fatigue behavior of sintered ferrous materials with particular reference to their use in automobiles, Zw.F. Vol. 73, p. 451-456, (1978).

17. ASTM Standard Specifications for:
    Copper infiltrated sintered carbon steel structural parts   B303
    Iron-copper sintered metal powder structural parts   B222
    Sintered aluminum structural parts   B595
    Sintered austenitic stainless steel structural parts   B525
    Sintered brass structural parts   B282
    Sintered carbon steel structural parts   B310
    Sintered copper-steel structural parts   B426
    Sintered nickel-silver structural parts   B458
    Sintered nickel steel structural parts   B484
    Sintered bronze structural parts   B255

18. P. E. Mathews, The mechanical properties of brass and developmental non-ferrous P/M materials, Int. J. of Powder Metallurgy, Vol. 5, No. 4, p. 59-69, (1969).

19. P. E. Mathews, Cubraloy, a new development in aluminum bronze powder metallurgy, 1971 Fall Powder Metallurgy conference, proceedings, p. 205-216, (1972).

20. F. R. Hensel, E. I. Larsen and E. F. Swazy, Notes on copper base compacts and certain compositions susceptible to precipitation hardening, Trans, AIME, Vol. 166, p. 533-547, (1946).

21. E. Fetz, D. R. Hollingsbery, R. L. Cavanagh, Precipitation hardening sintered copper-base alloys, Mod. Dev. in Powder Metallurgy, Vol. 7, ed. by H. H. Hausner and W. E. Smith, Princeton, p. 537-583, (1974).

22 P. W. Taubenblat, W. E. Smith and R. L. Bieniek, Production and properties of Incramute, a noise damping powder alloy, Mod. Dev. in Powder Metallurgy, Vol. 10, ed. by H. H. Hausner and P. W. Taubenblat, Princeton, p. 241-258, (1977).

23. J. L. Holman and L. A. Neumeier, Powder metallurgy consolidation of manganese-copper damping alloys, ibid., p. 259-283.

24. P. W. Taubenblat, Importance of copper in P/M, Int. J. of Powder Metallurgy, Vol. 10, No. 3, p. 169-184, (1979).

25. H. H. Hirsch, Improving conductivity of copper P/M parts by pretreatment of green compacts, Powder Metallurgy, Vol. 22, p. 49-61, (1979).

26. F. I. Zaleski and R. A. Powell, The effect of lithium stearate additions and various atmospheres on sintered brass compacts, 15th annual meeting MPIF, Proceedings, p. 28-40, (1954).

27. H. D. Ambs and A. Stosuy, The powder metallurgy of stainless steel, ch. 29 in Handbook of stainless steel, ed. by D. Peckner and I. M. Bernsteim, N.Y., (1977).

28. L. W. Kempf, Properties of compressed and heated aluminum alloy powder mixtures, in "Powder Metallurgy," ed. by J. Wulff, Cleveland, p. 314-316, (1942).

29. G. D. Cremer and J. J. Cordiano, Recent developments in the formation of aluminum and aluminum alloys by powder metallurgy, Trans. AIME, Vol. 152, p. 152-165, (1943).

30. C. Durdaller, Copper infiltration of iron-base P/M parts, Technical bulletin from Hoeganaes Corporation, (1969).

31. F. V. Lenel, Steam treatment of porous iron, in "Powder Metallurgy" ed. by J. Wulff, Cleveland, p. 512-519, (1942).

32. R. S. Jamison and E. Geijer, Machinability of sintered steel, 15th annual meeting of MPIF, Proceedings, p. 82-93, (1959).

33. S. A. Kvist, Turning and drilling of some typical sintered steels, Powder Metallurgy, Vol. 12, No. 24, p. 538-565, (1969).

34. A. P. Pignon, Secondary machining operations for P/M products, Mod. Dev. in Powder Metallurgy, Vol. 8, ed. by H. H. Hausner and W. E. Smith, Princeton, p. 163-186, (1974).

35. O. W. Reen, The machinability of P/M materials, Mod. Dev. in Powder Metallurgy, Vol. 10, ed. by H. H. Hausner and P. W. Taubenblat, Princeton, p. 431-452, (1977).

18

Hot Forging and Cold Forming of Preforms

The basic method of producing P/M structural parts described in chapter 17 involves conventional cold pressing and sintering. In order to obtain improved mechanical properties the parts may be subsequently repressed and resintered or infiltrated. They may also be heat treated. However, parts made by these techniques are still deficient in ductility, impact energy and fatigue strength compared with wrought materials. This deficiency is due to residual porosity. To obtain the same properties as in wrought materials, the porosity must be eliminated. This may be done by producing preforms from metal powders and hot forging these preforms to obtain parts of closely controlled dimensions and completely or near completely dense. This chapter will be mainly concerned with hot forging of preforms. In a final section cold forming of preforms, a technique which is less widely used, is discussed.

Hot forging of preforms was suggested as early as 1940,[1] when Tormyn[2] produced preforms by compacting hammer milled steel turnings and sintering the compacts. These preforms were hot forged into sleeves which were than cold trimmed and finally cold coined. However, industrial P/M hot forging started only in the 1970's based on development work in the late 1960's. In the proceedings of the 1970 International Powder Metallurgy Conference, nine papers on forging of P/M preforms were published,[3]

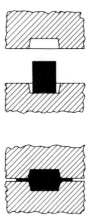

FIGURE 18-1   Schematic of presicion forging with flash of solid metal coupon.

including a complete description of a pilot plant process for a differential side pinion gear produced by forging of P/M preforms. Since then the output of literature on this subject has increased enormously with symposia included in several powder metallurgy conferences and with numerous reviews. A very comprehensive review with 228 literature references by W. J. Huppmann and H. Hirschvogel entitled "Powder Forging" is found in International Metals Review 1978.[4] The term "Powder Forging" for both the process and the product of forging preforms produced from metal powder has the advantage of being short, even if it is not quite accurate, and will be used in this chapter. The industrial production of powder forgings has increased considerably slower than the literature output in the field and is still small compared with the production of structural parts by conventional techniques. The economics of the production of conventional P/M structural parts as well as powder forgings will be discussed in chapter 19.

In conventional forging a fully dense blank is forged. The type of conventional forging closest to powder forging is precision forging in which the blank is forged in a closed die. The weight and geometry of the forging blank as well as the preheating and forging cycle are closely controlled. Nevertheless, even in precision forging, illustrated schematically in Figure 18-1, a flash is formed which must be trimmed and the forging must generally be machined. In powder forging the blank is a more precise preform produced from metal powders by compacting and sintering. Three different approaches have been taken in forging these preforms.

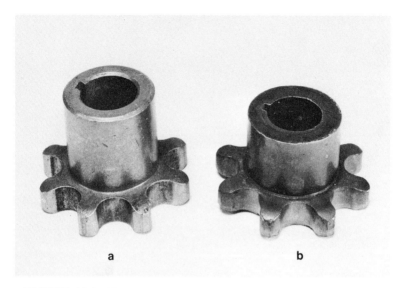

a   b

FIGURE 18-2 Hot repressed sprocket, preform and forging (Courtesy: Gould, Inc.). (a) sintered preform, (b) final forging.

The first of these approaches is hot repressing, sometimes called hot densification, in which the shape of the preform is close to that of the final piece except for its length in the forging direction. In this process the friction between die and preform during hot forging is high and therefore the pressure necessary to get complete densification would also be high, which would involve rapid wear of the forging tools. Hot repressing is therefore generally used in applications where densities on the order of 98% of theoretical are satisfactory. An example of a hot pressed component, both the preform and the final forging, shown in Figure 18-2, is a sprocket for a chain saw.[5]

The second approach is the one most widely used industrially. It is a precision forging process without a flash in which the shape of the preform is simpler than that of the final part, so that the desired final shape is produced to closely controlled dimensions in the hot forging step. As an example, Figure 18-3 shows the preform and the forged shape of the differential side pinion gear whose pilot plant production was developed in the late 1960's[6,7] The figure illustrates that this is not a hot repressing operation but that the preform is upset (and extruded) in hot forging. The advantages of hot forging by upsetting (and extrusion) over hot forging by repressing may be seen on the basis of Figures 18-4 and 18-5.[8] Figure 18-4

447

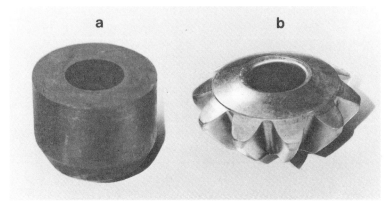

FIGURE 18-3 Differential pinion gear, preform and forging (a) sintered preform, (b) finished part.

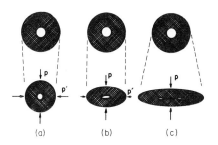

FIGURE 18-4 Schematic of void deformation under (a) isostatic pressure, $p^1 = p$; (b) repressing $p^1 < p$ (no lateral deformation); (c) true forging, $p^1 = 0$ (lateral flow).

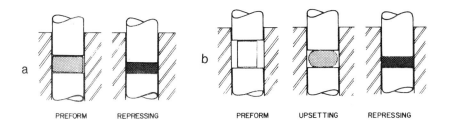

PREFORM   REPRESSING   PREFORM   UPSETTING   REPRESSING

FIGURE 18-5 Schematic of overall deformation in (a) repressing, (b) forging (upsetting in trap die).

shows schematically the deformation of a porous solid represented by a material element with a void in it under:

    a.   condition of hydrostatic pressure,

    b.   repressing condition with little lateral flow and

    c.   true forging condition involving considerable lateral flow.

In Figure 18-5 the overall deformation of a preform during

    a.   repressing and

    b.   hot forging by upsetting

are illustrated. The latter involves much more lateral flow especially in the beginning of deformation. This leads to more rapid initial densification and also involves more shear stress at pore surfaces, producing relative motion between opposite sides of the collapsed pore. Mechanical rupturing of any oxide film present at the pore surface exposes the metal and insures a sound metallurgical bond across collapsed pore surfaces. Upset forging also produces fibering of inclusions in the lateral direction. Toward the end of the forging stroke by upsetting when the preform has reached the diewall, the mode of deformation becomes the same as in repressing.

A third approach to hot forging of P/M preforms considers the process more from the point of view of the conventional forge shop. It has been described by Huseby and Scheil[9] and by Hoefs.[10] In this view the results of the forging step must not necessarily be a product with dimensions as closely or nearly as closely controlled as those of conventional P/M structural parts. The forging may even have a flash which must be trimmed after forging. In conventional hammer forging the stock to be forged is heated to temperatures in the range of 1150 to 1300°C. The heated stock is manipulated through a die with a series of 3 to 4 impressions with multiple hammer blows in each impression for a total of 10 to 15 blows. The success of conventional forging depends very much on the skill of the hammerman. Hammer forging a P/M preform would make it possible to lower the forging temperature, to replace the several impressions of the hammer forging die by a single impression and the multiblows by a single blow.

The following is primarily concerned with the second approach to powder forging and includes discussion of preform design, of the individual steps involved, i.e. choice of powder, compacting, sintering, forging, and of the resulting properties.

I. Preform Design

A principal problem in this type of powder forging is the design of the preform. Its deformation during forging should lead to rapid densification and at the same time avoid strains which cause fracture. Kuhn and his associates[8] have developed a theory of deformation processing porous materials which has been of considerable help in the design of preforms and has made it possible to avoid some of the pitfalls involved in design merely by trial and error. A detailed presentation of the theory[11] and, in particular, the development of a yield criterion during plastic flow of a porous body, paralleling the classical yield criterion of the von Mises theory would lead too far. However, Kuhn's use of axial compression (upsetting) tests will be presented to illustrate two important aspects in deformation processing of porous materials, the plastic Poisson's ratio and a fracture criterion.

When a cylinder is compressed axially, it is strained in compression in the axial direction and strained in tension in the vertical direction. The ratio of the transverse strain to the axial strain is Poisson's ratio for plastic deformation. For a fully dense material this ratio is 0.5, which can readily be shown as a result of the fact that the volume of the cylinder remains constant. During compressive deformation of a porous cylinder some material flows into the pores, the volume decreases and the density increases. The ratio of expansion of the cylinder in the radial direction to its reduction in height, Poisson's ratio for plastic deformation for porous bodies, will be less than the value of 0.5 for solid bodies. Densification and lateral flow occur simultaneously rather than in sequence. From experiments on frictionless plastic deformation of porous bodies both at room and at elevated temperature, Kuhn[8] derived a relationship between Poisson's ratio ν for plastic deformation and ρ, the relative density of the porous body:

$$\nu = 0.5\, \rho^{a}$$

The exponent a was found to be equal to 1.91 for cold deformation of sintered bodies from iron powder and 2.0 for deformation at 700° F for sintered bodies from the aluminum alloy 601 AB.

Axial compression of a cylinder, the upset test, may also be used to develop a fracture criterion for porous compacts. During this test, friction at the top and bottom surfaces of the cylinder where the die contacts the cylinder retards radial outflow and leads to barreling, as illustrated in Figure 18-6. The greater the friction because of absense of lubrication and the greater the aspect ratio, i.e. the ratio of height to diameter, the more will the circumferential free surface bulge. This bulging causes a secondary tensile stress in the circumferential direction, as seen in Figure 18-6. The more the surface bulges, the greater will be the circumferential tensile stress. One type of fracture observed in upset forging, free surface fracture, is a direct consequence of these tensile circumferential stresses. Figure 18-7

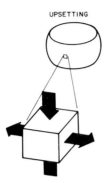

FIGURE 18-6 Circumferential tensile stress and axial compressive stress at the bulge surface of cylinders compressed axially with high contact friction.

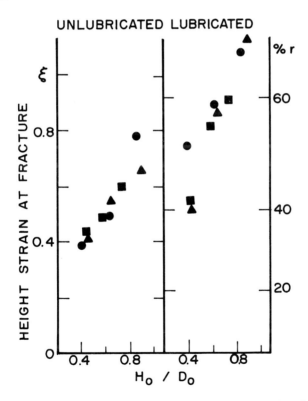

FIGURE 18-7 Overall height strain at fracture for hot upsetting of 601 AB aluminum alloy powder compacts, sintered at 700° F (370° C); ■ 76%, ● 88%, ▲ 93% relative density.

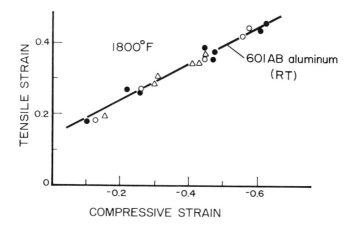

FIGURE 18-8 Locus of surface strain at fracture during upsetting at 1800° F (980° C) of sintered 4620 alloy steel powder cylinders.
● Induction sintered at 2350° F (1290° C) for 3 minutes
O Induction sintered at 2050° F (1120° C) for 3 minutes
Δ Conventionally sintered at 2050° F (1120° C) for 30 minutes; sintered cylinders of 601 AB aluminum alloy powder were upset at room temperature.

gives the height strain at fracture as a function of the aspect ratio of sintered aluminum alloy powder 601 AB compacts during hot upsetting. The figure indicates that the height strain at fracture is considerably larger for upsetting in a lubricated than in an unlubricated die and that it increases with increasing aspect ratio. Surprisingly the height strain at fracture appears to be little affected by original porosity of the compacts. This may be explained by two opposing effects which compensate each other. As initial pore volume increases, the ability of the material to withstand circumferential tensile stress decreases, but, at the same time, the Poisson's ratio for plastic deformation decreases with increasing porosity and therefore the amount of lateral spread also decreases.

The technique for establishing a fracture criterion consists of measuring the axial and circumferential strain on small grid marks at the equator of the bulge surface of the upset cylinder, as shown in Figure 18-6. The ratio of these strains during compression is affected by the friction conditions, the aspect ratio of the cylinder and also the temperature of the test. When the compressive strain is plotted vs the tensile strain, a straight line with a slope of one half is obtained. This is illustrated in Figure 18-8 in which the locus of surface strains at fracture during upsetting at 1800° F of 4620 low alloy

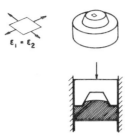

FIGURE 18-9   Schematic of hub extrusion, indicating tensile strain at top free surface.

steel powder compacts is plotted. In this experiment the method of sintering the cylinders prior to upsetting does not appear to affect the locus of fracture strains. It so happens that the locus of fracture strains during upsetting at room temperature of aluminum alloy 601 AB powder compacts fits on the same straight line.

The fracture strain loci can be considered as fracture criteria for evaluating the deformation to fracture in more complex deformation processes. Kuhn[8] suggests that "the progressing deformation strain paths in potential fracture regions of the process under consideration are first determined through plasticity analysis or measurements on a model material. These strain paths are then compared with the fracture locus of the material. Any strain path crossing the fracture locus before the deformation process is complete indicates that fracture is likely and alterations of the process may be utilized to change the strain paths so they do not cross the fracture locus."

Hot forging of preforms often involves not only upsetting, but also extrusion. In extrusion forging a hub, as illustrated in Figure 18-9, compression of the flange section involves radial flow inward and up into the hub section. The top surface of the hub is a free surface which undergoes bulging and tensile strains. Free surface fracture occurs on this surface when the strains reach the critical amount for fracture. These strains may be altered by changing the draft angle of the hub or by using a preform that partially fills the hub section of the die. Not only free surface fracture, but also contact surface fracture may occur during extrusion forging of a hub on a cylinder. Again Kuhn and associates[12] analysed the conditions under which such cracks will be found and how they may be eliminated by design changes. The problem of fracture during extrusion forging of a hub was approached in a somewhat different manner by Fischmeister and

FIGURE 18-10   Partially formed pinion gear showing cracks forming on the teeth.

associates.[13,14] A third type of crack formation during hot forging, which fortunately is much less frequent than the other types of fracture is internal fracture which cannot be detected by surface observation; it has also been investigated by Kuhn and associates.[12]

Kuhn[8] has applied his concept of workability analysis to the design of preforms for actual parts to be forged. Only one example will be mentioned here, which involves the forging of differential side pinion gears, the prototype of hot forging of preforms illustrated in Figure 18-3. The preform shown in this figure is flat-topped with a beveled bottom surface. When this preform is forged, cracks are developed in the partially formed gear during tooth formation, as shown in Figure 18-10. These cracks are closed up in the final stages of the forging stroke as the material is pressed against the die surface. However, oxidation of the crack surface and trapping of lubricant in the cracks may lead to structural weakness of the tooth surfaces. Kuhn showed that a modification of the preform contour shown in Figure 18-11 involving a slightly tapered top surface and a rounded instead of a beveled bottom surface will eliminate crack formation. Such a preform may be produced by isostatic compression.

II.   Powder Selection

The powder forging technique is used for structural parts with mechanical properties superior to those attainable with conventional P/M techniques. For this reason, mixtures of prealloyed atomized steel powders and graphite have been widely chosen as the raw material for preforms. The most common compositions are those containing nickel and molybdenum as alloying elements, such as compositions with 0.4% nickel and 0.6% molybdenum and those with 2% nickel and 0.5% molybdenum. The

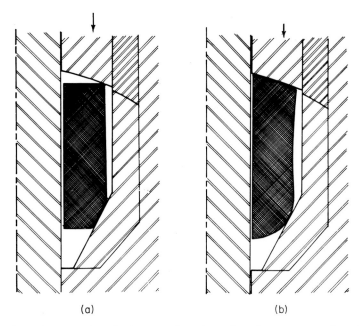

(a)   (b)

FIGURE 18-11 Longitudinal section view of preforms and tooling for forging pinion gear; (a) original preform which led to crack formation at teeth (Figure 18-10), (b) modified preform which led to defect-free gears.

advantage of these compositions is that they contain low contents of alloying elements, which have a tendency to oxidize, e.g. only 0.2-0.3% Mn and less than 0.1% Cr. The objection to the nickel-molybdenum compositions and particularly the 4600 composition (2% Ni, 0.5% Mo) has been that they are relatively expensive and may not have sufficiently high hardenability for applications which require combinations of high strength and toughness. Several approaches have been suggested to remedy this situation. One approach[15] has been to add copper powder to a mixture of a prealloyed atomized steel powder with 0.4% Ni and 0.6% Mo and of graphite powder in order to increase hardenability of the sinter forged steel. After compacting, sintering and forging specimens from mixtures with up to 2.1% copper, the forgings were shown to have hardenability much higher than those of the copper-free alloys, particularly when they were sintered at 1230°C in dry hydrogen. Another approach[16] has been to start with laboratory prepared atomized alloy powders, with manganese contents up to 1.4% and up to 1% of molybdenum, nickel and copper. Molybdenum had a beneficial effect on the hardenability of the P/M steels, but the effect of manganese was strongly dependent on the residual oxygen content of the steels. As was pointed out by Lindskog and Grek,[17] water atomized prealloyed powders inevitably show surface oxidation. During subsequent annealing in hydrogen-rich atmospheres such oxides as FeO and NiO with a moderate free energy of formation are reduced, but more stable oxides, like MnO and Cr_2O_3 remain virtually unchanged. In order to reduce the oxygen content of prealloyed powder compacts containing Mn and Cr to values under 1000 ppm sintering in very dry atmospheres at temperatures as high as 1300°C may be necessary. Residual oxides affect not only hardenability but also impact energy as will be discussed in the section on properties.

As was pointed out in chapter 17, structural parts are conventionally produced from mixtures of iron powder and the powders of the alloying ingredients. This approach when used for conventional P/M structural parts as well as for powder forgings has the great advantage of a much wider choice of composition than when prealloyed powders are used, but invariably leads to alloys which are inhomogeneous. The use of master alloys which produce a liquid phase during sintering has been repeatedly suggested as a method to hasten homogenization when preforms for powder forging are sintered. Among the master alloys are those of manganese and copper, as suggested by Fischmeister and Larsson,[18] master alloys of 10% of either nickel, molybdenum or manganese with approximately 5% carbon coated with a thin layer of copper, as suggested by Kaufman,[19] and masteralloys of ferromolybdenum containing 1, 4 or 5% carbon, as suggested by Y. E. Smith and H. Watanabe.[20] A requirement for rapid homogenization during sintering is wetting of the iron by the liquid, so that the iron particles are uniformly and completely coated. This reduces the distances over which diffusion must take place to about one half of the austenite grain size of the iron particles.

III. Compacting and Sintering

The same methods used for conventional compacting and sintering of P/M structural parts may be used in the fabrication of powder forgings. After sintering the preforms would be cooled in the cooling zone of the sintering furnace and later reheated to the forging temperature. As was pointed out in the previous section, powder forgings should have low oxygen contents, preferably in the range of a few hundred ppm, in order to give them good hardenability and good impact properties. Control of sintering temperature and time is necessary to obtain these low oxygen contents. On the other hand, when powder forgings requiring less stringent mechanical properties are to be produced, omitting a separate sintering step has been suggested. The green preforms are heated to the forging temperature and forged immediately. G. Bockstiegel[21] expressed the opinion that the concept of powder forging without a separate sintering step is "likely to be useful in the production of non-power transmitting automobile parts and under certain conditions perhaps also in the production of power transmitting parts." He recently described in detail experiences with an automatic powder forging line based on this principle.[22] It contains a heating furnace with two parallel muffles through which the preforms are transported by means of a pulling beam with pins at regular distances. The preforms are heated in this furnace to 1100° C in an atmosphere of nitrogen with controlled additions of endogas. Powder forging without a separate sintering step has also been advocated by Japanese authors, among them Morimoto et al.,[23] for transmission spur gears for motorcycles and by Nishino et al.,[24] who claim that forging of powder preforms without sintering gives mechanical properties comparable with those attained by the usual sinter forging method.

A variation of the compacting and sintering steps in powder forging would substitute dry bag isostatic pressing in place of compacting in rigid dies and sintering in a high frequency induction coil in place of conventional sintering. Dry bag isostatic pressing of preforms was discussed by Sellors;[25] high frequency induction sintering by Vernia.[26] A process for producing pinion gears using both of these steps was described by Eloff and Wilcox.[27] They sintered the isostatically pressed preforms in a 3 KHz induction coil in an atmosphere of nitrogen with 10% hydrogen. No preheating was necessary since the compact did not contain any lubricant. The sintering times were 1 to 6 minutes and the sintering temperatures either 2100 or 2350°F. These sintering treatments were sufficient to produce oxygen contents below 200 ppm in speciments of 4600 prealloyed steel with 0.2% carbon. After induction sintering the specimens were cooled to approximately 1850° F in the induction coil, removed from the coil and forged.

IV. Forging

Forging machines for powder forging have to fulfill certain requirements which were outlined by Huppmann and Hirschvogel:[4]

The force-displacement characteristics must match deformation characteristics of the preforms.

Workpiece-tool contact times should be as short as possible.

The machines shoud be stiff and the ram should have good guidance to obtain good tolerances in the powder forgings.

Mechanisms for ejection of the parts are necessary.

Because of these requirements forging hammers which do not have sufficiently accurate guidance and hydraulic presses which are too slow are inappropriate for powder forging. Instead mechanical presses, in particular crank presses with short, fast strokes and short contact times, are used for powder forging. This applies mainly to precision hot forging by upsetting to closely controlled dimensions without a flash. When the conventional forging approach is applied to forging P/M preforms, the same equipment as in conventional forging may be used. Flow stress and the force for forging preforms are initially lower than for conventional forging, but rise toward the end of the forging stroke, as density increases. Tooling for powder forging will be quite different from that for conventional forging and will more closely resemble tooling for powder compacting. On the other hand, the provisions necessary in tooling for powder compacting to obtain uniform density in multilevel parts will not be required. An example of a relatively simple tool design for powder forging is shown in Figure 18-12, in which a powder preform is upset during forging and ejected after forging.

Only limited data are available for tool life for forging tools for preform forgings. Since the preforms are heated in a protective atmosphere, tool wear due to scale formation is less of a problem then in conventional forging. Also the flow stress in preform forging is lower and no flash need be formed which adds to tool life. On the other hand, the dimensional tolerances in preform forging are tighter than in conventional forging, which means that tools, or at least those parts of the tooling subjected to wear, must be replaced more often. A tool life of 5 to 10,000 forgings for readily replaceable high-wear components of the tooling and of 10,000 to 20,000 forgings for other components has been quoted.[4]

Before a preform is forged it must be heated to the forging temperature. For preforms having low alloy steel compositions the forging temperature is in the same range as for conventional forgings, i.e. from 800 to 1200°C. With increasing temperature the flow stress decreases and the formability increases, but higher forging temperature generally means decreased tool life in spite of the lower force. Preforms may be heated in a continuous furnace or by high frequency induction heating, in both cases with protective atmosphere. In order to produce forgings to close dimensional tolerances, the forging temperature must be carefully controlled. This requires[22]

1. automatic temperature controls of the equipment in which the preforms are heated to the forging temperature,

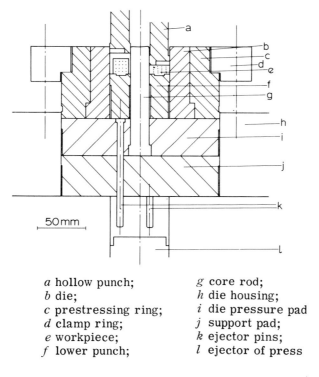

a hollow punch;	*g* core rod;
b die;	*h* die housing;
c prestressing ring;	*i* die pressure pad
d clamp ring;	*j* support pad;
e workpiece;	*k* ejector pins;
f lower punch;	*l* ejector of press

FIGURE 18-12 Example of powder forging tool construction.

2. automatic control of the temperature of the forging tools which are generally operated at an elevated temperature,
3. automated operation of the transfer unit by which the preheated preforms are transported from the heating equipment to the preheated die.

The pilot production system developed for forging differential side pinion gears, which includes elaborate temperature control equipment, was described by Halter.[6] Bockstiegel[21] has given examples of variations in dimensions and of out-of-roundness which may occur when the temperatures of preforms and of tooling are not carefully controlled.

Another important requirement in preform forging is lubrication. Systems in which the preform is lubricated by dipping it into or spraying on a colloidal suspension of graphite in a water carrier have been used successfully. They provide not only lubrication, but also protection against oxidation. On the other hand, spray lubrication of the forging tools has the

459

advantage of keeping the lubricant from penetrating into the structure of the porous preform.

One problem peculiar to preform forging is the avoidance of local porosity near the surface of the forging. Surfaces which are in contact with the diewalls may be chilled and may not be completely densified exhibiting surface porosity. Development of surface porosity may be overcome by rapid deformation rate, by minimum contact time between preform and tooling, by using relatively high forging temperatures and by providing adequate lubrication.

V. Mechanical Properties

Powder forgings are used for applications where they compete with wrought materials. Therefore, the design engineer will compare the mechanical properties of wrought and powder forged materials. One important difference in their properties is, that those of wrought materials tend to be anisotropic while those of powder forged materials are more isotropic. As was pointed out by Brown and Smith,[28] the anisotropy in wrought materials is due to the working operations by which ingots are transformed into wrought billets or bars. The working produces banding of interdendritically microsegregated alloying elements and elongation and microstringering of non-metallic inclusions. For this reason the mechanical properties and, in particular, the fatigue strength of wrought steels are considerably better in the longitudinal than in the transverse direction. The anisotropy is less pronounced in powder forgings as was shown in a comparison of fatigue curves for wrought and powder forged steels.

In chapter 17 on P/M structural parts the effect of porosity on mechanical properties was discussed. Porosity has a similar effect upon the properties of powder forgings as upon those of conventional P/M structural parts. The small amount of porosity in powder forgings, generally less than 2%, will decrease tensile and yield strength only moderately. Two percent porosity will lower these properties by approximately 20% in powder forgings of normalized low alloy steels, but the ductility will be lowered 50-60%, as was shown by Kaufman and Mocarski[29] in tests on normalized low alloy steel samples. On the other hand, even very small amounts of porosity have a strong effect upon the toughness of powder forgings. Figure 18-13 from the work of Bockstiegel and Blände[30] shows the effect of porosity on the Charpy V-notch impact energy of quenched and tempered powder forgings of a 4640 type material with a Vickers hardness level of 330-360. Even when the porosity is so low that it can no longer be detected metallographically, the impact energy is

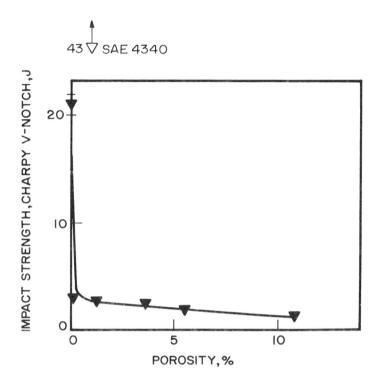

FIGURE 18-13 Impact energy of hot repressed and of conventional compacts of SAE 4340 steel composition, heat treated to hardness level HV$_5$ 320-350 as function of residual porosity.

only one half of that of the corresponding wrought material and a porosity of 0.2% further lowers the impact energy drastically.

In powder forgings from atomized steel powder the effect of inclusions is of principal interest. These inclusions may originate from slag entrapment from the melt or deterioration of melting furnace refractories or they may be present because oxides are not completely reduced during sintering of the preforms, particularly those of alloying ingredients, such as manganese or chromium. Bockstiegel and Blände[30] have investigated the influence of deliberately added slag particles upon the toughness of fully dense powder forgings of the same type as those used in their work on the effect of porosity. Figure 18-14 shows the effect of coarse and fine inclusions upon

461

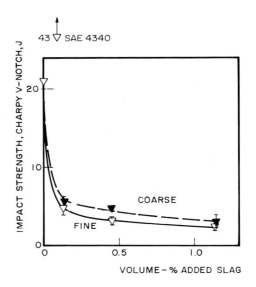

FIGURE 18-14 Impact energy of hot repressed compacts of SAE 4340 steel composition, heat treated to hardness level HV_5 320-350, as a function of varying volume percentages of added "coarse" and "fine" slag particles. Residual porosity less than 0.1%.

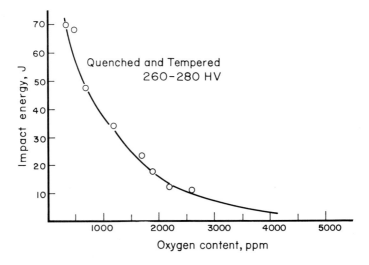

FIGURE 18-15 Room temperature Charpy V-notch impact energy of forged alloy powder specimens (0.29% Ni, 0.62% Mo, 0.58% Mn, 0.35% Cr, 0.4% C), quenched and tempered to Vickers hardness 260-280 as a function of residual oxygen content.

the Charpy V-notch impact energy of quenched and tempered powder forgings with a Vickers hardness level of 320-350.

Figure 18-15 from the work of Lindskog and Grek[17] illustrates the effect of oxide content upon impact energy. In this figure the impact energy of dense powder forgings with a composition of 0.58% Mn, 0.62% Mo, 0.29% Ni and 0.4% C, quenched and tempered to a Vickers hardness of 260-280 is plotted vs. their oxide content, which is undoubtedly due to the presence of manganese oxide. The oxide content can be drastically lowered by sintering the preforms at high temperatures in a highly reducing atmosphere.

Not only porosity and non-metallic inclusions affect the impact energy of powder forgings, but also the amount of lateral flow during the forging operation according to the investigation of Kuhn and Downey.[31] In order to obtain impact energy values approaching or exceeding those of conventional wrought material, height strains on the order of 60-80% during hot upsetting are necessary.

In more recent investigations of the toughness of powder preforms not only the Charpy impact energy, but also the fracture toughness (K_{Ic} values)[32-33] and data on the dependence of toughness on the forging direction.[33 34] have been determined.

In the discussion of the fatigue strength of P/M structural parts the possible effect of pores acting as cracks and contributing to crack growth was included. Crack propagation may be the reason for the wide variability of the endurance limit of conventional P/M structural parts of higher tensile strength. In the paper in which C. Razim[35] discussed this effect he also pointed out that the endurance limit of fully dense powder forgings is not only much higher, but also less variable than that of porous structural parts. Recent values, published by T. Krantz et al,[16] for the fatigue ratio, i.e. the ratio of fatigue strength to ultimate tensile strength of powder forged low alloy steels with 0.3% combined carbon, quenched and tempered to a hardness of R_c 30-34 were in the range from 0.37 for an 0.8% Mn steel to 0.50 for a type 4630 steel. The authors did not find any correlation between the oxygen content of the powder forged steels and their fatigue ratio, but the possibility of such a relationship is still debated. Investigations of Ferguson et al.[36] indicate that the amount of lateral flow in powder forging has a strong effect upon the endurance limit of the forging in axial fatigue tests, i.e. cycling between a low and a high tensile stress. The endurance limit increases with increasing lateral flow. On the other hand, in rotating bend fatigue tests, i.e. cycling between a tensile and a compressive stress, lateral flow during forging does not effect the fatigue limit. They explain this difference by an analysis of internal stress concentration of collapsed pores in the powder forgings.

VI. Cold Forming of P/M Preforms

A process in which blanks cut from steel bar stock are cold formed by upsetting and by forward and backward extrusion has been of particular interest to the automotive industry. It is a fast process for producing stronger parts to close dimensional tolerances with little or no scrap loss. In this respect it resembles the process of hot forging P/M preforms and competes with it as will be discussed in chapter 19.

The growth of the technique of cold forming parts from wrought steel blanks has been delayed in the United States because the proper design of tooling for cold forming is still somewhat of an art and the requirements regarding surface finish, cleanliness and control of microstructure of the bar stock are quite rigid.

In the first part of this chapter the hot forging of parts from sintered P/M preforms instead of from wrought steel blanks was discussed. Similarly, parts may be cold formed from sintered P/M preforms instead of from wrought steel blanks. The principal advantages of sintered P/M preforms over wrought steel blanks in cold forming parts are the critical material factors mentioned:

Good surface finish, cleanliness and control of microstructure together with the facts that the weight of the preforms can be controlled to ± ½ percent, the squareness of preform end surfaces with respect to the diameter can be held to 0.001 cm per cm of diameter and the length of the preforms to ± 1 mm.

On the other hand, an understanding of deformation processing characteristics of porous materials is even more critical in cold forming than in hot forging of P/M preforms, since preforms will fracture at lower strains during cold forming than during hot forging. Cold forming of preforms has been repeatedly suggested and investigated in the laboratory.[37, 38] A number of automotive parts, e.g. diesel engine tappets and automotive valve caps and also certain soft magnetic parts are being produced from P/M preforms by cold forming.[39] As in cold forming generally, the extrusion punch face pressures are high, on the order of 150 tons/sq.in. For ferrous powder preforms, cemented carbide lined dies, solid carbide punches and guide sleeves are needed and blanks must be given a zinc phosphate soap conversion coating for lubrication. The densities of cold formed ferrous parts from P/M preforms are 7.5g/cm^3 or higher.

Applying the orbital forging process developed in Poland to P/M preforms has also been investigated.[40]

Cold forming of preforms is not limited to those with steel compositons.

The cold forming of sintered preforms from copper powder for components in electrical and plumbing applications has been investigated in pilot plant operation. Detailed data on the fabrication and mechanical properties of aluminum alloy parts cold formed from P/M preforms in the as-formed and the heat treated conditions has been published.[41] Data on the fatigue strength of cold formed aluminum alloy P/M connecting rods is included. The tests show that these rods have better fatigue limits than those from the diecast alloy usually used for the application. Cold formed aluminum alloy P/M parts would be used to take advantage of the characteristics of aluminum and its alloys, light weight, corrosion resistance, electrical and thermal conductivity. Cold formed parts would take the place of castings, extrusions and screw machine parts whenever the manufacturing economics favor the cold forming process and the good mechanical properties of cold formed parts, as-formed or as heat treated, are needed.

References

1. R. P. Koehring, Some experiments in hot forging of iron powder briquettes, in "Powder Metallurgy," ed. by J. Wulff, Cleveland, (1942), p. 304-309.
2. A. H. Allen, Steel, May 26, 1941, p. 74.
3. Nine papers on "Forging of P/M preforms," in Mod. Dev. in Powder Metallurgy, vol. 4, ed. by H. H. Hausner, N.Y., (1971), p. 369-473.
4. W. J. Huppmann and M. Hirschvogel, Powder forging, Int. Metals Reviews, No. 233, Vol. 23, No. 5, p. 209-239, (1978).
5. S. Altemeyer, Density/Economy relationships of P/M parts, Machine Design, June 29, 1972.
6. R. F. Halter, Pilot production system for hot forging P/M preforms, Mod. Dev. in Powder Metallurgy, vol. 4, ed. by H. H. Hausner, N.Y., (1971), p. 385-394.
7. G. Lusa, Differential gear by P/M hot forging, ibid., p. 425-430.
8. H. A. Kuhn, Deformation processing of sintered powder materials, ch. 4, p. 99-138 in "Powder Metallurgy Processing," ed. by H. A. Kuhn and A. Lawley, Academic Press, N.Y., (1978).
9. R. A. Huseby and M. A. Scheil, Forgings from P/M preforms, Mod. Dev. in Powder Metallurgy, vol. 4, ed. by H. H. Hausner, N.Y., p. 395-417, (1971).
10. R. H. Hoefs, Manufacturing and economic aspects of hammer forging of P/M preforms, Mod. Dev. in Powder Metallurgy, vol. 7, ed. by H. H. Hausner and W. E. Smith, Princeton, p. 457-484, (1974).
11. H. A. Kuhn and C. L. Downey, Deformation characteristics and plasticity theory of sintered powder metal materials, Int. J. of Powder Metallurgy, vol. 7, No. 1, p. 15-25, (1971).
12. S. K. Suh and H. A. Kuhn, Three fracture modes and their prevention in

forming P/M preforms, Mod. Dev. in Powder Metallurgy, vol. 9, ed. by H. H. Hausner and P. W. Taubenblat, Princeton, (1977), p. 407-426.

13. G. Sjöberg, Material flow and cracking in powder forging, Powder Metallurgy Int., vol. 7, No. 1, p. 30-33, (1975).

14. H. Fischmeister, G. Sjoberg, B. O. Elfström, K. Hamberg, and V. Mironov, Preform ductility and transient cracking in powder forging, Mod. Dev. in Powder Metallurgy, vol. 9, ed. by H. H. Hausner and P. W. Taubenblat, Princeton, (1977), p. 437-448.

15. S. Mocarski and D. W. Hall, Properties of hot formed Mo-Ni-Mn P/M steels with admixed copper, Mod. Dev. in Powder Metallurgy, Vol. 9, ed. by H. H. Hausner and P. W. Taubenblat, Princeton, (1977), p. 467-489.

16. T. Krantz, J. C. Farge, and P. Chollet, Hardenability and mechanical properties of hot forged Mn-Mo steels made from prealloyed powders, Mod. Dev. in Powder Metallurgy, vol. 10, ed. by H. H. Hausner and P. W. Taubenblat, Princeton, (1977), p. 15-41.

17. P. Lindskog and S. E. Grek, Reduction of oxide inclusions in powder preforms prior to hot forging. Mod. Dev. in Powder Metallurgy, vol. 7, ed. by H. H. Hausner and W. E. Smith, Princeton, (1974), p. 285-301.

18. H. F. Fischmeister and L. E. Larsson, Fast diffusional alloying for powder forging using a liquid phase, Powder Metallurgy, vol. 17, No. 33, p. 227-240, (1974).

19. S. M. Kaufman, The use of master alloys for producing low alloy P/M steels, Mod. Dev. in Powder Metallurgy, vol. 10, ed. by H. H. Hausner and P. W. Taubenblat, Princeton, (1977), p. 1-13.

20. Y. E. Smith and H. Watanabe, Premix ferromolybdenum in alloy sintered compacts with the aid of a liquid phase, Mod. Dev. in Powder Metallurgy, vol. 9, ed. by H. H. Hausner and P. W. Taubenblat, Princeton, (1977), p. 277-300.

21. G. Bockstiegel, Some technical and economic aspects of P/M hot forging, Mod. Dev. in Powder Metallurgy, vol 7, ed. by H. H. Hausner and W. E. Smith, Princeton, (1974), p. 91-127.

22. G. Bockstiegel, Experiences with an automatic powder forging line, Powder Metallurgy Int., vol. 10, No. 4, p. 176-180, (1978).

23. K. Morimoto, K. Oyata, T. Yamamura, T. Yukawa, N. Yamada, and N. Sekiguchi, Transmission spur gear by powder forging, Mod. Dev. in Poder Metallurgy, vol. 7, ed. by H. H. Hausner and W. E. Smith, Princeton, (1974), p. 323-339.

24. Y. Nishino, H. Nishie, K. Obara, H. Doi and M. E. Offerhaus, The elimination of the sintering steps in P/M hot forging process, Mod. Dev. in Powder Metallurgy, vol. 9, ed. by H. H. Hausner and P. W. Taubenblat, Princeton, (1977), p. 509-523.

25. R. G. Sellors, Factors affecting isostatic pressing of ferrous preforms, Powder Metallurgy, vol. 13, No. 26, p. 85-99, (1970).

26. P. Vernia, Short cycle sintering by induction heating, Mod. Dev. in Powder Metallurgy, vol. 4, ed. by H. H. Hausner, N.Y., (1971), p 479-486.

27. P. C. Eloff and L. E. Wilcox, Fatigue behavior of hot formed powder

differential pinions, Mod. Dev. in Powder Metallurgy, vol. 7, ed. by H. H. Hausner and W. E. Smith, Princeton, (1974), p. 213-233.

28. G. T. Brown and T. B. Smith, The relevance of traditional materials specifications to powder metal products, Mod. Dev. in Powder Metallurgy, vol. 7, ed. by H. H. Hausner and W. E. Smith, Princeton, (1974), p. 9-31.

29. S. M. Kaufman and S. Mocarski, The effects of small amounts of residual porosity on the mechanical properties of P/M forgings, Int. J. of Powder Metallurgy, vol. 7, No. 3, p. 19-30, (1971).

30. G. Bockstiegel and C. A. Blände, The influence of slag inclusions and pores on impact strength of powder forged iron and steel, Powder Metallurgy Int., vol. 8, No. 4, p. 155-160, (1976).

31. H. A. Kuhn and C. L. Downey, How flow and fracture affect designs of preforms for powder forgings, Int. J. of Powder Metallurgy, vol. 10, No. 1, p. 59-66, (1974).

32. R. J. Dower and W. E. Campbell. The toughness of P/M forgings as a function of processing route, Mod. Dev. in Powder Metallurgy, vol. 10, ed. by H. H. Hausner and P. W. Taubenblat, Princeton, p. 53-71, (1977).

33. W. J. Bratina, W. F. Fossen, R. D. Hollingsberry and R. M. Pilliar, Anisotropic properties of powder forged ferrous systems, ibid. p. 157-169.

34. P. Hanejko, Mechanical property isotropy of P/M hot formed material, ibid., p. 73-88.

35. C. Razim, Fatigue behavior of sintered ferrous materials with particular reference to their use in automobiles, ZwF, vol. 73, p. 451-456, (1978).

36. B. L. Ferguson, H. A. Kuhn and A. Lawley, Fatigue of iron-base P/M forgings, Mod. Dev. in Powder Metallurgy, vol. 9, ed. by H. H. Hausner and P. W. Taubenblat, Princeton, (1977), p. 51-75.

37. R. J. Dower and G. I. Miles, The production of mild steel rings by a combined powder metallurgy and cold forming process, Powder Metallurgy, vol. 19, No. 3, p. 141-152, (1976).

38. R. J. Dower and G. I. Miles, The cold forging of two automobile components from sintered powder preforms, 5th European Symp. on Powder Metallurgy, Stockholm, May, (1978), Preprints vol. 1, p. 59-64.

39. G. M. Waller and S. W. McGee, Cold formed parts from P/M preforms, Contribution to an ASM conference, Cleveland, O., April, (1972).

40. J. Kotschy, M. Lewicka, Z. Leszczyniak and W. Rutkowski, Rocking die method for cold forming P/M preforms, Mod. Dev. in Powder Metallurgy, vol. 9, ed. by H. H. Hausner and P. W. Taubenblat, Princeton, (1977), p. 427-436.

41. K. E. Buchovecky and H. L. Hurst, Fabrication and properties of cold formed aluminum P/M parts, Mod. Dev. in Powder Metallurgy, vol. 6, ed. by H. H. Hausner and W. E. Smith, Princeton, (1974), p. 189-207.

19

Applications of P/M Structural Parts and Powder Forgings Their Economics and the Energy Requirements to Produce Them

In the first section of this chapter the method of producing structural parts by powder metallurgy for automotive and other applications is contrasted with competing methods. Examples of structural parts designed for powder metallurgy fabrication are shown. In the second section methods of cost analysis for powder metallurgy structural parts are discussed and specific examples of comparative costs by powder metallurgy and by competing methods are presented. The third section is concerned with the energy requirements for producing powder metallurgy structural parts. Examples of powder forgings produced at present and the criteria for potential applications of powder forgings are shown in section four, while the economics of producing parts as powder forgings are discussed in the last section.

I. Applications and Economics of P/M Structural Parts

P/M structural parts found their first applications in automobiles.[1] Even today over half of the parts produced are used in the automotive industry,[2] but their use has rapidly spread into the fields of household appliances, farm and garden equipment,[3,4] business machines,[5] power tools[6] and many miscellaneous applications. P/M structural parts have replaced primarily cast iron castings and parts machined from steel forgings or bar stock or stamped from sheet stock. Powder metallurgy parts are produced to close tolerances at high production rates at relatively low labor costs. Little or no loss of material as machining scrap is involved compared with the competing methods.

An important aspect in the replacement of small sand or permanent mold castings by P/M structural parts have been changes in the structure of foundry technology in recent years. In many industries, particularly the farm equipment industry, foundries have been a part of larger plants. Some of these foundries have been closed because of environmental factors or the lack of skilled labor, others have been automated for producing large castings. Small precision sand or permanent mold castings have become less readily available.

Production on automatic screw machines is generally the lowest cost method of machining for components designed for this type of fabrication. With this method a percentage of the bar stock ends up as chips. Other components such as gears cannot be finished on screw machines, but subsequent hobbing or milling operations are necessary, which improve the competitive position of producing the parts as P/M structural parts. Even with the recent advances in gear production, powder metallurgy fabrication is often lower in cost. Specific examples will be discussed in a later section of this chapter.

Many parts, particularly small levers and gears, have been converted from stampings to powder metallurgy because in powder metallurgy fabrication such relatively expensive finishing operations as edge shaving are not necessary.

In comparing the cost of powder metallurgy fabrication with these competing methods, the cost of iron powder, the raw material for P/M structural parts in relation to the raw material for castings or machined or stamped steel parts, i.e. cast iron, bar or sheet stock, plays a decisive role. As Anderson and Winquist[7] pointed out, the cost of iron powder compared to the index price of steel decreased in the years between 1966 and 1971 so that conventional powder metallurgy parts became less costly than parts from steel forgings or machined stock in high volume production, provided that the required physical properties are attained. This competitive advantage still applies.

Parts produced by other methods of fabrication such as investment casting, die casting and cold forming have been less frequently replaced by P/M structural parts. Costs of investment castings are higher than those of P/M structural parts, but powder metallurgy can usually not match the variety of materials and the complexity of shape of this method of casting. Most die castings are produced from zinc, aluminum and magnesium alloys and cannot be readily replaced by ferrous P/M structural parts. Aluminum powder metallurgy parts have been substituted in a number of cases for aluminum die castings because of the better dimensional tolerances and ductility of powder metallurgy parts. Recent advances in cold forming have extended this method of fabrication, at least potentially, to many shapes where the process competes with P/M structural parts. A direct comparison of costs for specific components processed by powder metallurgy and by cold forming will be discussed in a later section of this chapter.

The discussion of P/M structural parts has so far centered on applications in which parts originally produced by some other method of fabrication are replaced by P/M structural parts. Such replacements were the rule in the early years of development of structural parts. These were also parts which were not highly stressed, such as the oil pump gears, crankshaft sprockets, ball joint suspension bearings, windshield wiper gears and cams and shock absorber components in automobiles. There are still many applications for structural parts, which are replacements for parts designed for fabrication by other methods, but more and more parts are now designed as P/M structural parts, in which full use is made of the advantages of this type of fabrication. This change in the character of applications started with the introduction of automatic transmissions and of power steering in many U. S. cars. One of the first examples, where the design of an automotive mechanism was based on the use of powder metallurgy parts, was the pump for the Ford power steering mechanism. The parts include the rotor, both upper and lower plate, actuator and cam insert, shown in Figure 19-1. Winquist[2] pointed out that "the plates are of a very unusual design and could not possibly be economically manufactured by any other process."

Recent examples of P/M structural parts which were designed for powder metallurgy fabrication rather than as replacements for assemblies and parts produced by other fabrication methods are shown in Figures 19-2 and 19-3. Figure 19-2 shows a complex conveyor labyrinth seal wheel

FIGURE 19-1 Components for pump of Ford automobile power steering mechanism including rotor, upper and lower plate, actuator and cam insert (Courtesy: Ford Motor Co.).

FIGURE 19-2 Conveyor labyrinth seal assembly consisting of a journal and outer seal cap, an inner seal cap and a wheel.

FIGURE 19-3   Commutator directing oil in the steering mechanism of off-
road and farm equipment.

assembly, consisting of a journal and outer seal cap, an inner seal cap and a
wheel. The parts have a combined weight of 11 lbs. 2 ounces. They are made
from copper steel with the journal and outer seal cap and the inner seal cap
from a material corresponding to the MPIF designation FC-0208-R (see
table 17-3) with 0.6 to 0.9% C, 1.0 to 3.0% Cu and a density in the range of
6.4 to 6.8 g/cm³. The material of the wheel contains an additional 2% Cu for
increased strength. The outer face of the wheel is case hardened for
increased wear resistance. The wheel was designed with mating ribs and
grooves with an 0.015" clearance between them, which is subsequently filled
with grease. The close dimensional control of this intricate design is
possible only through fabrication as a P/M structural part assembly.

Figure 19-3 represents a commutator which directs oil in a steering
mechanism in off-road and farm equipment. It is produced from two parts,
a cap and a commutator body. The pieces are compacted, presintered,
assembled, resintered and infiltrated with copper and, at the same time,
joined into one strong complex shape. The material in the infiltrated part
corresponds to MPIF designation FX-2008-T (see table 17-15) with 0.6 to
1.0% C, 15 to 25% Cu and a density in the range of 7.2 to 7.6 g/cm³.

II. Methods of Cost Analysis for Structural Parts and Comparative Costs for Producing Them

Methods of cost analysis for structural parts by powder metallurgy have been discussed repeatedly. D. S. Anderson and L. A. Winquist[7] have given an outline of the American method of cost analysis, in which the total cost is taken as the sum of

> material cost per part
> direct labor cost per part
> direct and indirect overhead and
> administrative cost, including profit.

In this analysis, the direct overhead includes indirect labor, tool maintenance cost and supplies, while indirect overhead is composed of depreciation allowances, salaried employees and service department costs. The direct and indirect overhead is included in total cost as a percentage of direct labor cost. The administrative costs include sales and marketing costs, organizational administrative expenses and profits. They are taken as a percentage of the sum of material cost, labor cost and direct and indirect overhead.

In his article on powder metallurgy, G. Zapf[8] describes his system of cost analysis for P/M structural parts, based on his experience as general manager of the largest German parts producing plant. It is quite detailed and differs markedly from the American system. Zapf includes the following items in his analysis:

> Raw material cost
> Cost of shaping, i.e. pressing and sizing costs
> Cost of sintering
> Tool costs
> Set-up costs
> Costs for the risk of producing scrap or having to rework parts and for quality assurance

Percentages for development and administrative costs and for profit are included in each of his items. In Zapf's raw material cost all costs necessary to produce a mix ready to feed into the press including the pertinent overhead costs are contained. In order to arrive at the cost of pressing, Zapf determines a "sales price" per hour for the use of the press needed for the part. This "sales price" comprises fixed costs, primarily depreciation of the press, variable machine costs including energy, repairs, etc. and variable labor cost including a certain overhead of the direct labor cost. Similarly, the cost of sintering is calculated on the basis of the cost per hour of furnace time used. The cost per piece is derived from the hourly costs of using presses and furnaces on the basis of the number of pieces pressed or sintered per hour. Tool costs and set-up costs depend on the complexity of the part and on the number of parts upon which these costs are based. The last item, costs for the risk of producing scrap and having to rework parts and for quality assurance, according to Zapf, amounts to 9 to 16% of the sum of the first five cost items. In the tables of his article Zapf presents data on the changes in costs, in particular raw material and shaping costs, during the years 1968 to 1978.

Two comparisons of the costs of producing parts by powder metallurgy and by competing methods have been published. One of them, by McGee and Burgess,[9] compares the cost of machining and of P/M fabrication for five parts, several of them gears, the other by H. G. Schuhbauer,[10] the cost of coldforming and of P/M fabrication for several types of shapes. McGee and Burgess emphasize in their study that cost effectiveness of P/M applications is based on optimum tooling methods and on market acceptance of a new method of fabrication, in addition to producing a functional, reliable part at a lower cost. The analyses for three of the parts which have found market acceptance are briefly discussed. The first is the type of automotive oil pump gear which was the initial P/M structural part produced. It is interesting to note that the requirements on mechanical properties of this part have increased considerably since 1940 when the P/M gear replaced one machined from cast iron. In the study, a gear machined from SAE 1045 bar stock and subsequently hardened is compared with a P/M gear from MPIF FC-0208-S material (2% Cu, 0.8% C, 7.0 g/cm^3 density) and also hardened. The comparison, which gives cost percentages for the raw material and the individual operations for producing both the machined and the P/M gear, shows that the P/M gear can be produced at a cost 68.27% lower than that of the machined gear. Not surprisingly, 80% of the 36 million automotive and truck oil pump gears produced annually in the U. S. are made by powder metallurgy.

A second example is a fluid power pump rotor. A high density, 7.2 g/cm^3, compact produced by pressing, presintering, repressing and final sintering from a high carbon nickel steel compositon hardened after sintering is compared with a rotor machined from 86L20 material carburized and hardened. Grinding after hardening is required for both the machined and the P/M rotor. Nevertheless, there is a 57% cost advantage of the P/M over the machined rotor. In spite of its high density the mechanical properties of the P/M rotor are marginal for certain applications. Its estimated use is 40% of all rotors produced.

The third example is a spur bevel differential pinion produced from a copper infiltrated high carbon steel compact, hardened and ground after infiltration. The competing machined gear from SAE 4620 steel is carburized and hardened. The gears are used in garden tractors. The P/M gears are quite satisfactory for tractors in the 6 to 12 HP class, but marginal with respect to impact and fatigue resistance for the 12 to 20 HP classes of tractors. The cost advantage is 26% and about 40% of the gears are produced by P/M.

Schuhbauer's cost comparisons are based on estimates at a plant producing both cold formed and P/M parts. The comparison is limited to the cost of parts having dimensions which allow production by both techniques. For each part relative material, forming, tool and scrap costs are included in the estimates; sales, development and administrative costs are not included. According to the estimates certain shapes, such as a can with a cruciform inside produced by backward extrusion, a flanged tube produced by forward extrusion and a cold formed gear preform could be produced at lower cost by powder metallurgy. The cost comparisons are based merely on the shape of the parts and do not take into account that cold formed parts have, in general, better mechanical properties than P/M parts. For certain applications the shapes produced by P/M may therefore not be adequate in mechanical properties.

III. Energy Requirements

While the supply of energy available to industry appeared to be virtually unlimited until a few years ago, the situation has drastically changed since the middle 1970's. Actual energy shortages have occurred and greater potential shortages in the future have been forecast. For this reason, the energy requirements for the P/M industry and, in particular, those for producing P/M structural parts have been critically assessed in several recent papers by Bocchini[11] in Italy, Kaufman[12] in the United States and Zapf[8] in Germany. These requirements are compared with those for parts produced conventionally, i.e. by machining from bar stock or from forgings.

All three authors give figures for the energy requirements of producing one ton of steel and one ton of iron powder (sponge iron, water atomized or both). The wide variation in these figures for powder - Zapf's 10 GJ (10^9 joules) per (metric) ton, Bocchini's 22 GJ/ton and Kaufman's 22 to 23 GJ/ton - and for finished steel - Zapf's 11.6 to 16.0 GJ/ton, Bocchini's 15 to 18 GJ/ton and Kaufman's 42 GJ/ton - are apparently due to the fact that the calculations are based on different premises. Even the question of whether producing 1 ton of steel or 1 ton of iron powder requires more

energy is answered differently, less energy according to Zapf and Kaufman, more energy according to Bocchini.

However, the most important saving in energy by P/M production of
parts, regardless of the relative energy requirements for producing powder
and producing steel, are due to the following reasons:

    1.   For each part to be produced less powder is needed than bar
stock because the gross weight of bar stock for each part includes the
percentage which becomes chips. The net weight of a part from powder is
almost the same as the gross weight of powder used.

    2.   The weight of the final piece machined from bar stock is
greater than that of the piece produced from powder because the density of
the sintered part is less than the solid density of the material.

Both Zapf and Bocchini present detailed data for energy requirements
for producing certain parts from powder and from bar stock or forgings.
Zapf includes only the energy requirements for the necessary operations to
produce the piece from bar stock or forgings, such as turning, milling,
broaching, grinding, deburring and in the case of forgings, rough and final
forging. In the case of P/M the operations include pressing, sintering,
repressing, resintering and for some of the parts turning, grinding and
deburring. Bocchini also includes the energy requirements for producing
the raw material, the amount of bar stock or of powder.

Bocchini's ratios of the energy required for the sintered part to that for
the machined parts for the 5 parts for which he makes calculations range
from 0.59 to 1.12. Zapf shows lower energy requirements for all the six
parts studied when produced from powder ranging from 41 to 60% of the
energy needed for production by machining.

Among the energy requirements for the various operations in powder
metallurgy production those for sintering are high. They could be lowered
considerably by changes in the present sintering practice. This was brought
out in papers by Burgess[13] and Tamalet[14] and is also discussed by Bocchini[11]
and Kaufman.[12] Possible changes would be:

      Shorter sintering times by high frequency induction or by infrared
        radiation sintering
      Decreased atmosphere consumption
      Use atmospheric nitrogen in place of partially combusted hydro-
        carbons as sintering atmospheres
      Make use of the heat content of the atmosphere exhausted from
        the furnace
      Sintering in vacuum instead of an atmosphere
      Redesign furnaces with lower mass and more efficient insulation.

Preform   Final Forging

FIGURE 19-4   Preform and final forging of the lower roller clutch cam for General Motors automatic transmission. (Courtesy: Federal Mogul Corporation)

FIGURE 19-5   Powder Forgings; inner and outer race of ball bearings at right. (Courtesty of Federal Mogul Corporation)

IV. Present and Potential Applications of Powder Forgings

Examples of powder forgings are shown in Figures 19-4, 19-5 and 19-6. Figure 19-4 shows both the preform and the final forging of a lower roller

FIGURE 19-6   Parts forged from P/M preforms from the United States,
Germany, Sweden and Japan.

clutch cam for an autombile automatic transmission produced from 4600
prealloyed steel powders. Figure 19-5 represents on the left various
transmission parts and on the right an inner and outer race for a ball
bearing. These parts are produced in the United States. Figure 19-6 shows a
group of forged parts made in the United States, Germany, Sweden and
Japan, indicating the range of size and the variety of shapes which can be
produced by the technique.

Pietrocini[15] has critically reviewed applications for powder forgings
which hold promise for the future, noting the conditions of stress under
which the parts reviewed operate. The review includes:

Gearing, such as the Receppa joint, bevel gears in agricultural use,
drive sprockets, sliding clutch components and sector gears

Hydraulic components, such as hydraulic pistons, Gerotor pump
components, hydraulic couplings, valve lifter bodies, cam rings for power
steering pumps

Magnetic parts, such as alternator rotor poles

Fasteners, such as valve spring retainers

Bearing components, such as taper roller bearing cones.

Bockstiegel[16] has shown, in the form of a schematic drawing, examples of
European passenger car parts which he considers potentially suitable for
P/M hot forging. Of particular interest is the use of powder forgings for

FIGURE 19-7   Connecting rod for Porsche Model 928 automobile forged from P/M preforms.

automobile engine connecting rods which has been repeatedly discussed.[17, 18] The connecting rod for the Porsche Model 928 automobile is being produced by powder forging in England. Figure 19-7 shows this part.

V.   The Economics of Producing Parts as Powder Forgings

In spite or perhaps because of the relatively slow development of powder forging as a production technique, several economic analyses and discussions of the future potential of the process have been published.[16, 19, 20, 21, 22] Some of the analyses, e.g. the ones by Bockstiegel[16] and by Hoefs,[20] present detailed comparative cost figures for certain parts produced by conventional and by powder forging.

Among the parts analyzed by Bockstiegel[16] is the differential ring gear, which is produced by one U. S. automobile manufacturer as a powder forging. Bockstiegel establishes savings of 25% when the gears are produced by powder forging rather than by conventional open die forging and machining. In order to make the discussion of comparative costs more general, Bockstiegel suggests a formula for the conditions under which powder forgings can compete favorably with conventional forgings:

The powder price $\times$ the powder weight $+$ P/M hot forging cost $-$ saved machining cost should be smaller than the price of the billet per kg (for conventional forging) × slug weight + conventional forging cost.

Other analyses, e.g. the one by Huppmann and Hirschvogel,[21] attempt a more general economic analysis of powder forging comparing it with competing processes, i.e. conventional P/M processing, die forging, cold forming and precision casting. The authors come to the conclusion that the principal competition of powder forging is cold forming.

Those who assess the future of powder forging as a production technique emphasize two points, the cost of powder with respect to that of steel and the necessity of writing off a large capital investment when a part now being produced by conventional methods is to be replaced by a powder forging.

References

1. F. V. Lenel, Oil pump gears, in "Powder Metallurgy" ed. by J. Wulff, Cleveland, p. 502-511, (1942).
2. L. A. Winquist, Automotive applications of standard P/M parts, Mod. Dev. in Powder Metallurgy, vol. 6, ed. by H. H. Hausner and W. E. Smith, Princeton, p. 87-100, (1974).
3. T. L. Burkland, Use of P/M parts in the farm equipment industry, ibid., p. 37-57.
4. N. L. Ward, P/M applications in garden equipment, Mod. Dev. in Powder Metallurgy, vol. 11, ed. by H. H. Hausner and P. W. Taubenblat, Princeton, p. 197-209, (1977).
5. P. V. Schneider, Applications of powder metal parts on business machines, ibid., p. 181-196.
6. M. Feir, The use of structural powder metal parts in power tools, Mod. Dev. in Powder Metallurgy, vol. 6, ed. by H. H. Hausner and W. E. Smith, Princeton, p. 59-69, (1974).
7. D. S. Anderson and L. A. Winquist, The economics of powder metallurgy, Int. J. of Powder Metallurgy, vol. 9, No. 2, p. 111-121, (1973).
8. G. Zapf, Pulvermetallurgie in "Handbuch der Fertigungstechnik," Carl Hanser Verlag, Munich, in print.
9. S. W. McGee and F. K. Burgess, Identifying cost effective powder metallurgy applications, Mod. Dev. in Powder Metallurgy, vol. 11, ed. by H. H. Hausner and P. W. Taubenblat, Princeton, p. 143-162, (1977).
10. H. G. Schuhbauer, Technological and economic comparison of sintered and cold extruded parts, Powder Metallurgy, 3rd European P/M symposium conference supplement, part II, p. 469-489, (1971).
11. G. F. Bocchini, Comparison of energy consumption between conventional technologies and powder metallurgy, Powder Metallurgy Int., vol. 10, No. 2, p. 78-82, (1978)

12. S. M. Kaufman, Energy consumption in the manufacture of precision metal parts from iron powder, Int. J. of Powder Metallurgy, vol. 15, No. 1, p. 9-20, (1979).

13. W. C. Burgess, Sintering furnace and development, Powder Metallurgy Int., vol. 8, No. 1, p. 28-32, (1978).

14. M. Tamalet, How to decrease the energy demand of powder metallurgy furnaces, Powder Metallurgy Int., vol. 9, No. 3, p. 119-126, (1977).

15. T. W. Pietrocini, Hot formed P/M applications, Mod. Dev. in Powder Metallurgy, vol. 7, ed. by H. H. Hausner and W. E. Smith, Princeton, p. 395-410, (1974).

16. G. Bockstiegel, Some technical and economic aspects of P/M hot forming, ibid., p. 91-135.

17. S. Corso and C. Downey, Preform design for P/M hot formed connecting rods, Powder Metallurgy Int., vol. 8, No. 4, p. 170-173, (1976).

18. C. Tsumuki, I. Niimi, K. Hasimoto, T. Suzuki, T. Inukai, K. Ushitani and O. Yoshihara, Connecting rods by P/M hot forging, Mod. Dev. in Powder Metallurgy, vol. 7, ed. by H. H. Hausner and W. E. Smith, Princeton, p. 385-394, (1974).

19. J. W. Wisher and P. K. Jones, The economics of powder forging relative to competing processes - present and future, ibid., p. 33-50.

20. R. H. Hoefs, Manufacturing and economic aspects of hammer forging P/M preforms, ibid., p. 457-484.

21. W. J. Huppmann and M. Hirschvogel, Powder Forgings, Int. Metals Reviews, vol. 23, No. 5, p. 209-239, (1978).

22. L. W. Lindsay-Carl, The possibility of using powder forgings in highly stressed automobile components, Powder Metallurgy, 3rd European P/M symposium conference supplement, Part II, p. 517-529, (1971).

20

Cermets

The term "cermet" is being used for heterogeneous bodies which consist of an intimate mixture of "ceramic" and metallic components. "Ceramic" components in this sense are inorganic materials with high melting points and include, besides the conventional silicate ceramics, oxides, nitrides, carbides, borides, silicides and also carbon in the form of graphite or of diamonds. The term was first employed for materials intended for use at elevated temperatures having a combination of high strength and of thermal and mechanical shock resistance. These materials which, at present, are of only limited use for high temperature applications have been briefly described in an appendix to chapter 16 on cemented carbides. In the sense given above, the term "cermets" includes, of course, all of the cemented carbides discussed in chapter 16, although the term did not yet exist when cemented tungsten carbides were developed. There are, however, "cermets" which are neither cemented carbides nor high temperature materials. Three groups of these "cermets" will be discussed in this chapter. They are:

> Metallic friction materials
> Dispersion type fuel elements and control rods for nuclear
> > reactors
> Metal bonded diamond tool materials

Other materials to which the term "cermets" may be applied are oxide dispersion strengthened alloys to be discussed in chapter 21, certain contact materials, such as silver-graphite and silver-cadmium oxide and the so-called metallic brushes for motors and generators combining copper and copper alloys with graphite. The contacts and brushes will be discussed in chapter 24.

I. Metallic Friction Materials

Metallic friction materials[1,2] include brake linings and clutch facings produced from mixtures of metal powders and friction producing ceramic ingredients by compacting and sintering. They should be distinguished from conventional, so-called "organic" friction materials, consisting of asbestos linings bonded with an organic resin and the recently developed so-called "semi-metallic" friction materials in which metal powders, friction producing ingredients and organic resins are compounded. Neither of these latter two types of materials are powder metallurgy products. The principal advantage of metallic friction materials over the other two types is their ability to absorb energy at a higher rate and have higher wear resistance. They can withstand higher temperatures and have higher thermal conductivity; they also have a more constant coefficient of friction over a range of temperatures and pressures and are less affected by heat, cold, oil, grease, salt water or fungi. Because of their higher cost they are not yet used in automobiles, but in heavy duty applications for aircraft and tanks as brake linings and on tractors, heavy trucks, busses, heavy earth-moving equipment, heavy presses, etc., as clutch facings. There are two principal types of applications for metal powder friction materials: "dry" application such as airplane brakes and standard clutches and "wet" applications in oil immersed clutches in powershift transmission, brakes, etc.

Metallic friction materials consist of a dispersion of a friction producing ingredient in a metallic matrix. Originally, a copper-tin alloy was used as metallic matrix in which case copper powder, tin powder, the friction producing ingredient in powder form, and other ingredients, which modify the frictional behavior, are mixed, compacted and sintered. This matrix material is still widely used, but in the last ten years copper-zinc matrix materials have been developed mainly for oil immersed applications. The copper-zinc materials have the ability to maintain a strong, yet porous, matrix which retains oil. The more porous Cu-Zn materials have higher friction coefficients and higher energy absorption capacity than Cu-Sn materials.

The original method of producing metallic friction materials was developed by S. K. Wellman in the 1930's. It is widely used particularly for "wet" type friction materials and is distinguished by fine particle size grit of the friction producing material. Later on, a different method of producing friction materials was developed by Bendix Corporation, in which a much coarser grit of friction producing material is used. This method is primarily used for dry clutch applications and aircraft "buttons." Originally the methods differed both in the way the friction materials were fabricated and attached to their backing and in their compositon. However, with the original patents for friction compositions expired and with the numerous modifications in composition developed for specific applications, the original sharp differentiation in the composition of the two types of friction materials has gradually disappeared.

In the original Wellman compositions the friction producing ingredient is very fine silica powder. Less than 10% of silica is added to the mixture of metal powders. Iron powder, lead powder and graphite powder are used to modify the frictional behavior. A typical composition would be: 60-75% Cu, 5-10% Sn, 5-10% Fe, 5-15% Pb, 2-7% SiO_2, 5-10% graphite. The iron is believed to increase friction and exert a scouring action to retard seizing, the lead to form a lubricating film to prevent "grabby action", the graphite to give smooth engagement and enhance the frictional characteristics by breaking up the metallic matrix.

A typical composition for a material for wet powershift transmission with a Cu-Zn metallic matrix is 60-70% Cu, 5-10% Zn, 0.4% Sn, 5-1% graphite and 2-7% less than 200 mesh grit consisting of SiO_2, Al_2O_3, etc. for producing friction.

Wellman's method of manufacturing metallic friction materials is quite unique[3] and based on the facts that the linings generally have a low and uniform section thickness, large surface area, low green strength and need to be bonded to a steel backing. The linings often referred to as "cookies" are compacted at pressures of 138 to 414 MPa (20 to 60,000 psi). The sintering of the "cookies" is done simultaneously with bonding them to the steel backing. This is a sort of hot pressing operation at low pressures of 0.7 to 2.8 MPa (100 to 400 psi). This low pressure is mainly intended to insure perfect contact between the "cookie" and the steel backing during sintering and bonding. A stack is built upon a base; the stacking consists of alternate layers of the copper plated steel backing plates and the green cookies. To keep them from sticking, the cookies are coated on one side with graphite. The stack is covered by a "furnace can," inside of which a reducing atmosphere is maintained. The top of the can rests on the stack, its open bottom rests on an oil or water seal, which surrounds the base upon which the stack is built. A bell type furnace is lowered over stack and can. Through the top of the furnace the stack is subjected to pressure by means of an air diaphragm connected to a compressed air supply. Figure 20-1 is a diagrammatic cross sectional view of the furnace showing the stacked disks with the bell-type heating unit. The stack is heated to the sintering temperature of 540-980°C (1000-1800°F) in a protective atmosphere, such as natural gas, and held for 30 to 60 minutes. The lower sintering temperature is used where retention of steel core hardness is required. After

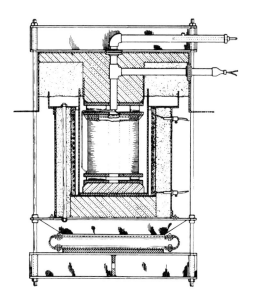

FIGURE 20-1 Schematic cross-sectional view of sintering furnace for friction materials showing the arrangement of stacked disks covered by can, bell-type heating unit and the diaphragm applying pressure to the stack.

sintering, the bell furnace is removed and the stack allowed to cool within the can. While the stack cools, the bell furnace can be lowered over another furnace base with its stack and can on it. The friction elements are finished by such operations as grinding parallel and flat, forming to shape, drilling and counter boring of rivet holes and notching. Disks for oil immersed applications are grooved to provide a path for cooling oil. Type and size of grooves are tailored to the requirements of the application. An assembly of friction facings produced by the Wellman method for dry clutch applications is shown in Figure 20-2. It includes "buttons" for aircraft brakes and an assembly of the buttons attached to a backing. Figure 20-3 shows bronze powershift disks for a military tank.

In the Bendix compositions the friction producing ingredient is commonly "mullite," an aluminum silicate produced by calcining the mineral "kyanite." The particle size of mullite used is much coarser than Wellman's silica. Mullite particles are friable and therefore shatter during brake or clutch applications, rather than rounding off. This makes mullite a good friction producing ingredient.

The amount of mullite and also of graphite added to the metal powders

FIGURE 20-2 Assembly of friction facings for dry clutch applications (Courtesy: S. K. Wellman, Inc.).

depends upon the application. In aircraft brakes the amounts of mullite and graphite are high and may be as much as 50% by weight. In clutches, the amount of non-metallic ingredients is considerably lower; as much as 85% by weight of the mixture may be metal powders. Because of the large amount of non-metallic materials in the mixture, the green strength and also the sintered strength of the compacts is low. One technique developed by Bendix[4] to provide the necessary strength for operation is to insert the compacts into steel cups. The cups are attached to the brake or clutch backing plate. In friction material for airplane brakes, the compacts are pressed ("preformed") at a low pressure, inserted in the steel cups and simultaneously repressed ("compacted") and crimped into the steel cup. The steel cups are spindled to keep them flat and then sintered. After sintering, the compacts are coined to densify the friction materials and to dove-tail the steel cup walls into the friction material. In friction materials for heavy duty clutches ("clutch buttons"), the compacts are sintered before being inserted into steel cups. They are repressed and the cup walls simultaneously crimped into the material. Alternatively, the friction material may be brazed into the cup coincident with sintering prior to repressing and cup wall crimping. Ten percent by volume of thin steel fibers

FIGURE 20-3 Bronze power shift friction disks for military tank (Courtesy: S. K. Wellman, Inc.).

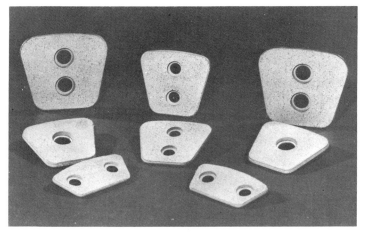

FIGURE 20-4 Buttons for airplane brake in steel cups (Courtesy: Bendix Corp., Friction Materials Div.).

of uniform length may be added to the powder mixture. While the buttons shown in Figure 20-2 have no side wall, buttons in which the friction material is held in steel cups are shown in Figure 20-4. A plate with the buttons attached to it by riveting is shown in Figure 20-5.

FIGURE 20-5 Plate with airplane brake buttons riveted to it (Courtesy: Bendix Corp., Friction Materials Div.).

II. Dispersion Type Nuclear Fuel Elements

The cores of nuclear fuel elements contain a fissionable compound, i.e. a compound of the metals uranium or plutonium. Because of the instability of metallic uranium during fission, fuel elements made of compounds, very often the compound UO_2, are generally used. Even with these compounds the cores of nuclear fuel elements undergo radiation damage during use, which limits the life of the fuel elements and the percentage of the available uranium which has undergone fission (total burn-up) before the fuel elements have to be taken out of service. One method of minimizing the amount of radiation damage in the cores of nuclear fuel elements is to disperse the fissile compound, commonly UO_2, in a metal matrix using powder metallurgy methods to produce the fuel elements.[5] There are two types of radiation damage, i.e. dynamic damage due to the motion of fission fragments and static damage due to the effects of fission fragments as

489

impurities in the lattice. The range over which radiation damage occurs in the metallic matrix of the dispersion type fuel element depends upon the distance over which the fragments travel from the surface of the fissile particles into the metallic matrix. In order to minimize damage, the distribution of the dispersed phase should be uniform, the size of the dispersed phase particles should be uniform and large because with large particles the volume percentage of the undamaged matrix metal will be the largest and the matrix can play its role as structural material which gives the fuel element its strength. The dispersed phase should, however, have a particle size not larger than 1/4 to 1/3 the minimum cross section of the finished fuel element. The volume percentage of undamaged matrix will, of course, also depend upon the total relative volume percentage of matrix and dispersed phase. This ratio will be primarily governed by nuclear design considerations, but is also limited by the size of the fissile particles and the ductility of the matrix; generally 35 volume % is the maximum of fissile material included in the fuel element core. The uranium in the dispersed phase of dispersion type nuclear fuel elements is generally enriched. The choice of the matrix metal for dispersion type nuclear fuel elements is based upon considerations of minimum capture cross section for neutrons, good heat conductivity and strength properties. The best known matrix metals are aluminum and stainless steel. The method of producing dispersion type nuclear fuel elements with an aluminum matrix widely used in test and research reactors will be briefly described.[6,7] For these fuel elements, -325 mesh aluminum powder and UO_2 particles in the range from 45 to $105\,\mu$m are blended, the mixture is pressed into compacts 5 x 6 x 0.6 cm under a pressure of 350 MPa (50,000 psi). The green compacts are assembled into a "picture frame" in which the UO_2 containing compact is surrounded by pure aluminum on all four sides and on top and bottom, as shown in Figure 20-6. The composite fuel plate is hot rolled at 590°C from the starting thickness to 0.18 cm in six passes. The plates are then annealed and finally cold rolled to the final thickness and assembled into a fuel assembly by brazing. Hot rolling of the fuel elements has been shown to minimize forming stringers of the dispersed phase, which are undesirable.

The techniques for producing nuclear fuel elements with a stainless steel matrix are generally similar to those with an aluminum matrix. The UO_2 should be uniform in size, density and chemistry with equiaxed, preferably spherical particle shape. Stainless steel powders of many types, in particular those with 302, 304, 347 and 316 compositions have been used. Instead of prealloyed powders, mixtures of elemental powders have also been used to produce fuel elements. The mixture is pressed to high green densities, preferably 90% of theoretical, since this gives less stringering of

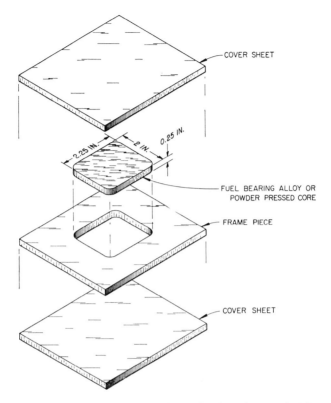

COVER SHEET

0.25 IN.

2.25 IN.

2 IN.

FUEL BEARING ALLOY OR
POWDER PRESSED CORE

FRAME PIECE

COVER SHEET

FIGURE 20-6 Exploded view of "picture frame" for dispersion type fuel element
consisting of UO_2 in an aluminum matrix.

UO_2 during rolling. Both cold and hot compacting have been used. The
compacts are sintered in hydrogen at 1150 to 1250°C (2100 to 2300°F) and
rolled using the same picture frame technique as for aluminum matrix fuel
elements. By using rolling reductions of 20% per pass, good bonding is
assured. An alternate technique for producing fuel elements with a circular
cross section would be hot extrusion of hot or cold pressed compacts in a
stainless steel can.

Besides UO_2, other fissile compounds, such as uranium nitride and
uranium carbide, have been used in dispersion type fuel elements.

Considerable development work has been done on dispersion-type
nuclear fuel elements using refractory metals, such as tungsten,
molybdenum, chromium and niobium as matrixes. Such fuels are
applicable to very high temperature gas cooled reactors, to thermionic

converter reactors with fuel cladding at thermionic emitter temperatures (about 1800°C) and to nuclear rocket auxiliary power and propulsion. Special methods of fabrication, such as high energy compaction, hot isostatic pressing and vacuum hot pressing have been developed for these fuel elements.

Not only fissile materials, but also neutron absorbing materials may be dispersed in a metallic matrix. Such dispersions of neutron absorbing materials can be used as control elements in nuclear reactors.

III. Metal Bonded Diamond Tool Materials

The third type of "cermets" discussed in this chapter are metal bonded diamond tool materials used for grinding. Most grinding tools use either aluminum oxide or silicon carbide as refractories. For use as grinding wheels aluminum oxide and silicon carbide are generally bonded with other ceramics. Metals or alloys as bonds have been suggested, but are not used commercially. In order to grind very refractory materials, such as sandstone, granite, concrete, fused alumina, glass and certain cemented carbides, conventional grinding wheels based on alumina or silicon carbide are unsatisfactory; instead, wheels based on diamonds are used. Grinding wheels, saws and cut-off wheels are produced from diamond powder held in a matrix of either resin or metal. Those with a metal matrix are produced by the methods of powder metallurgy.

The grit size of the diamonds used in metal bonded diamond wheels depends on the particular application. Unfortunately, very fine diamond powder, which is lowest in cost, has not been found satisfactory. Both natural and man-made diamonds are used.

A number of alloys are used as "bond" in metal bonded diamond tool materials. Probably the most common is a bronze high in tin with 15-20% tin, balance copper. This relatively brittle bond will allow the diamonds to do the cutting, while tougher bonds, e.g. the common 90 Cu - 10 Sn, have a tendency to glaze. Bronze-iron combinations and steel powders are also used as bond material. Cemented carbide grit is occasionally used as a "filler" in diamond grinding wheels.

Several methods are used for fabricating the wheels. The mixture of metal powders and diamonds may be cold pressed, often as a rim around a core consisting of steel or of bond material without diamonds, and then sintered.

Bond materials which have to be sintered at high temperatures where there is danger of chemical decomposition of the diamond into graphite may be hot pressed in resistance heated graphite molds. The short time necessary for this type of hot pressing minimizes the danger of decomposition. Hot pressing in graphite molds is primarily used for producing segments, which are then brazed to the rim of steel wheels.

Finally, hot pressing of metal powder-diamond mixtures in steel molds under higher pressure but at lower temperatures than in graphite molds is used.

References

1. B. T. Collins and C. P. Schneider, Sintered metal friction materials, Mod. Dev. in Powder Metallurgy, vol. 3, ed. by H. H. Hausner, N.Y., (1966), p. 160-165.
2. A. Jenkins, Powder metal based friction materials, Powder Metallurgy, vol. 12, No. 24, p. 503-518, (1969).
3. E. F. Cone, Sintered metal in friction devices. Metals and Alloys, vol. 14, No. 6, p. 843-850, (1941).
4. E. W. Drislane, Cerametallix clutch material, Society of Automotive Engineers, September 1965, Milwaukee Meeting, Publication No. 650683.
5. A. N. Holden, Dispersion fuel elements, Gordon and Breach, N.Y. (1967).
6. C. E. Weber and H. H. Hirsch, Dispersion type fuel elements, Proceedings First International Conference on Peaceful Uses of Atomic Energy, vol. 9, p. 196-202, (1953).
7. E. J. Boyle and J. E. Cunningham, MTR type fuel elements, ibid., p. 203-207.

21

Wrought Powder Metallurgy Products

"Wrought powder metallurgy products" is a convenient term for products in which the powder metallurgy process does not lead to a finished product with controlled dimensions, as in structural parts, selflubricating bearings and cemented carbide tool inserts. Many powder metallurgy products from refractory metals, which have been treated in chapter 14, belong into this category. As shown in that chapter, products from tungsten, molybdenum and tantalum are generally cold pressed and sintered and then worked. In contrast, other "wrought powder metallurgy products," those from beryllium, magnesium, aluminum and copper alloys, nickel and iron base supperalloys and special steels, as well as some from titanium alloys, are generally hot consolidated by the methods described in chapter 13. The result of the hot consolidation step are in many cases products which correspond to ingots produced by casting because they must undergo subsequent working steps, such as forging, rolling, extrusion and machining. In other cases, e.g. in hot isostatically pressed nickel base supperalloy turbine disks, the hot consolidated products have shapes near those of the final component. Since metals and alloys are generally more expensive in the form of powders than as castings or products wrought from castings, and since hot consolidation of powders is a high-cost process, there must be reasons why the "wrought powder metallurgy products" are produced from powder. In some cases, as in the turbine disks, fabrication from powder may mean a saving in expensive raw material.

Similar reasons apply to certain hot consolidated titanium alloy products, which have been discussed in chapter 14. However, more commonly, the "wrought powder metallurgy products" have properties, particularly mechanical properties, superior to those of conventionally produced products. The following wrought powder metallurgy products will be discussed with the reasons why they are made from powder:

1. Metals and alloys which can be produced in fine-grained form only by powder metallurgy. This applies particularly to beryllium, but is also of importance for certain magnesium alloys. Both beryllium and magnesium are hexagonal and exhibit pronounced plastic anisotropy.

2. Highly alloyed aluminum base and nickel base alloys and alloy steels produced by hot consolidation of prealloyed powders. These products may have superior properties because

   a. No macrosegregation which is often difficult to avoid in cast ingots of highly alloyed materials is encountered.

   b. The dispersion of carbides in the microstructure of steels and of second phase particles in that of non-ferrous alloys can be more readily controlled. This also may make extrusion and forging of hot consolidated powder compacts easier than that of cast ingots.

3. Dispersion strengthened materials in which a second phase having very little or no solubility in the matrix at the solidus temperature of the alloy is finely dispersed in the matrix.*

*The argument presented regarding the conomic reasons why wrought products are produced by powder metallurgy may not be valid in all cases. Recently a process for producing seamless stainless steel tubing by hot extruding preforms consisting of spherical stainless steel powder cold isostatically pressed in mild steel molds has been developed in Sweden[1] as described in chapter 13. Although the properties of this stainless steel tubing, which are briefly described in a appendix to section V of this chapter, are somewhat superior to the conventional product, this does not appear to be the main reason for the development of the process. It is probably related to the relatively poor yield in producing seamless stainless steel tubing conventionally.

I. Beryllium

Practically all beryllium is commercially produced by powder metallurgy since this used to be the only method of producing a fine grained product. A method of making cast ingot source beryllium relatively fine grained and non-textured by three-directional upset forging was recently discovered,[2] but it is still in the development stage. Although very pure beryllium single crystals oriented for basal slip can be extensively deformed, slip along basal planes generally interacts with barriers such as grain boundaries and causes the development of kink bands, according to a model suggested by Stroh,[3] which result in the nucleation and propagation of cracks without measurable elongation. On the other hand, beryllium single crystals which are plastically deformed along prismatic planes have a higher critical resolved shear stress, but deformation continues to much higher strains. Because of this plastic anisotropy cast coarse-grained beryllium is completely brittle. Fine grained polycrystalline beryllium produced from powder has a certain degree of ductility. Its strength increases with decreasing grain size.

The methods of producing beryllium powder were discussed in chapter 2. In order to get the best mechanical properties in P/M beryllium the grain size of the beryllium powder should be as small as possible and its purity as high as possible.[4] These requirements are difficult to combine because the disintegration into powder, the size classification of the powder and its tendency to react with the atmosphere are apt to introduce oxide impurities. The best results are obtained by impact attriting the beryllium swarf, obtained by machining, in an inert gas stream against a beryllium target and by classifying the attrited powder in cyclone separators, producing a powder which after hot pressing has grain sizes in the range of 8-12 μm for -44 mesh powder and 3.5-5μm for specially classified powders. Metallic impurities in the beryllium are also detrimental because, by forming eutectics of aluminum, silicon and magnesium which are liquid at the thot pressing temperature, they promote growth and agglomeration of the oxide particiles at the grainboundaries which causes a loss in ductility and impact strength.

Beryllium powder may be consolidated by loose powder sintering as described in chapter 12. Commerical purity beryllium is commonly produced by vacuum hot pressing in graphite molds[5] as described in chapter 13. Vacuum hot pressed beryllium, depending on the grade, has yield strengths of 200-250 MPa (30-36,000 psi) with 2-4% elongation or yield strengths of 400 MPa (60,000 psi) with 1-2% elongation.

Better mechanical properties have been obtained from high purity impact attrited powder which is cold isostatically pressed in evacuated elastomeric bags at 410 MPa (60,000 psi) and then hot isostatically pressed in evacuated and sealed steel cans at 105 MPa (15,000 psi) pressure and peak temperatures of 915 to 1160°C.[4] Typical mechanical properties of this hot isostatically pressed beryllium are:

Yield strength	280-460 MPa (41-66,000 psi)
Tensile Strength	470-600 MPa (67-87,000 psi)
Elongation	4-6½%

The higher strength values are obtained with the finer grain-size beryllium.

Hot pressed beryllium can be extruded and rolled. During extrusion the basal planes (0001) line up perpendicular to the extrusion direction, while the prism planes ($10\bar{1}0$) are in the extrusion direction. For this reason, the extrusions have good ductility in the extrusion direction (10% elongation with 900 MPa (130,000 psi) tensile strength and 550 MPa (80,000 psi) yield strength). Because of this preferred orientation during plastic working it is also possible by cross rolling to produce beryllium sheet which is relatively ductile in the plane of the sheet, but very brittle in the direction perpendicular to the plane of the sheet.

Beryllium is finding increasing use in aircraft applications, particularly jet engine parts. It is used in gyroscopes and other guidance instruments. Among the aerospace applications of beryllium were heat sinks for the Apollo space capsules. In nuclear applications the principal advantage of beryllium is its outstanding ability to slow down neutrons to thermal velocities and at the same time not react with the neutrons.

II. Magnesium Alloys[6,7]

Magnesium and magnesium alloys, which also have a hexagonal close packed structure, show another type of plastic anisotropy. When stressed in tension, the alloys deform by slip, but when stressed in compression they deform by twinning. For this reason, coarse grained magnesium alloys have a lower compressive yield strength than tensile yield strength. When the magnesium alloys are sufficiently fine-grained, they have about the same yield strength in tension and compression. Such fine grained alloys can be produced from coarse magnesium alloy powders by the method described in chapter 2. The magnesium alloy powders, or "alloy pellets," have particle sizes in the range from 70-450 μ m. As described in chapter 13, the powder is loaded directly into the container of an extrusion press, the ram is advanced and the pellets extruded through the extrusion die. The coarse pellets do not readily bridge and entrap air between particles. Any shape which can be conventionally extruded can be extruded from pellets. Pellet extrusions have a considerably finer grain size than extrusions from chill-cast magnesium alloy ingots and therefore a ratio of compressive to tensile yield strength near one. The effect of the smaller grain size upon the mechanical properties in the longitudinal direction of a powder extrusion as compared

Table 21-1   Properties of AZ31 Magnesium Alloy as Extruded from Billets and from Powder

Type Billet	Pct. Elong.	Tensile Yield Strength kpsi	Tensile Yield Strength MPa	Compressive Yield Strength kpsi	Compressive Yield Strength MPa	Tensile Strength kpsi	Tensile Strength MPa
Solution heat treated direct chill billet	14	20	138	12	83	35	241
As cast direct chill billet	19	25	172	21	145	39	269
Atomized powder	21	30	207	28	193	43	297

with conventional billet extrusions for AZ 31 magnesium alloy (3% Al, 1% Zn, 0.2% Mn) extruded from a 4″ container into a ½ x 3″ rectangle at 5 feet per minute at 700° F is shown in Table 21-1.

III.   Aluminum Alloys

Development work on aluminum alloys, hot consolidated from atomized prealloyed powder, goes back 20 years. The early work[8] showed that unusual tensile properties could be obtained in heat treated extrusions from alloy powders higher in zinc, magnesium and copper than the conventional 7000 series wrought alloys. As an example a tensile strength of 113,800 psi, a yield strength of 113,300 psi and an elongation of 4% were reported for a heat treated extrusion from a powder containing 9.7% Zn, 3.24% Mg, 1.98% Cu and 0.2% Cr. However, no data on resistance to stress corrosion cracking, one of the most important properties of aluminum alloys, were given. Recent extensive work[9,10,11] on 7000 series aluminum alloys hot consolidated from powder has been directed towards developing optimum processing conditions and optimum combinations of properties, including resistance to stress corrosion cracking and to exfoliation, high fatigue strength and good ductility and toughness both in the extrusion and in the transverse directions.

The compositions investigated correspond in general to the 7000 series of precipitation hardened aluminum alloy, but several of them had additions of iron and nickel (0.8% Fe and 0.8% Ni up to 3.25% Fe and 5.0% Ni) or of cobalt (0.75 to 1.6% Co). Such additions cannot be readily made in castings because of the appearance in the ingots of segregated coarse intermetallic compound phases which cause ingot cracking.

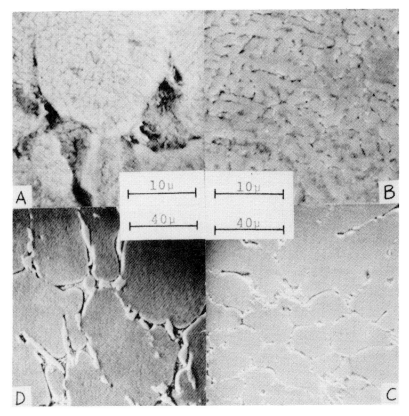

FIGURE 21-1 Microstructures of aluminum alloy (5.9% Zn, 2.6% Mg, 1.8% Cu, 0.8% Fe, 0.8% Ni, bal Al) from atomized powders and from ingots
A   25 μm powder, average cell size 0.9 μm
B   115 μm powder, average cell size 1.8 μm
C   6″ ingot, average cell size 15 μm
D   16″ ingot, average cell size 42 μm

The starting material for the powder metallurgy alloys are atomized powders with a particle size distribution depending on the atomizing conditions. The finer the powder, the faster is the solidification rate, which, in turn, determines the cell size in the microstructure of the powder particles. In Figure 21-1 the microstructures of powder particles and of cast ingots are shown, as they depend on average particle size and ingot size. The cell size in powder and ingot structures determines, in turn, the grain size and the inter-particle spacings in extrusions.

The steps in the fabrication of the powder metallurgy materials are:

Cold compacting in rigid dies or cold isostatic compacting to densities of 70-80% of theoretical.

Preheating, typically for one hour at 480°C (900°F), in flowing argon to remove the hydrogen. The hydrogen results from a reaction between aluminum and H_2O which is produced by the decomposition of hydrated aluminum oxides at the particle surfaces.

Instead of preheating in argon, the compacts may be encapsulated and preheated and hot-pressed in vacuum.

Hot compacting at pressures of 200 to 400 MPa (30-60,000 psi).

Hot working at 290 to 370°C (550-700°F). In extrusion the same die as in hot pressing may be used by replacing the dummy block with the extrusion die. For forging, compacts should be scalped after hot pressing.

Solution heat-treating and aging.

The principal advantage of the high-strength powder metallurgy alloys compared to conventional alloys is that they can be aged to considerably higher values of yield strength without losing their resistance to stress corrosion cracking. As an example, the powder metallurgy alloy MA67 with 8.0% Zn, 2.5% Mg, 1.0% Cu and 1.5% Co was compared with the conventional ingot metallurgy, "I/M", alloy 7075. In a stress corrosion test, in which samples of both the P/M and the I/M alloys were exposed to the industrial atmosphere of New Kensington, PA, for 4.1 years under a stress of 40,000 psi (276 MPa), the maximum transverse yield strength to which the I/M alloy could be heat-treated without stress corrosion cracking failure was 66,000 psi (455 MPa) while the P/M alloy could be heat-treated to 76,000 psi (523 MPa) without failure. The P/M alloys also have higher fatigue resistance than I/M alloys. Both stress corrosion cracking and fatigue resistance of the P/M alloys is related to their structure having finely divided second phase particles. The principal advantage of P/M alloys preheated and hot pressed in vacuum compared with those preheated in argon and hot pressed in air is their better transverse fracture toughness (higher ratio of notched tensile strength to yield strength).

Parts hot consolidated from aluminum alloy powder may not only have better properties than conventionally produced parts, but may sometimes also be lower in cost[12]. Preforms from aluminum alloy powder of the 7075 composition (1.6% Cu, 2.5% Mg, 6.0% Zn, balance Al) were cold isostatically pressed at 276 MPa to the shape of a howitzer component with a rib height to thickness ratio larger than 10:1 and a rib thickness of 5 mm. The preforms were sintered in nitrogen at 570°C and then isothermally forged in air at 400°C with a graphite-based lubricant. The properties of these forgings were equivalent to those of forgings produced conventionally, which had to be forged in two sets of dies and machined much more extensively than the powder forgings. A comparison showed the powder metallurgy process to be lower in cost because of a saving in raw material, the simplified forging procedure and the reduced amount of machining.

Potential applications of the high strength powder metallurgy aluminum alloys are in ordnance (receiver forgings in rifles not subject to exfoliation corrosion, cartridge cases with maximum combination of fracture toughness and yield strength) and in aerospace applications (for aircraft components subjected to predominantly compressive stresses).

IV. Powder Metallurgy Superalloys

The therm "superalloys" is applied to alloys of iron, nickel and cobalt which have high strength at temperatures of 600°C and above. They are of primary interest for components in jet aircraft engines and for aerospace applications. A large amount of work has gone into the development of superalloys by powder metallurgy. Two extensive reviews of this work have recently appeared.[13,14] The development work is concerned, on the one hand, with nickel-base superalloys having compositions near those of alloys produced by casting or by casting and working and, on the other hand, with dispersion strengthened alloys. In this section only the first group will be discussed, while the second group will be treated in the section on dispersion strengthened alloys.

The high temperature strength of nickel base superalloys is based on the presence of coherent precipitates, which are nickel-aluminum and nickel titanium intermetallic compounds, produced by solution treatment and aging. In addition to high temperature strength the alloys must have corrosion resistance, which requires sufficient chromium in the analysis. The advanced alloys developed to obtain balanced stress rupture and corrosion resistance exhibit in the cast condition gross segregation and structural inhomogeneity, which is a principal reason why these alloys are being produced by powder metallurgy methods. Segregation and inhomogeneity are then limited to the dimensions of individual powder particles. A principal application of powder metallurgy superalloys has been for turbine disks in jet engines. In this application stresses up to 480 Mpa (70,000 psi) and temperatures up to 760°C (1400°F) are encountered. Up to this temperature a relatively fine grained material has better strength than coarse grained material. When superalloys are needed for use at higher temperatures, alloys with a larger grain size are needed which can be obtained by thermo-mechanical treatments. The principal methods of producing supperalloy powders, argon atomization, vacuum atomization and the rotating electrode process, have been presented in chapter 2. The general aspects of hot consolidating powders have been discussed in chapter 13, but not the specific details applying to superalloy powders. As was pointed out, one method of fabricating jet engine components is by hot isostatic pressing of powder to "sonic shape." In order to obtain this goal, thermoplastic processing has been developed. It depends on the phenomenon that cold straining a superalloy powder greatly reduces the grain size upon recrystallization and softens the powder at elevated temperature. The spherical superalloy powder particles may be passed through a vertical feed rolling mill producing disk shaped cold strained particles. The rolling operation may be integrated into the fabrication process of producing turbine disks by hot isostatic pressing as indicated in Figure 21-2. Other methods of fabrication for turbine disks include forging steps. The so-called HIP + Forge route illustrated in Figure 21-3 includes hot isostatic pressing to a billet, followed by two forging steps. Because of the relatively fine grain size of the hot isostatically pressed billet, the loads necessary for forging these billets are much lower than for cast ingots. Another patented method, which involves the so-called "gatorizing" process,[15] is shown schematically in Figure 21-4. The atomized powder is consolidated by hot extrusion of the canned powder. The canned billets are

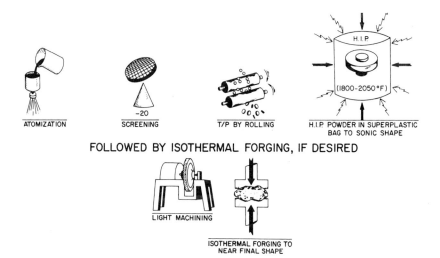

FIGURE 21-2 Schematic flow chart showing use of thermoplastic processing for producing turbine disks.

503

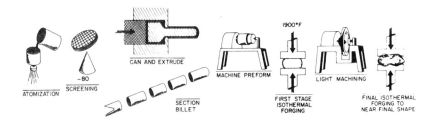

FIGURE 21-3 Schematic flow chart showing isothermal forging ("gatorizing") for producing turbine disks.

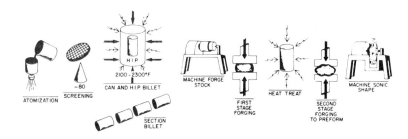

FIGURE 21-4 Schematic flow chart using hot isostatic pressing followed by forging for producing turbine disks.

heated to a temperature just below the recrystallization temperature of the alloy so that during extrusion recrystallization to a very fine grain size, less than 5 μm, takes place. This very fine grained material can be isothermally forged, i.e. "gatorized," at quite low loads to near final shape because it is super-plastic.

Producing superalloys by powder metallurgy involves certain problems. The powders have to be very low in both oxygen and nitrogen in order to avoid films on the powder particles interfering with bonding. Another problem is the formation of carbides with the formula MC on prior powder particle boundaries in the consolidated material, which has a deleterious effect on the forgeability and the mechanical properties of the alloys. The precipitation of these carbides may be minimized by controlling the carbon content of the alloys and by hot isostatic pressing at low temperature. The composition of several nickel-base powder metallurgy superalloys is

Table 21-2   P/M vs Cast and/or Wrought Compositions

	C	Cr	Ni	Co	Mo	W	Cb	Ti	Al	Ta	B	Zr
Astroloy												
C/W	.06	15	Bal.	15	5.25	—	—	3.5	4.4	—	.03	—
P/M	.03	15	Bal.	17	5.0	—	—	3.5	4.4	—	.03	.05
IN-100												
Cast	.18	10	Bal.	15	3.0	—	—	4.7	5.5	—	.01	.06
P/M	.06	10	Bal.	15	3.0	—	—	4.7	5.4	—	.01	.06
IN-792												
Cast	.21	12.7	Bal.	9	2	3.9	—	4.2	3.2	3.9	.02	.1
P/M	.04	12.5	Bal.	10	2	3.9	—	4.3	3.2	3.5	.02	.1
Rene 95												
C/W	.15	14	Bal.	8	3.5	3.5	3.5	2.5	3.5	—	.01	.03
P/M	.05	12.7	Bal.	8	3.5	3.5	3.5	2.5	3.5	—	.01	.03

compared with that of cast and/or wrought alloys in Table 21-2. The lower carbon content of the powder metallurgy alloys is evident. As in conventional alloys, the structure and the properties of powder metallurgy nickel-base superalloys may be improved by heat-treating and thermomechanical processing techniques, such as producing serrated grain interfaces and a necklace structure.

V.   Tool Steels

Tool steels are another example of applying the technique of producing an alloy powder by inert gas atomizing and hot consolidating the powder. The process was introduced in the U. S. by Crucible, Inc. in 1970[16] and at about the same time in Sweden by Stora Kopparberg and ASEA.[17] According to P.Hellman et al,[17] the fabrication of these steels involves the following steps: the spherical powder is charged into large cylindrical sheet-steel capsules which are vibrated. The capsules are first cold isostatically pressed at 400 MPa (58,000 psi) to increase the density of the powder and thereby its thermal conductivity so that it can be more readily hot isostatically pressed. Before being hot isostatically pressed the capsules are evacuated at temperatures of 400-500°C and then sealed. The powder is hot isostatically pressed at pressures of 100 MPa (14,500 psi) and temperatures of 1100°C

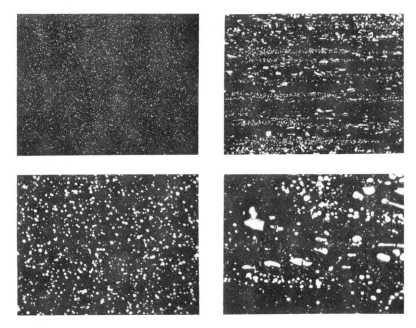

FIGURE 21-5 Longitudinal microstructures of T-15 tool steel; on the left; by hot isostatic pressing of powder on the right: by convention processing. Magnification: upper row 136X, lower row 543X (Courtesy: Crucible Inc., Specialty Metals Division).

(2000° F). Quite large compacts weighing up to 640 kg (1400 lbs) are produced.

By this technique, carbide segregation can be avoided and compacts with a uniform macro- and microstructure produced, having fine evenly sized and distributed carbide particles and a fine and uniform grain size. The size and distribution of the carbide particles is illustrated in Figure 21-5, in which the structures of 2″ diameter round hot rolled bars at 136X and 543X magnification produced from a conventionally cast ingot and from a P/M compact are compared with each other. The controlled microstructure provides improved impact and bend fracture strength, important in tools for intermittent cutting and improved grindability without impairing wear resistance. It also minimizes distortion of the material during heat treatment. The powder metallurgy technique permits modification in the composition of tool steels which could not be produced by conventional processing.[18] Two of these compositions are shown in table

Table 21-3   Compositions of CPM Rex 76 and CPM-10V Tool Steels

	CPM Rex 76	CPM-10V
C	1.50	2.40
Cr	3.75	5.3
V	3.0	9.8
W	10.0	0.3
Mn	—	0.5
Mo	5.25	1.3
Si	—	—
Co	9.0	—

21-3. The first of these CPM Rex 76 (Crucible Particle Metallurgy of Crucible, Inc., Specialty Metals Division, Colt Industries) is a cobalt rich high speed tool steel with exceptional hot hardness and wear resistance and greatly increased tool life in difficult cutting operations. It would be unforgeable if produced by conventional processing. The second composition CPM 10V (also of Crucible, Inc.) is obtained by producing a melt of the well known tool steel composition H-11. In this melt additional vanadium and carbon are dissolved before it is atomized. The atomized powder is hot isostatically pressed, hot worked and heat treated. After heat treatment, the microstructure of the material consists of small, spherical, uniformly distributed carbides in a matrix of a tough air-hardening medium alloy steel. It has an outstanding combination of wear resistance and toughness at temperatures up to 480°C (900°F). It has been successfully used for powder metallurgy compacting dies and in many other tool and die applications. In both cases, it replaced considerably more expensive cemented carbide tooling.

Another example of a tool steel composition which can be produced only by powder metallurgy techniques is the alloy NM 100 (Nuclear Metals Corp.)[19] whose composition and properties are shown in table 21-4. As cast

Table 21-4   Composition and Properties of NM-100 Alloys
Composition: 17.5 Cr, 10.5W, 9.5Co, 1.25C, 0.75V

Temp.		Yield Strength		Tensile Strength		Elong.
°F	°C	kpsi	MPa	kpsi	MPa	%
70	20	246	1700	277	1910	1.5
1000	538	182	1255	242	1670	3.7
1200	650	79.5	548	111	765	11.9

the alloy is unforgeable. However, when produced by canning and hot extruding a prealloyed powder, it can be forged and heat treated. It combines the wear, corrosion and oxidation resistance of a martensitic stainless steel and the resistance to softening of a tungsten bearing tool steel. It has found use for bearings in the hot section of gas turbines.

Appendix

As mentioned in the footnote to the introduction of this chapter, a process for hot extruding seamless stainless steel tubing from powder was developed in Sweden.[1] The sequence of operations in this process were described in chapter 13. Five grades of stainless steel were included in this development work, four of which were austenitic stainless steels, grades 316L, 316Ti, 304L and 304Ti and one a low interstitial ferritic grade (0.017% C, 17.2% Cr, 2.4% Mo, 0.63% Ti, 90 ppm N and 160 ppm O). The mechanical properties (yield strength, tensile strenght, elongation and Charpy impact energy) of the steels correspond to those of conventionally produced seamless stainless steel tubing. However, the ductile-brittle transition temperature, particularly for the ferritic steel, is much lower than for conventional steel, which is attributed to titanium carbonitride inclusions at the grain boundaries, which inhibit grain growth. Intergranular pitting and crevice corrosion of the steels produced from powder are equivalent or superior to conventional material.

VI. Dispersion Strengthened Alloys

The third group of wrought powder metallurgy alloys are dispersion strengthened alloys. The important characteristic of these alloys is high strength at elevated temperature. The strengthening mechanism is related to that in precipitation strengthened alloys, in which a submicroscopically fine dispersed second phase, whose crystal structure is generally coherent with that of the matrix, is dispersed in the matrix and produces the strengthening effect. However, the strength of precipitation strengthened alloys is limited with regard to the temperature range in which it applies. The dispersed alloy phase is produced by quenching from a temperature where the alloy is a single phase solid solution and then heating to an intermediate temperature to cause precipitation. When the alloy is further heated to a temperature above that at which precipitation took place, the precipitated phase will coarsen ("overaging of the alloy") and eventually redissolve. On the other hand, in dispersion strengthened alloys the dispersed phase should be stable up to the solidus temperature of the alloys. The mechanisms by which fine dispersions alter the yield strength, the work

508

hardening, the creep and the fracture behavior of an alloy have been studied in a large number of investigations.[20,21,22] High temperature strength properties, in particular creep rate, are affected by parameters related to dispersion geometry, such as particle spacing, and by those related to grain size and to grain shape (ratio of grain length to grain width, called grain aspect ratio). The results indicate that creep both by dislocation climb over second phase particles and by grain boundary sliding are important in explaining the properties of the alloys. A generally accepted model for creep in dispersion strengthened alloys has not yet been developed. For this reason, mechanisms of dispersion strengthening will not be discussed in spite of the fact, that powder metallurgy is by far the most important method of producing dispersion strengthened materials.

The dispersed phase in dispersion strengthened alloys is commonly an oxide, but may also be a stable intermetallic compound or even a pure metal. A fine dispersion of an oxide in a metal matrix can generally be produced only by powder metallurgy since it is very difficult to disperse solid oxide particles in a molten metal. The only fusion metallurgy method of producing an oxide dispersion is by internal oxidation. An alloy of a more noble matrix metal and a less noble alloying constituent is heated in an oxygen containing atmosphere whose oxygen pressure is low enough so that the matrix metal is not oxidized. The oxygen diffuses through the lattice of the matrix metal and reacts with the less noble alloying consitituent producing a fine dispersion of the oxide of this constituent in the matrix matal. The principal disadvantage of producing oxide dispersion strengthened alloys by internal oxidation is the fact that it depends upon the solubility of oxygen and its rate of diffusion through the matrix metal, which are generally low. Long times are therefore necessary to produce oxide dispersions of appreciable section thickness by this method. This difficulty may be overcome by internally oxidizing the alloy in powder form since in this case the distances over which the oxygen must diffuse are limited to the radii of the powder particles. The internally oxidized powder must be subsequently consolidated. This is, indeed, the method by which dispersion strengthened copper alloys are produced, as will be discussed later on in this section.

Dispersions of oxide in a metal matrix may, of course, be produced by mixing the matrix metal in powder form with a fine powder of the oxide and consolidating the mixture. However, very fine dispersions of oxide in the matrix with oxide particle sizes and spacings between oxide particles of not more than a few tens nm are necessary to obtain the desired strengthening effect. Both the size of the oxide particles and of the metal powder must be extremely fine to obtain these spacings. For this reason, conventional mixing of metal and oxide powders is not used for producing oxide dispersion strengthened alloys. Very small spacings between oxide particles have been achieved by a process called "mechanical alloying" which has been applied primarily to nickel base dispersion strengthened alloys, but also to iron and aluminum base alloys and will be discussed later in this section.

A. Dispersion Strengthened Aluminum[23,24]

The first of the dispersion strengthened alloys, SAP (sintered aluminum powder), was discovered more or less by chance in Switzerland.[25] It consists of a fine dispersion of aluminum oxide in an aluminum matrix. Its method of production is peculiar to this alloy system. The raw material for SAP is flake aluminum powder which has been produced from sheet aluminum in stamping mills for many decades. It is widely used for paints, printing inks and similar applications. Flake aluminum powder has a very low apparent density and it is therefore difficult to press. Aluminum flake powder may also be produced by milling atomized aluminum powder in ball mills. By suitable control of the milling process, in particular control of the stearic acid added to the powder during milling, the individual aluminum flakes produced in milling are permitted to weld together to produce a powder with a bulk density near 1 gm/cm³, which can be processed more readily than the very low density flake powder used for paints.

The flakes in the milled powder have thicknesses in the range from 1 μm down to 0.1 μm depending upon the details of the milling process. Each flake is covered on both sides by surface layers of aluminum oxide approximately 10 nm thick.

The process of consolidating the milled powder into a semifinished product is quite similar to that of hot consolidating aluminum alloy powder described in section III. The powder is cold compacted in the container of a vertical extrusion press under pressures of 90 to 200 MPa (13 to 29,000 psi). Several cold pressed compacts are wrapped in aluminum sheet, gradually heated to 600°C and maintained at this temperature. During the sintering the hydrated aluminum oxide on the surface of the powder particles gives off water vapor which reacts with the aluminum forming additional aluminum oxide and hydrogen. The sintered compacts are hot pressed at pressures up to 1000 MPa (145,000 psi) and then extruded. Impact extrusion, forging, drawing and hot and cold rolling are also possible, usually from material previously extruded.

The aluminum oxide layer on the original flake particles is broken up during working into submicroscopic particles finely dispersed in the

Table 21-5   Tensile Properties of two SAP Alloys

Oxide Content in Weight %	Test Temp °C	0.2% offset Yield Strength		Tensile Strength		Elong. % on 5 times diameter
		kpsi	MPa	kpsi	MPa	
5 - 8	20	23.5	162	36.3	250	17
(Grade M257)	427	10.8	74.5	12.9	89	5
	482	10.3	71	10.8	74.5	4
15 - 17	20	33.0	228	52.0	359	7
(Grade M286)	427	14.7	101	52.0	109	3
	482	11.9	82	12.8	88	3

aluminum matrix. Since the thickness of the original oxide layer on the flake particles is more or less constant, the amount of oxide in the final product depends primarily upon the thickness to which the flake particles are reduced during the milling process. Oxide dispersion strengthened aluminum with oxide contents from 1 to more than 15% by weight have been produced. The amount of oxide determines the spacing between oxide particles, which in turn determines the strength properties of the product.

Since the solubility of oxygen in aluminum up to the melting point of aluminum is extremely small, SAP alloys have remarkable strength up to this temperature. Also, in contrast to precipitation hardened aluminum alloys, their mechanical properties are not affected by prolonged soaking at elevated temperatures.

The short time tensile properties of two SAP alloys at a series of temperatures are shown in table 21-5.[24] Even more important than the short time tensile properties of SAP are their creep properties. They are shown in Figure 21-6[24] for the alloys SAP 865 (13.5 wght % Al_2O_3), SAP 895 (10.5 wght % Al_2O_3) and SAP 930 (7 wght % Al_2O_3). The times necessary to cause creep elongations of 0.1, 0.2 and 0.5% are plotted as a function of the stress.

In spite of their good strength properties at temperatures up to 500°C, not many applications of SAP have been found because the material is more expensive than other materials which have similar strength-weight ratios at the temperatures in question. It is being used as canning material for neclear fuel elements in organic moderated and cooled reactors.

B.   Dispersion Strengthened Lead

The high strength of oxide dispersion strengthened aluminum near the aluminum melting point has led to intensive research and development

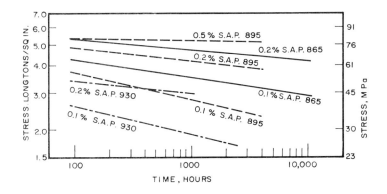

FIGURE 21-6 Creep properties of SAP 865 (13.5% Al_2O_3), 895 (10.5% Al_2O_3) and 930 (7% Al_2O_3) at 400°C. Stress for creep of 0.1, 0.2 and 0.5% as a function of time.

work on other oxide dispersion strengthened systems. The only one consisting of a combination of base metal and its oxide is lead, dispersion strengthened with PbO, produced by surface oxidizing lead powder under controlled conditions, pressing and hot extruding. The material is called DS Lead by St. Joseph Lead Company and has much better creep resistance than other Pb alloys which have a tendency to deform under their own weight. Table 21-6 shows data for DS lead. Suggested uses are sound attenuation, chemical construction, radation shielding and batteries.

Table 21-6 Comparison of properties of Chemical lead, Antimonial Lead and Dispersion Strengthened Lead.

	Chemical Pb	Pb with 6% Sb	DS Lead
Tensile Strength MPa	17	34	57
Yield Strength MPa	8	20	43
Elongation %	50	35	18.5
Creep rate in % per year at room temp. at stress in MPa	3.0 at 2.3	0.9 at 3	0.1 at 14
Stress to rupture in 1000 hours, MPa	6.5	8	24

C. Dispersion Strengthened Copper

In oxide dispersion strenthening of copper and nickel matrix material it was found that the oxide of the matrix metal is not satisfactory as a dispersed phase because it is not stable. Considerable work has been done to find suitable oxides. Aluminum oxide is a suitable dispersoid in a copper matrix. Such an alloy is being commercially produced under the name of "Glidcop" for applications in which the high electrical conductivity of unalloyed copper is combined with elevated temperature strength.[26] The material is produced by atomizing a copper-aluminum alloy, mixing it with cuprous oxide and aluminum oxide and internally oxidizing the copper-aluminum alloy producing a fine dispersion of aluminum oxide in copper. The cuprous oxide acts as a source of oxygen; any excess is reduced with hydrogen. The powder is enclosed and sealed in copper containers and extruded at 925°C (1700°F) into the desired shape.

In Figure 21-7[26] the tensile properties and the electrical conductivity of copper dispersion strengthened by Al_2O_3 are plotted as a function of the aluminum oxide content. Relatively small amounts of oxide, less than 1 volume %, produce large increases in hardness and tensile strength, while the decrease in electrical conductivity is quite modest. The strength and hardness of dispersion strengthened copper may be further enhanced by

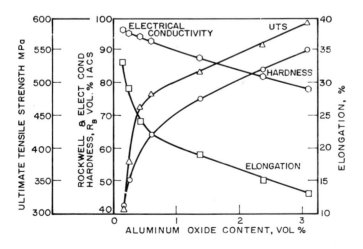

FIGURE 21-7 Properties of as-extruded copper dispersion strengthened with Al_2O_3 ("Glidcop") as a function of aluminum oxide content in volume percent.

cold reduction. The cold worked alloys show excellent resistance to softening after exposure to temperatures up to 930°C. The drop in tensile strength upon annealing lessens with increasing oxide content. The yield strength of a cold worked alloy with 0.45 vol. % Al_2O_3 decreases 30% from 470 PMa to 330 MPa upon annealing one hour at 930°C, while that of an alloy with 2.68 vol. % oxide decreases only 17% from 580 MPa to 480 MPa with the same annealing treatment. This resistance to softening at elevated temperature distinguishes the dispersion strenthened copper alloys from alloys precipitation strengthened with chromium or zirconium, which soften at much lower annealing temperatures. Also, the tensile strength at elevated temperatures of dispersion strengthened copper is higher and its stress rupture life is longer than those of precipitation strengthened alloys. The principal uses are for resistance welding electrodes and for lead wire in incandescent lamps.

D. Dispersion Strengthened Superalloys

By far the greatest amount of research has been done on oxide dispersion strengthened superalloys. The aim in this work is to produce alloys which retain their strength at higher temperatures than the nickel-base superalloys strengthened by precipitation strengthening with the gamma prime phase. The oxides in the fine dispersion in nickel, iron and cobalt are generally thorium or yttrium oxide which have been found to be stable while aluminum oxide is not stable.[27] In addition to much work on oxide dispersion strengthened nickel alloys primarily of a fundamental character, three commercial approaches to producing the alloys based on wet methods were developed. The best known of these approaches is "TD-Nickel" (Thoria dispersed nickel)[28] originally developed by the E. I. DuPont Company.[29] The others are the so-called "DS-Nickel" of Sherritt Gordon Mines Ltd.[30][31] and oxide dispersion strengthened nickel and cobalt base alloys developed by Sylvania Electric Products Inc.[32]

"TD-Nickel"[28] is a nickel alloy with a dispersion of 2% ThO_2 which is produced by precipitating nickel hydroxide from a nickel salt solution upon a colloidal suspension of thorium oxide in the solution. The resulting precipitate is dried and reduced with hydrogen in the solid state to a fine powder of nickel and thorium oxide. The powder is isostatically pressed into compacts. The compacts are sintered, canned and extruded. The extrusions may be rolled into sheet. The process was later extended to matrix alloys of nickel and chromium and nickel and molybdenum and finally to ternary matrix alloys of nickel, chromium and aluminum with a dispersion of thorium oxide.

In the Sherritt Gordon Mines process,[30,31] a solution of nickel ammine ammonium carbonate is boiled at atmospheric pressure to produce a fine suspension of basic nickel carbonate. To the agitated slurry of this compound the desired amount of thoria in the form of a sol is added, which is adsorbed on the carbonate particles. The suspension is reduced in an autoclave with hydrogen at 135°C (275°F) and a pressure of 2.4 MPa (350 psi) to a nickel-thoria powder. The powder is compacted and the compacts hot rolled into strip.

The Sylvania process[32] is based on selective reduction of a spray dried mixture of salts. An alloy of nickel with 15% molybdenum and 4% ThO_2 is produced from an aqueous ammonium oxalate solution in which appropriate amounts of nickel oxalate, molybdenum trioxide and thorium nitrate are dissolved. The solution is atomized to a droplet size of less than 10 μm. The droplets are brought into contact with hot gases and dried to a powder which is heated in air to produce the oxides of the metals. The nickel and molybdenum oxides in the mixture are reduced to the metals at temperatures near 1000°C. The resulting powder is consolidated by powder rolling followed by hot rolling at 800°C or by cold hydrostatic compacting, heating to 1200°C and rolling. The high temperature properties of the strip produced are critically dependent upon the amount of room temperature deformation following the hot consolidation.

At the present time, the principal interest in dispersion strengthened superalloys is based on alloys fabricated by a new method, "mechanical alloying," rather than on alloys produced by the wet methods. "Mechanical alloying" was first suggested by Shafer et al. in 1961.[33]

The metal powder and oxide to be dispersed in the metal matrix are introduced into an attritor grinding mill which is a high energy driven ball mill with the powder and balls held in a stationary tank and agitated by rotating impellers. During milling the ingredients of the powder mixtures are reduced in size and brought into intimate contact by flattening and crushing the particles, welding them together and repeating the process again and again.

Mechanical alloying is being used by the International Nickel Company to produce a type of nickel alloy which can be both precipitation and dispersion hardened.[34,13,14] A mixture of metal and alloy powders, typically carbonyl nickel powder, chromium powder, refractory metal powders, nickel-aluminum-titanium, nickel-zirconium and nickel-boron master alloy powders and yttrium oxide powder are processed into composite powder particles. The powders are consolidated by methods similar to those for other superalloys involving packing the powder in cans and extruding the canned powder. The extrusions may be rolled into sheet.

Table 21-7[35]   Nominal Composition of INCO "Mechanical Alloys."

Alloy	Cr	Y₂O₃	Al	Ti	C	Fe	Ta	Mo	W	Ni	Zr	B
Incoloy MA754	20	0.6	0.3	0.5	0.005	-	-	-	-	Bal	-	-
Incoloy MA956E	20	0.5	4.5	0.5	-	Bal	-	-	-	-	-	-
MA 6000E	15	1.1	4.5	2.5	-	-	2.0	2.0	4.0	Bal	0.1	0.01

The alloys contain aluminum and titanium in the matrix and can therefore be strengthened by precipitation of the gamma prime phase. They also contain very finely divided yttrium oxide which imparts dispersion strengthening at temperatures where conventional supperalloys overage. Some of these mechanically alloyed materials are now available commercially, e.g. the alloy Incoloy Ma 754. Its composition together with those of two newer Experimental alloys, Incoloy MA 956 E and MA 6000 E is shown in table 21-7[35]. Alloy MA 754 has properties somewhat better than those of thoriated nickel-chromium. Y₂O₃ as a dispersoid has the advantage over ThO₂ of not giving any radiation problems. Incoloy MA 754 is intended primarily as a vane material for jet engines.

Incoloy alloy MA 956 E is ferritic and based on the Fe-Cr-Al alloys. Because of its Al and Cr content it has sufficient oxidation and corrosion resistance so that it does not need protective coatings.

One of the newest experimental alloys is MA 6000E. Its composition is similar to some of the new high strength nickel-base superalloys, but also includes 1.1% Y₂O₃ to provide dispersion strengthening. It is fabricated by extrusion and rolling and may be subsequently zone annealed to provide a grain structure with a grain aspect ratio greater than 10. The stress for 1000 hour stress-rupture life of this alloy is compared in Figure 21-8[35] with that of TD-nickel, of the superalloy IN 792, whose composition was given in Table 21-2 and of a directionally solidified alloy DS-MAR M200 + Hf. Its outstanding stress rupture properties over the temperature range from 800 to 1150°C are evident. Above 1000°C it is vastly superior to the common nickel-base superalloys. It has good oxidation, sulfidation and carburization resistance and is intended for rotating, centrifugally loaded parts, such as jet engine blades.

Successful development work has been undertaken by the International Nickel Company in producing dispersion-strengthened mechanically alloyed aluminum alloys by attrition grinding mixtures of aluminum powder, metallic alloying elements and aluminum oxide.[36] Properties superior to those of the SAP alloys are claimed.

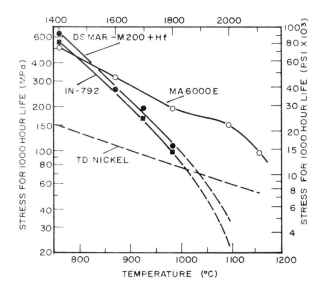

FIGURE 21-8 Stress for 1000 hours life to rupture as a function of temperature for alloys MA 6000 E, DS MAR M200 (a directionally solidified alloy) and TD nickel.

E. Dispersion Strengthened Materials with Dispersed Phases Other than Oxides

In dispersion strengthened materials a metallic phase, i.e. a pure metal or an intermetallic compound may serve as dispersoid, if the metallic phase is practically insoluble in the matrix metal near the melting point of the matrix. Two powder metallurgy methods for producing such a metallic dispersoid phase have been suggested. One is called interference hardening. An example is a magnesium alloy containing aluminum and zirconium as alloying elements.[37] Individually, aluminum and zirconium are soluble in both solid and molten magnesium. If they are added to the same melt they form the intermetallic compound $AlZr_4$ which is insoluble in molten magnesium and settles out. A dispersion of $AlZr_4$ in magnesium can be produced by solid state precipitation. A magnesium-zirconium alloy powder is produced by atomizing. The powder is coated by flake aluminum powder, hot pressed to a dense compact and heat-treated to diffuse the aluminum. $AlZr_4$ precipitates throughout the matrix. The dispersion is very fine, thermally stable and effectively inhibits growth of the grains which are 2 to 3 μm in diameter. The resulting alloy which has tensile and compressive yield strengths of 300 MPa with 10 percent elongation can be extruded into complex shapes. In this respect, it is superior to most aluminum and magnesium alloys which are cast or extruded.

Another method of producing dispersion strengthened alloys with an intermetallic compound as dispersoid is to air atomize a liquid alloy of a base metal with a solute, which is soluble in the liquid metal, but practically insoluble in the solid metal, even near the melting point.[38] Examples are alloys of aluminum with Fe (7.8%) and with Fe and Ni in which the dispersed phases are $FeAl_3$ and $FeNiAl_9$. A powder is produced by atomizing a liquid Al-Fe or Al-Fe-Ni alloy. The powder is compacted, hot pressed and extruded. These dispersion strengthened alloys are not quite as stable as SAP, but easier to produce. Research on aluminum alloys dispersion strengthened by intermetallic compounds has recently been revived.

A dispersion strengthened alloy with a pure metal as the dispersed phase is somewhat of a curiosity. Such an alloy was investigated at Rensselaer Polytechnic Institute.[39] Its composition is 95% Pb, 5% Cu produced by atomizing the molten alloy and extruding the resulting powder. It has quite good creep resistance.

References

1. C. Aslund, A new method of producing stainless steel tubes from powder, Fifth European Powder Metallurgy Symposium, Stockholm, Preprints, Vol. 1, p. 278-283, June, (1978).

2. D. Webster, Keynote paper - physical metallurgy, Fourth International Conference on Beryllium, London, Oct., (1977)

3. A. N. Stroh, Phil. Mag., Vol. 3, p. 597, (1958).

4. G. J. London, G. H. Keith and N. P. Pinto, Grain size and oxide content affect beryllium properties, Metals Engineering Quarterly, Vol. 16, No. 4, p. 45-57, (1976).

5. W. W. Beaver and H. F. Larson, The powder metallurgy of beryllium, in "Powder Metallurgy," ed. by W. Leszynski, N.Y., p. 747-772, (1961).

6. R. S. Busk and T. E. Leontis, The extrusion of powder magnesium alloys, Trans. AIME, Vol. 188, p. 297-306, (1950).

7. R. S. Busk, The pellet metallurgy of magnesium, Light Metals, Vol. 23, p. 197-200, (1960).

8. S. G. Roberts, An exploratory investigation of prealloyed powders of aluminum, in "Powder Metallurgy," ed. by W. Leszynski, N.Y., p. 799-817, (1961).

9. J. P. Lyle, Jr., and W. S. Cebulak, Fabrication of high-strength aluminum products from powder, in "Powder Metallurgy for High-Performance Applications," ed. by J. J. Burke and V. Weiss, Syracuse, p. 231-254, (1972).

10. J. P. Lyle, Jr., and W. S. Cebulak, Properties of high-strength aluminum products, Metals Engineering Quarterly, Vol. 14, No. 1, p. 52-63, (1974).

11. W. S. Cebulak, E. W. Johnson and W. Markus, High-strength aluminum P/M mill products, Int. J. of Powder Metallurgy, Vol. 12, No. 4, p. 299-310, (1976).

12. S. Bhattacharya and K. M. Kulkarni, Development of high strength aluminum P/M isothermally forged component, Fifth European Powder Metallurgy Symposium, Stockholm, Preprints, Vol. 1, p. 19-25, June 1978.

13. G. H. Gessinger and M. J. Bomford, Powder metallurgy of superalloys, Internat. Metall. Reviews, Review 181, Vol. 19, p. 51-76, (1974).

14. J. S. Benjamin and J. M. Larson, Powder metallurgy techniques applied to superalloys, J. of Aircraft, Vol. 14, p. 613-623, (1977).

15. J. B. Moore, J. Tequesta and R. L. Athey, Fabrication method for the high temperature alloys, U. S. Patent 3,519,503, (1970).

16. E. J. Dulis, High-speed tool steels by powder metallurgy, in "Powder Metallurgy for High-Performance Applications," ed. by J. J. Burke and V. Weiss, Syracuse, p. 317-329, (1972).

17. P. Hellman, H. Larker, J. B. Pfeffer and I. Stromblad, The ASEA/Stora process, a new process for the manufacture of tool steel and other alloy steels from powder, Mod. Dev. in Powder Metallurgy, Vol. 4, ed. by H. H. Hausner, N.Y., p. 573-582, (1971).

18. W. T. Haswell, Presentation at Westec, (1978).

19. E. F. Bradley, R. A. Sprague and W. B. Tuffin, A new stainless steel from Powder, Metal Progress, Vol. 88, No. 3, p. 83-85, (1965).

20. G. S. Ansell, The mechanism of dispersion strengthening: a review, in "Oxide Dispersion Strengthening," ed. by G. S. Ansell, T. D. Cooper and F. V. Lenel, N.Y., p. 61-141, (1968).

21. See also the contributions of M. F. Ashby, L. L. J. Chin and N. J. Grant, R. L. Jones and A. Kelly, I. G. Palmer and G. C. Smith, M. C. Inman and P. J. Smith, B. A. Wilcox and A. H. Clauer, R. Grierson and L. J. Bonis, R. W. Fraser and D. J. I. Evans, P. Guyot in Part II, Mechanism of dispersion strengthening, ibid., p. 143-427.

22. B. A. Wilcox and A. H. Clauer, Grain-size in dispersion hardened nickel alloys, Acta Met., Vol. 20, p. 743-757, (1972).

23. J. P. Lyle, Aluminum powder metallurgy products, Proc. Metal Powder Assoc., 12th meeting, p. 93-105, (1956).

24. E. A. Bloch, Dispersion strengthened aluminum alloys, Metallurgical Reviews, Vol. 6, No. 22, p. 193-239, (1961).

25. R. Irmann, Techn, Rundschau, Vol. 41, No. 3, p. 19, (1949).

26. A. V. Nadkarni, E. Klar and W. M. Shafer, A new dispersion-strengthened copper, Metals Eng. Quart., Vol. 16, No. 3, p. 10-15, (1976).

27. J. A. Dromsky, F. V. Lenel and G. S. Ansell, Growth of aluminum oxide particles in a nickel matrix, Trans. AIME, Vol. 224, p. 236-239, (1962).

28. F. J. Anders, G. B. Alexander and W. S. Wartel, A dispersion strengthened nickel alloy, Metal Progress, Vol. 82, No. 6, p. 88-91 and 118 and 122, (1962).

29. F. J. Anders, Dispersion strengthened metal product and process, U. S. Patent 3,159,908, Dec., 1964.

30. R. W. Fraser, B. Meddings, D. J. I. Evans and V. N. Mackiw, Dispersion strengthened nickel by compaction and rolling of powder produced by pressure hydrometallurgy, Mod. Dev. in Powder Metallurgy, Vol. 2, ed. by H. H. Hausner, N.Y., p. 87-111, (1966).

31. L. F. Norris, R. W. Fraser and D. J. I. Evans, Cold rolling of dispersion strengthened nickel, in "Powder Metallurgy for High Performance Applications," ed. by J. J. Burke and V. Weiss, Syracuse, p. 257-280, (1972).

32. R. F. Cheney and J. S. Smith, Oxide strengthened alloys by the selective reduction of spray dried mixtures, in "Oxide Dispersion Strengtheneing," ed. by G. S. Ansell, T. D. Cooper and F. V. Lenel, N.Y., p. 637-674, (1968).

33. M. Quatinetz, R. J. Shafer and C. Smeal, The production of submicron metal powders by ball milling with grinding aids, Trans. AIME, Vol. 221, p. 1105-1110, (1961).

34. J. S. Benjamin, Dispersion strengthened superalloys by mechanical alloying, Metall. Trans., Vol. 1, p. 2943-2951, (1970).

35. Anonymous, Mechanically alloyed high temperature materials, J. Metals, Vol. 29, No. 12, p. 9-11, (1977).

36. J. S. Benjamin and M. J. Bomford, Dispersion strengthened aluminum made by mechanical alloying, Metall. Trans., Vol. 8A, p. 1301-1205, (1977).

37. G. S. Foerster, Dispersion hardening of magnesium, Metals Engin. Quart., Vol. 12, No. 1, p. 22-27, (1972).

38. R. J. Towner, APM alloys, Metals Engin. Quart., Vol. 1, No. 2, p. 24-40, (1961).

39. F. V. Lenel, Preparation and selected properties of certain dispersion strengthened lead-base alloys, Powder Metallurgy, Vol. 5, No. 10, p. 119-131, (1961).

Powder Metallurgy Applications Involving Sintering with a Liquid Phase

In this chapter the following powder metallurgy applications are discussed:

> Non-porous bearings by powder metallurgy
> High speed tool steel products by cold pressing and liquid phase sintering
> Silver-tin amalgams for dental restorations

The common characteristic of the products described is the fact that they are sintered with a liquid phase, although in other respects they are quite dissimilar. The mechanisms involved in sintering with a liquid phase have been discussed in chapter 11. Some of the products sintered with a liquid phase have been discussed in chapters 14 (Heavy metal), 16 (Cemented carbides) and 17 (Structural parts). Others will be described in chapter 23 (Alnico and cobalt-rare earth permanent magnets). The approach in developing the products described in this chapter has been mainly empirical. In their discussion emphasis is therefore put on engineering aspects and no attempt is made to present in any detail the sintering mechanisms involved.

I. Non-porous Bearings by Powder Metallurgy

This section will be primarily concerned with steel-backed precision automotive main and connecting rod bearings in whose fabrication powder metallurgy has played a fairly important role. Main and connecting rod bearings were originally produced by coating the bearing housing for main bearings and the big end of connecting rods with molten babbit metal and then machining the coating to the correct size as a bearing. Eventually precision bearings were developed consisting of a stell shell lined with the babbit metal. The composite bearing is fitted into the housing or the big end of the connecting rod. Two types of babbit metal were used primarily, a tin-based composition, SAE 12, containing 89% Sn, 7½% Sb and 3½% Cu and a lead-base babbit, SAE 15, containing 83% Pb, 15% Sb, 1% Sn and 1% As. These babbit bearings fulfilled certain of the requirements for a precision type bearing quite well:

They had sufficient deformability to permit flow at points of insufficient clearance resulting from rod distortion,

They were sufficiently deformable to permit imbedding of dirt and metal particles carried into the bearing with oil,

Their friction characteristics allowed contact with the shaft at high speed and load without galling or seizing,

They resist attack by the fatty acids present in oils or formed during oxidation of oil in use.

However, they were inadequate with respect to:

Sufficient strength at operating temperature to resist fatigue failure due to pounding.

The approach made in the last 40 years to improve the fatigue strength of automotive precision main and connecting rod bearings involved both changes in methods of fabrication and in composition. Three of these approaches and, in particular, the role which powder metallurgy plays in them will be discussed.

In the earliest one, which goes back to the early 1940s, a lead base babbit was retained as the bearing material, but powder metallurgy was used to produce a better bond to the steel shell and provide better fatigue resistance.[1]

Among the changes in composition, one very widely used was to substitute a copper-lead alloy as a bearing material for the traditional tin or lead base babbit. A relatively low cost powder metallurgy method of producing strip consisting of a steel backing with a copper-lead alloy coating was developed in the late 1940s.[2] The insufficient corrosion resistance of these bearings to acids in the oil was overcome by electroplating the bearing surface with a lead-tin-copper alloy.

Aluminum alloys have also been widely used as a material for the lining of precision main and connecting rod bearings. Alloys of aluminum with lead provide a particularly good lining material, but cannot be readily produced by fusion metallurgy because of the tendency of these alloys to segregate in the liquid state. Again an ingenious method based on powder metallurgy was developed in the 1970s to produce aluminum-lead bearing linings.[3]

The approach to improve the strength of the babbit metal layer made use of an intermediate sponge-like layer consisting of a porous copper-nickel or copper-iron alloy which was sintered from metal powders and simultaneously bonded to the steel backing.[1] The layer was infiltrated by the babbit metal with a pure babbit layer on top of the sponge layer, which after final machining is only 50 μm (0.002″) thick. With its anchor in the sponge layer the babbit metal proved to have much better fatigue resistance than the conventional 500 μm (0.020″) thick babbit lining, but is sufficiently deformable for use in the bearing.

The manufacture of the bearing involves the following steps:

1. Cold-rolled steel strip is cleaned and leveled

2. A layer of copper and nickel powders is applied loose to the strip to a given thickness. The original composition was 60% Cu, 40% Ni, but the nickel content was later lowered. It was also found that a mixture of copper and iron powder could be used. Other sponge compositions have been successfully used.

3. The strip with the applied loose powder is run continuously through a roller hearth furnace and heated near the melting point of Cu; the copper alloys partially with the Ni, and bonds it in the form of a sponge to the steel strip.

4. The strip with the sponge is lightly rolled to produce a layer of uniform thickness and to consolidate the sponge. The strip is then annealed to relieve the cold work in the sponge.

5. The strip with the sponge is run through a continuous vacuum-infiltration machine in which the sponge is infiltrated with a lead base babbit (3% Sn, 3% Sb, bal. Pb). As it emerges from the machine the babbit freezes in a layer on top of the sponge; most of the excess babbit is milled off so that about 0.25 mm (0.010″) of babbit remain above the sponge.

6. The strip is cut into blanks which are bent into bearing halves and finished in a series of operations, the last of which is a broaching operation in which only a thickness of 50 μm (.002″) above the copper nickel sponge is left over.

In this context, a bearing material should be mentioned which is also fabricated using an intermediate sponge-like layer, although it is not employed for automotive main and connecting rod bearings. It has been developed in England[4] and is a dry bearing using neither oil nor grease as a lubricant but relies for lubrication upon the low coefficient of friction of polytetrafluoroethylene plastic. The sponge-like layer in these bearings is produced from spherical bronze powder particles which are sintered to a copper plated steel backing in a process quite similar to that described for spherical bronze filters in chapter 5. The sponge is impregnated with a mixture of lead powder and the polytetrafluoroethylene plastic. It has a thin surface layer of the lead-plastic mixture over the top of the bronze spheres.

The bearing makes use of the high wear-resistance and low friction of the polytetrafluoroethylene on the one hand and the low thermal expansion and the high thermal conductivity of the bronze over the steel backing on the other hand. It may be used over a wide temperature range from $-200°$ C ($-328°$ F) to as high as $280°$ C ($536°$ F). The material is used in the form of bushings, thrust washers or strip in applications where the use of dry bearings is essential.

The sequence of operations to produce bearings with a copper-lead lining by powder metallurgy[2] involves the following steps:

1.   A copper-lead alloy, generally with 60 to 75% copper, balance lead is melted in an induction furnace which provides excellent dispersion of the lead in the copper. The melt is poured into a tundish and flows through a nozzle where it is atomized with water. Careful control of the oxide content of the powder is important.

2.   A layer of the atomized copper-lead powder is applied loose to cold-drawn steel strip to a certain thickness.

3.   The strip with the applied loose powder is sintered in a continuous furnace at a temperature between 1550 and 1650° F, where the powder sinters and bonds to the strip.

4.   As the strip emerges from the furnace, it runs through a rolling mill where it is rolled sufficiently to increase the density of the copper-lead layer close to that of the cast alloy.

5.   The strip continues from the rolling mill through a second continuous furnace where a thorough bond of the copper in the bearing layer to the steel backing is established. The temperature of the second sintering operation is the same as that of the first.

6.   The strip is cut into blanks, bent into bearing halves and finished in a series of mechanical operations.

7.   After the bearing halves have been mechanically finished they are electroplated with a thin overlay of a lead-tin-copper alloy to improve the resistance of the bearing to attack by acids in the oil.

In the fabrication by powder metallurgy of bearings with an aluminum-lead alloy lining,[3] it is not feasible to start with a mixture of aluminum and lead powder since the softer lead would squeeze between the harder aluminum particles when the powder mixture is consolidated. Instead, an alloy having the desired composition is air atomized so that the lead is located in the interior of the aluminum alloy particles where it does not interfere with the bonding of the aluminum particles. Also, the aluminum-lead alloy does not roll bond satisfactorily to the steel strip; therefore an intermediate layer of pure aluminum must be provided. Consolidation of the powder into strip and bonding of the alloy composition to the pure aluminum is achieved by powder rolling. The composition of the bearing linings is 8.5% Pb, 4.0% Si, 1.5% Sn, 1.0% Cu, bal. Al.

The production of the bearings involves the following steps:

1. An induction melting furnace is charged with aluminum, copper and silicon and these ingredients are melted. The temperature is then raised to 1800° F and the lead and tin added. A homogeneous liquid alloy is formed by the stirring action of the furnace.

2. The molten alloy is poured into a tundish furnace and metered through a ceramic nozzle into the atomizing chamber, 21 feet high by 4 feet in diameter, where it is atomized. The oversize and undersize particles are removed. The particles consist of a lead-tin alloy phase dispersed in a hypoeutectic Al-Si-Cu matrix.

3. A bimetallic strip is formed by powder rolling as discussed in chapter 6 and illustrated in Figure 6-28. One surface of the strip is pure aluminum, the other consists of the Al-Pb prealloyed powder. Actually a third layer on the opposite side of the prealloyed layer consisting of a blend of aluminum and lead-tin particles is also formed to facilitate rolling, but this layer is later machined off. The compacted strip has theoretical density.

4. The coils of composite strip are sintered. During sintering the fine eutectic of the aluminum-silicon coarsens and discrete particles of Si are formed.

5. The sintered strip is roll bonded to the steel strip. To obtain a satisfactory bond the thickness of the strip is reduced 50%.

6. The strip is cut into blanks and the blanks finished into bearings. A finished bearing is shown in Figure 22-1.

II. High Speed Steel Products by Cold Pressing and Liquid Phase Sintering

In chapter 11 the so-called "heavy alloy" sintering process was discussed in which complete densification of cold pressed compacts of a powder mixture

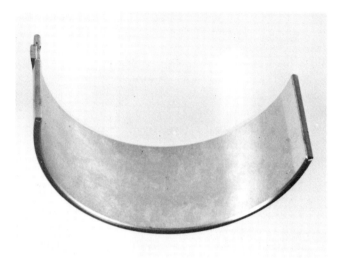

FIGURE 22-1 Steel back engine bearing lined with aluminum-lead alloy produced by powder rolling (Courtesy: Gould, Inc.).

is achieved. The mixture contains a minor low melting constituent—nickel-copper or nickel-iron in heavy alloy (chapter 14), cobat in cemented carbides (chapter 16)—and a major high melting constituent, W and WC respectively. Such mixtures with a lower melting constituent also play a role in sintering alnico and rare earth-cobalt type permanent magnets. The liquid phase necessary for densification may, however, also be formed during sintering compacts from a single metal powder, rather than a mixture which contains a lower melting constituent. The compacts are sintered at a temperature slightly above the solidus temperature of the alloy where a small amount of liquid phase is formed at grain-boundaries. This type of liquid phase sintering was investigated by Westerman[5] on compacts from a series of nickel base superalloys, but no commercial use of liquid phase sintered superalloys developed. A process based on this type of sintering has, however, been developed for high speed steel alloy compositions in the "Fuldens" process.[6] It involves cold compacting of a finely divided high speed tool steel powder and sintering, was originally developed at Aerojet General Corporation (Patent No. 3,744,993) and is practiced by Consolidated Metallurgical Industries. It is quite different from the process for producing tool steels by hot consolidating inert gas atomized spherical powder described in Section V of chapter 21. The powder used in the Fuldens process is produced by atomizing melts of the required tool steel composition with water rather than with inert gas. The powder particles with a particle size of approximately 100 μm are extremely hard, having a microstructure characteristic for high speed steel ingots. The particles are further pulverized by impacting them in a stream of gas against a hard target (cold-stream process, see chapter 2). The resulting powder has particle sizes corresponding to Fisher subsieve sizer numbers between 6 and 12 μm and a Gaussian particle size distribution. In order to reduce the oxide on the particle surface and to make the powder compactible it is annealed in hydrogen. The annealed powder is pressed at pressures of 410 to 690 MPa (60 to 100,000 psi) into the desired shape in rigid dies or by isostatic pressing. The green compacts are sintered under conditions of sintering time, temperature and atmosphere which must be carefully controlled. Sufficient liquid phase must be produced so that the compacts are densified to near theoretical density. On the other hand, excessive grain growth and the development of a coarse carbide structure must be avoided. After sintering the compacts are annealed, before they are given the usual heat treatment for high speed steels: hardening and multiple tempering. Parts with several different high-speed tool steel compositions, such as M2, T15, M3 type 2, M42 and M4 have been produced and their performance as cutting tools has been found equivalent and sometimes superior to those of conventional high speed steels. Details on the metallurgy have not yet been published.

The process is used for cutting tools of complex shapes, such as hobs, pipe taps and reamers by cold isostatic pressing in flexible envelopes. Some of these shapes are shown in Figure 6-12. Quite good control of the dimensions of these compacts is possible even though they are pressed isostatically and shrink to near theoretical density during sintering. Producing such shapes from powder rather than by machining means considerable materials saving. High speed steel indexable tool steel inserts may also be produced from powder by compacting in rigid dies and sintering, similar to the inserts produced from cemented carbides.

A process for producing high speed tool steel parts which appears to be somewhat similar to the Fuldens process was developed in England by Powdrex,[7] Ltd., but even fewer details are available for it.

III. Silver-tin Amalgams for Dental Restorations[8,9]

When the intermetallic compound Ag_3Sn in finely divided form is mixed or, as the dentist calls it, triturated with liquid mercury, a plastic amalgam is formed which the dentist uses for dental restorations. He packs or "condenses" the amalgam into the tooth cavity. During a short intermediate period the amalgam exhibits "carvability" so that the dentist can cut it to the contours required, but after about 10 minutes the amalgam begins to harden and to solidify into the desired restoration. The reaction between the liquid mercury and the finely divided silver-tin intermetallic compound should be considered a process of liquid phase sintering at room temperature with a transient liquid phase (see chapter 11) in which the mercury is used up through reaction with the intermetallic compound.

The reaction is commonly written:

$$\gamma + Hg \rightarrow \gamma + \gamma_1 + \gamma_2$$

in which γ is the rhombohedral silver-tin compound Ag_3Sn, γ_1, a body centered cubic phase of the composition Ag_5Hg_8 and γ_2 a hexagonal phase of the composition Sn_6Hg. Both γ_1 and γ_2 are solid at room temperature. The reaction starts during "trituration" and "condensation." The condensed mass hardens as the mercury continues to react with the tin-silver compound particles. The particles become coated with crystals of the γ_1 and γ_2 intermetallic compounds. The final restoration consists of the remaining unreacted Ag_3Sn compound embedded in a matrix of γ_1 and γ_2.

The important variables in producing the amalgam are the exact composition of silver-tin alloy, its particle size and particle size distribution, the ratio of the compound and mercury and the method of mixing them and the possible additions of other alloying ingredients to the silver-tin compound. They determine the rate of setting, i.e. the rate at which the silver-mercury and tin-mercury intermetallic compounds are formed, the dimensional changes, generally a small expansion, during their formation, the corrosion resistance and the mechanical properties of the amalgam restoration, in particular its resistance to creep under compressive stresses.

Commonly a silver-tin compound with 25 to 27 weight percent Sn is used which is within the homogeneity range of the Ag_3Sn phase γ. Small amounts of Ag-Sn β phase, richer in Ag, accelerate the setting reaction. Two methods are used to produce the finely divided Ag_3Sn compound. In one method, the compound is cast into a ingot and the ingot homogenized and cleaned. Particles are cut on a lathe, the shavings pulverized in a hammer mill and the powder classified into particle sizes. A fine-cut powder will have a median particle size of 50 μm, a microcut of 30 μm. Coarse cut powders with a median particle size of 70 μm are no longer used. Another method of producing the compound powder is by atomizing into spherical particles using argon or nitrogen as the atomizing medium. Atomized powders are pickled, lightly pressed into pellets and given a low temperature stress relief treatment.

Silver-tin compound and mercury used to be mixed with mortar and pestle, but this method has been superseded by mixing in mechanical triturators for periods of 10 to 20 seconds. With this method it is also possible to lower the ratio of mercury to alloy from 1:1 to 1:1 ¼ by weight.

Small additions of 6% copper and 0.2% zinc to the silver-tin alloy castings were commonly made, with the copper facilitating comminution and the zinc acting as a scavanger for oxygen, but with modern casting methods the zinc additions are no longer necessary.

A more recent variation in dental amalgams has been the addition of 20% of fine particles of the silver-copper eutectic with 28.1% Cu to the Ag_3Sn alloy filings. During amalgamation the silver is replaced in the eutectic by tin and a dispersion of a copper-tin intermetallic compound is formed. The trade name of this alloy is "Dispersalloy." The new amalgam contains less of the γ_2 tin-mercury phase, which has the lowest corrosion resistance among amalgam constituents. It also has a higher compressive strength and a lower creep rate than conventional amalgams.

References

1. A. L. Boegehold, Copper-nickel-lead bearings, chapter 47 in "Powder Metal-lurgy," ed. by J. Wulff, Cleveland, p. 520-529, (1942).
2. E. R. Darby, Copper-lead bearings from metal powder, Proc. 3rd Ann. Spring Meeting, Metal Powder Assoc., p. 52-56, (May 1947).
3. M. L. MacKay, Innovation in powder metallurgy, an engine bearing material, Metal Progress, vol. 111, No. 6, p. 32-35, (1977).
4. Brochure of Garlock, Inc.: du, the new Glacier high performance non-lubricated bearing material.
5. E. J. Westerman, Sintering of nickel base superalloys, Trans. AIME, vol. 224, p. 159-164, (1962).
6. T. Levin and R. P. Hervey, P/M alternative to conventional processing of high speed steel, Metal Progress, vol. 115, No. 6, p. 31-34, (1979).
7. Anonymous, The Powdrex Process, Metals and Materials, p. 36-37, (Sept. 1977).
8. J. F. Bates and A. G. Knapton, Metals and alloys in dentistry, Review 215, International Metals Reviews, vol. 22, p. 39-60, (1977).
9. J. B. Moser, Metal powder applications in dentistry, Mod. Dev. in Powder Metallurgy, vol. 11, ed. by H. H. Hausner and P. W. Taubenblat, Princeton, p. 303-314, (1977).

23

Powder Metallurgy Products for Magnetic Applications

Powder metallurgy is used in the production of soft magnetic materials for direct current applications and for permanent magnets. In many, but by no means all, of these applications parts are produced by powder metallurgy because this method permits production to final shape with a minimum of subsequent machining and grinding and, at the same time, achieving desirable magnetic properties. Before the individual applications are discussed, the magnetic properties which are important in ferromagnetic materials will be briefly reviewed.

Soft ferromagnetic materials for direct current applications and permanent magnets are characterized by their hysteresis curves in which the magnetizing force H in Oersteds is plotted vs the magnetic induction B in Gauss. A typical hysteresis curve for a soft magnetic material is shown in Figure 23-1. It consists of the normal induction curve which shows how the induction varies as the magnetizing force is increased starting at zero, i.e. the demagnetized state. The slope of this curve, i.e. the ratio B/H, is the permeability of the material. With increasing magnetizing force the permeability increases, goes through a maximum and then decreases again. The maximum permeability is one of the characteristics of soft magnetic materials. The magnetic saturation, i.e. the value of induction or flux density when the magnetizing force has reached a specified value, e.g. 15 Oersteds, is another important characteristic of the material. When the

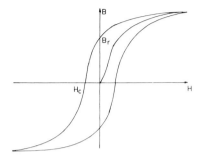

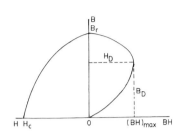

FIGURE 23-1 Schematic of magnetic hysteresis curve.

FIGURE 23-2 Schematic of demagnetization curve for permanent magnet alloy.

magnetizing force is decreased from this value, the curve of induction does not follow the original normal induction curve, but lags behind, as seen in Figure 23-1. The value of magnetic induction in Gauss, when the magnetizing force has been reduced to zero, is called the residual induction or remanence B_r. When a magnetizing force in the opposite direction is applied, the magnetic induction decreases further and reaches a value of 0 with a given magnetizing force in Oersteds, called the coercive force H_c. Remanence and coercive force are additional characteristics of soft magnetic materials.

Figure 23-2 shows the demagnetizing portion of the hysteresis curve from the residual induction B_r to the coercive force H_c for a permanent magnet material. In addition to the values B_r and H_c, the product of magnetizing force and induction at any point of the curve, the so-called energy product in gauss-oersted, is important for permanent magnets. In Figure 23-2 the value of this energy product is plotted as a function of the induction; it reaches a maximum at a certain value of induction.

In the first section of this chapter soft magnetic applications, such as pole pieces, armatures, relay cores and computer printers from iron, iron-silicon, iron-phosphorus and iron-nickel compositions, compacted from powder to shape and sintered, are described. They have adequately high permeability and saturation and low residual induction and coercive force. As described in the second section, powder metallurgy is also used to a limited extent to produce wrought soft magnetic nickel-iron-molybdenum and nickel-iron alloys, in the form of strip, rod and wire, where superior magnetic properties, in particular higher maximum permeability, can be obtained starting with very pure powders instead of cast ingots.

One type of permanent magnet material, the so-called Alnico magnets, is generally produced by casting, but small Alnico magnets are also often produced from powder. Their production and properties are discussed in section III. Another type of permanent magnets, recently developed, the cobalt-rare earth magnets, are always produced from powder because in the fabrication of the magnets the particles of the alloy powder must be magnetically oriented before they are pressed and sintered. The production and the properties of these magnets are described in section IV.

A third type of permanent magnet material, the elongated single domain (ESD) permanent magnets, are also produced from powder. The fabrication of these magnets involves the steps of electrodeposition of powder particles on a mercury cathode, magnetic alignment, compacting, embedding in a matrix, grinding into a coarse powder, realigning and compacting into the final magnet shape, but no sintering step. Since the sintering operation is the distinguishing characteristic in the fabrication of P/M products, the ESD magnets do not belong among them, strictly speaking. The same applies to another group of magnetic materials produced from alloy and metal powders. These are core materials for radio frequency alternating current applications produced from carbonyl iron powder and audio frequency core materials from permalloy powder. In these core materials the powder particles are covered by an insulating film which is to prevent metallic contact between the particles. Even when the cores are treated at elevated temperatures, the insulating film prevents sintering. The ESD permanent magnets and the cores will be briefly described in the last section of this chapter.

The methods of fabrication of certain non-metallic magnetic materials, the so-called ferrites, which include both soft magnetic materials and permanent magnets, are often quite similar to those used in powder metallurgy. However, oxide powders, not metal powders, are the raw materials for ferrites and they will therefore be excluded from the discussion. A monograph on "Oxide Magnetic Materials" by K. J. Standley, published by Clarendon Press, Oxford, second edition in 1972, describes these materials.

I. Soft Magnetic Materials for Parts Pressed to Shape

Four types of compositions are used for soft magnetic parts pressed to shape:
Pure iron
Iron-silicon alloys commonly with 3% Si
Iron-phosphorus alloys generally with 0.45 to 0.75% P
Iron-nickel alloys with 50% Fe and 50% Ni

FIGURE 23-3 Soft magnetic components from iron powder (Courtesy: Remington Arms Co., Inc.).

Many of these alloys in wrought form are difficult to machine because of their softness. Production by such methods as investment casting or powder metallurgy where machining is not necessary have definite advantages.

Soft magnetic parts from pure iron powder are the type most widely used. Examples of such parts are shown in Figure 23-3. For these parts atomized iron powder low in oxygen, carbon and nitrogen is used. Before the pure atomized iron powder became available, electrolytic iron powder was used, but the slightly better magnetic properties for parts from electrolytic compared to those from atomized iron powder generally do not warrant the use of the more expensive electrolytic powder. The parts are produced by compacting, sintering and, if necessary, repressing and annealing. The magnetic properties are a function of the sintered density, the sintering temperature and atmosphere and the particle size of the iron powder.[1] They are generally determined by the ballistic test method on ring shaped specimens.[2]

The effect of sintered density and of sintering temperature upon saturation flux density at a magnetizing force of 15 Oersteds, residual induction, maximum permeability and coercive force of specimens sintered 30 minutes in hydrogen is shown in Figures 23-4 to 23-7.[3] The improvement in these properties (higher saturation, residual induction and maximum permeabil-

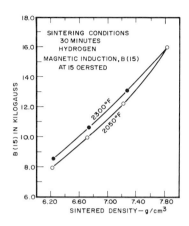

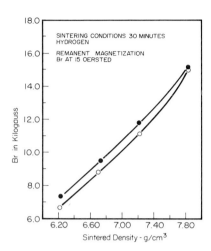

FIGURE 23-4 Magnetic induction at a magnetizing force of 15 Oersted for cores from iron powder sintered 30 minutes at 2050° F (1120° C) and sintered 30 minute at 2300° F (1260° C), cooled at a rate of 10° F (5.5° C) per minute as a function of sintered density.

FIGURE 23-5 Remanent magnetization (B$_r$) of cores from iron powder sintered under the same conditions as those of Fig. 23-4.

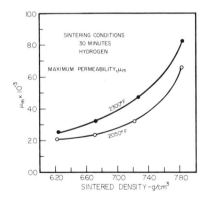

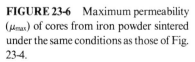

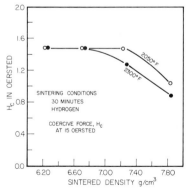

FIGURE 23-6 Maximum permeability (μ_{max}) of cores from iron powder sintered under the same conditions as those of Fig. 23-4.

FIGURE 23-7 Coercive force (H$_c$) of cores from iron powder sintered under the same conditions as those of Fig. 23-4.

535

ity and lower coercive force) with increasing density and increasing sintering temperature is evident. Sintering temperatures as high as 1370° C (2500° F) are not uncommon in the fabrication of magnetic parts. On the graphs the properties of fully dense material (density 7.83 g/cm^3) are shown. They were produced by hot repressing, a technique not yet generally used in commercial practice. The maximum permeability and coercive force for this fully dense material from powder are higher than those found in wrought annealed ingot iron ($\mu = 4000$, $H_c = 1.30$ Oersteds) and wrought annealed 1010 steel ($\mu = 3800$, $H_c = 1.80$ Oersteds). Pure iron magnetic parts are generally sintered in dissociated ammonia or hydrogen which, in porous compacts, lower the residual carbon and oxygen content. Parts sintered in endothermic and exothermic atmospheres have somewhat inferior magnetic properties.

Pure iron magnetic parts produced from coarse particle size fractions have pores with a larger mean pore area and a larger mean free path between pores than those from fine powder. Moyer[4] has shown that the pore characteristics are reflected in the values of maximum permeability and coercive force. As an example, specimens from a –60 +100 mesh fraction of atomized iron powder fabricated to a sintered density of 7.4 g/cm^3 and sintered at 1260° C (2300° F) have a mean pore area of 102 μm^2 and a free path between pores of 143 μm. Their maximum permeability is 4900 and their coercive force 0.9 Oersteds. Specimens from the –400 mesh fraction of the powder fabricated identically have a mean pore area of 43 μm^2 and a mean free path between pores of 62 μm. Their maximum permeability is 3200 and their coercive force 1.7 Oersteds.

Iron-silicon alloys for magnetic applications[5] are produced from powder by mixing the same pure atomized iron powder used for pure iron parts with ferrosilicon powder which may contain from 17 to 33% silicon. The ferrosilicon powder is hard and brittle, but since it constitutes only a minor constituent of the powder mixture for a 3% silicon alloy, the mixture can be compacted without major difficulties. In sintering compacts from the mixture, control of the sintering atmosphere, which is often dissociated ammonia, is critical. A low dewpoint is necessary in order to keep the atmosphere reducing to the ferrosilicon particles. Otherwise the oxide coatings on these particles remain during sintering and prevent interduffision between the iron and ferrosilicon and the formation of a homogeneous iron-silicon solid solution alloy. Under suitable reducing conditions such a homogeneous alloy is readily formed when compacts are sintered one hour at 1260° C (2300° F), as was pointed out in chapter 10. Because of the shift in the equilibrium constant for the oxidation-reduction reaction with increasing temperature, higher dewpoints are permissible at higher sintering tempera-

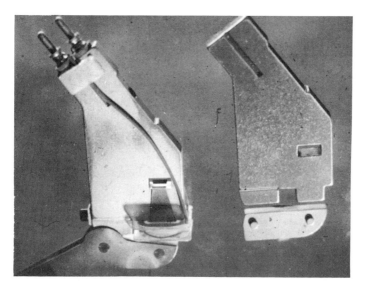

FIGURE 23-8 On the right: Actuator base pressed and sintered from 3% silicon steel (top) and armature from pure iron powder (bottom). On the left: assembly for use in computer printers (Courtesy: Remington Arms Co., Inc.).

tures, as discussed in chapter 7. At these higher sintering temperatures of 1260 to 1315° C (2300 to 2400° F) iron-silicon compacts shrink linearly 2 to 3% which makes the dimensional control during sintering more difficult. Sintering in vacuum is a satisfactory alternative to sintering in dry dissociated ammonia.

Iron-phosphorus alloys are produced by mixing ferrophosphorus powder with pure atomized iron powder, as was discussed in chapter 17. Alloys with phosphorus contents of more than 0.6% exhibit considerable shrinkage during sintering.

The magnetic properties of iron-silicon and iron-phosphorus alloys differ from those of pure iron.[5,6] Increasing silicon and phosphorus contents lower the saturation induction slightly. On the other hand, the maximum permeability is increased and the coercive force is decreased. Typical values of maximum permeability and coercive force for specimens with a density of 7.3 g/cm^3 are $\mu = 3000$, $H_c = 1.4$ Oersteds for pure iron, $\mu = 5000$, $H_c = 0.9$ Oersteds for a 3% silicon alloy and $\mu = 6000$, $H_c = 1.3$ for a 0.75% phosphorus alloy. Even more important is the increase in electrical resistivity with increasing silicon and phosphorus content from 10-20 μohm-cm for

FIGURE 23-9   Pole piece from 50% iron - 50% nickel prealloyed powder used in aerospace guidance control unit (Courtesy: Supermet division of Stenadyne Corp.).

pure iron to 41-52 μohm-cm for a 3% silicon-iron alloy and 27-28 μohm-cm for a 0.75% phosphorus alloy.

These alloys are used where a rapid decay of the magnetic field is important when the current producing the magnetizing force is switched off. The rapid decay is due to the higher reluctance and is related to the higher electrical resistivity and lower coercive force of the materials compared with pure iron.

An example for an application of this kind is the actuator base in high speed printers for computers produced from a 3% silicon alloy. It is shown in Figure 23-8, together with the armature from pure iron.

The 50% iron-50% nickel powder metallurgy alloys for magnetic applications[7] are produced from a prealloyed atomized powder. They are used for applications requiring high permeability. In order to obtain the desired high density of 7.5 g/cm^3, the powder is compacted at a pressure of 690 MPa (100,000 psi) and sintered at temperatures above 1200°C (2200°F) in vacuum. With this processing a maximum permeability of 40,000 and a coercive force of 0.2 Oersteds may be achieved, which are comparable to those of wrought nickel base magnetic alloys. A photograph of a pole piece from this material is shown in Figure 23-9.

II.   Wrought Permalloy from Powder

"Permalloy" is the name of a group of soft magnetic nickel-base alloys with very high permeabilities. They are generally produced from cast ingots, but powder metallurgy has been occasionally used for producing the alloy with 77% nickel, 14% iron, 4% molybdenum and 5% copper from pressed and sintered powder compacts.[8,9] The raw materials used are carbonyl nickel powder, carbonyl iron powders, molybdenum powder produced by hydrogen reduction of pure oxide and copper powder produced by atomizing oxygen free high conductivity copper. The powders are thoroughly mixed and pressed without the addition of a lubricant to the powder mixture at pressures between 310 and 520 MPa (45 and 75,000 psi) into compacts weighing about 7 kg (15 lbs), presintered several hours at 600° C and finally sintered at 1200° C for several hours. The sintered compacts are forged and then rolled into strip, rod or wire, taking precautions to avoid contamination. Using powder metallurgy, it is considerably easier to control the exact composition of the alloys, upon which the magnetic properties critically depend. No additions of small amounts of silicon and maganese, used in the cast alloys for deoxidation purposes, are necessary in the powder metallurgy alloys. For this reason, the powder metallurgy alloys may be annealed in wet hydrogen without impairing initial permeability. Also the powder metallurgy alloys do not develop a layer of fine grains on the surface of the bars, which is also detrimental to high values of initial permeability. Both of these phenomena in alloys from cast ingots are due to internal oxidation of residual silicon and manganese in the alloys.

In addition to permalloy, a 50% iron, 50% nickel alloy strip treated to give a cube texture and therefore a rectangular hysteresis loop has been produced from sintered compacts from mixtures of carbonyl iron and carbonyl nickel powder.[10] The cube structure is produced by cold rolling 97-99% and then annealing. Before the cube structure is produced, the grain size of the coarse-grained sintered compact has to be refined by means of light reductions with intermediate annealing at 700 to 800° C. In strip produced from powders, the development of the cube texture is not critically dependent upon the temperature of the final anneal. This is in contrast to strip with cube structure from cast ingots.

III.   Alnico Permanent Magnets

The excellent permanent magnet properties of alloys of iron, nickel and aluminum, often with additions of Co, Cu and Ti, were discovered by Mishima in 1932.[11] In contrast to the older permanent magnet alloys, which were steels which were magnetically hardened by quenching, the "Alnico" alloys are precipitation hardened alloys, which obtain their properties by cooling the alloy, which is isotropic at high temperatures, at a controlled rate with or without a subsequent aging treatment. The high coercive force of these alloys is connected with the spinodal precipitation of elongated single domain particles of a strongly magnetic iron-rich phase (α) in a weakly magnetic or non-magnetic (aluminum-rich) matrix (α' phase).[12] The properties of the cobalt-rich Alnico alloys can be further improved by cooling the alloys through the Curie point in a strong magnetic field.

Alnico magnets are produced by casting, generally in sand molds. If necessary, they are shaped by grinding since the alloy is too brittle to shape by plastic deformation or by machining. The cast alloy is also mechanically weak. These conditions limit the manufacture of magnets of small size and intricate shape by casting. For this reason, production of small Alnico magnets by powder metallurgy was developed in the 1930s.[13] Sintered Alnico magnets are produced[14] from a mixture of powders, all of which pass through a 200 mesh (75 μm) sieve. The mixture consists of about 50% of soft elemental powders, iron, nickel, cobalt and copper. The remaining material is a master alloy powder containing combinations of Co, Ni, Ti, Fe and aluminum produced by grinding a cast alloy. The master alloy has a melting point below 1200° C, since a 50 wt. % Fe, 50 wt. % Al alloy has a liquidus temperature of 1165° C. The master alloy provides a liquid phase during sintering, but this liquid phase is transient. When the final alloy composition, which is homogeneous at the sintering temperature, has been produced by interdiffusion at the end of the sintering cycle, the compacts are completely solid.

The powder mixture is pressed at a very high pressure of 1100 MPa (160,000 psi) in order to produce compacts with sufficient green strength. Since the powder mixture does not flow well, attention must be paid to the proper filling of the die cavity. In general, Alnico compacts should have a maximum 1:1 diameter to length ratio. Most pressed and sintered Alnico magnets have relatively simple shapes.

Alnico magnets are sintered at 1300° C (2370° F) in an atmosphere of very dry hydrogen, gettered by a titanium alloy. Sintering Alnico magnets in vacuo has been reported from Germany.[15] The compacts shrink during sintering between 5 and 7% linearly and have a final density of 7.0-7.2 g/cm³ depending on geometry.

As in cast Alnico magnets, certain grades are produced oriented by cooling the material through the Curie temperature in a magnetic field. In Table 23-1 the composition and magnetic properties of a series of Alnico alloys produced by powder metallurgy are shown.[14] Sintered Alnico magnets have certain advantages over cast magnets. Because of their fine grain size they are stronger than the coarse grained cast magnets. They also do not exhibit the cracks, cold shuts and segregation of impurities at grain boundaries encountered in cast magnets and therefore have a more uniform flux distribution. Sintered magnets can be produced to considerably closer dimensional tolerances than cast magnets. On the other hand, because of

Powder Metallurgy Products for Magnetic Applications

Table 23-1 Composition and Properties of Sintered Alnico Magnets

Designation	Composition	Condition	Residual induction B_r Gauss	Coercive force H_c Oersted	Maximum energy product. Mega-Gauss-Oersted	Induction at max. energy product. B_D Gauss	Demagnetizing force at max. energy product. H_D Oersted	B_D/H_D
Alnico II	Al 10 Ni 19 Co 13 Cu 3 Ti 0.5 Fe Bal	unoriented	7100	550	1.5	4500	335	13.5
Alnico 5	Al 8 Ni 14 Co 24 Cu 3 Fe Bal	oriented	10,900	620	3.5	800	440	18
Alnico 6	Al 8 Ni 16 Co 24 Cu 3 Ti 1 Fe Bal	oriented	9400	790	3.0	6000	500	12
Alnico 8B	Al 7 Ni 13 Co 39 Cu 3 Ti 6 Fe Bal	oriented	8300	1650	5.0	5000	1000	5
Alnico 8H	Al 7 Ni 14 Co 37 Cu 3 Ti 8 Fe Bal	oriented	7500	1900	5.5	4500	1220	3.7

their slight porosity, the magnetic characteristics of sintered magnets are generally not quite as good as those of cast magnets. This does not apply to Alnico 8 alloys which easily form a non-magnetic gamma phase and require quenching from a normalizing temperature. The appearance of a non-magnetic phase can be more readily avoided in powder metallurgy fabrication than by casting, and sintered Alnico 8 grades have, therefore, magnetic properties equal to those of cast grades. Because of their excellent properties, Alnico 8 grade magnets have become the volume leaders among sintered Alnico magnets.

IV. Cobalt-Rare Earth Permanent Magnets

The development of cobalt-rare earth permanent magnet materials goes back to the early 1960s, when Nesbitt, Wernicke and coworkers[16,17,18] established the structure, magnetic moment and Curie temperature of intermetallic compounds of the type $ReCo_5$, in which Re is a rare earth metal. These compounds have hexagonal crystal structure. Their usefulness for producing permanent magnets is based on their extremely high magnetic crystalline anisotropy in addition to high saturation magnetization and Curie temperature. Among the $ReCo_5$ compounds the one containing samarium, $SmCo_5$, has the highest anisotropy constant equal to 17.5×10^7 erg/cm^3 at 300° K.[19,20,21,22] The first cobalt-rare earth magnets were produced by bonding crushed powder of the compounds with plastic.[23] Beginning in 1969, a method of producing magnets was developed in which compacts of the intermetallic compound powders were sintered after they had been magnetically aligned and pressed.[24,25] The steps in this method of production are shown schematically in Figure 23-10, based on the production of $SmCo_5$ magnets. Alloys of two compositions, one corresponding to the compound $SmCo_5$, the other a sinter addition richer in samarium, e.g. 60 wt. % Sm, 40 wt. % Co, are prepared by induction melting and chill casting. The alloys are reduced to a fine powder by crushing, pulverizing and finally milling, which may be ball milling under toluene or jet milling with dry nitrogen. During milling, oxidation of the compound should be avoided as much as possible. The average particle size of the powders should be comparable to the critical diameter for a single domain particle, which means 6-8 μm. The particle size distribution should be narrow.

The powders with the two compositions are blended by tumbling to give an average composition of 62.6% cobalt, slightly richer in samarium than the stoichiometric $SmCo_5$ composition. The milled powder is enclosed in a plastic envelope and magnetically aligned in a magnetic field on the order of 60 kOe (kilo-oersteds), so that the c-axis of the single crystal particles is

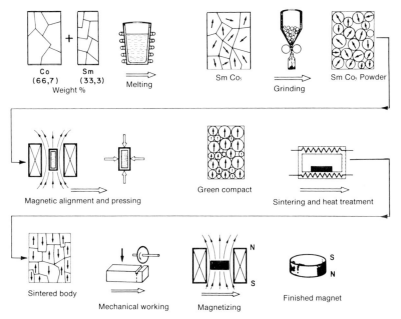

FIGURE 23-10 Flow chart for the production of cobalt-samarium permanent magnets.

parallel to the direction of the applied field. The powder is then hydrostatically compacted at a pressure of approximately 1400 MPa (200,000 psi) to a density of 6.9 g/cm^3, which is 80% of the solid density of the compound. The compacts are sintered in purified argon in the temperature range 1090 to 1150° C (1995-2100° F). They densify to approximately 7.7 g/cm^3, 90% of solid density. The porosity in the sintered compacts is non-interconnected. During sintering a liquid phase is formed from the two cobalt-samarium alloys. The densification of the compacts is probably not due to the heavy alloy liquid phase sintering mechanism (chapter 12), since the appearance of the liquid phase is only transient. Instead, densification must be primarily attributed to sintering within the composition range of the solid SmCo$_5$ phase but at an off-stoichiometric composition.[26,27] It is accelerated by the presence of vacant cobalt lattice sites in a composition containing more samarium than corresponds to stoichiometric SmCo$_5$. The structure of the sintered alloy consists of the SmCo$_5$ phase together with pores and inclusions of Sm$_2$O$_3$.

Following sintering a thermal treatment near 900° C (1650° F) is applied to the magnets to improve their coercive force, before they are magnetized in a field of 60 kOe. Because of their very high remanence and coercivity

Table 23-2 Magnetic properties of RE-Co$_5$ permanent magnets

	Sm Co$_5$	Sm$_{0.4}$ Pr$_{0.6}$ Co$_5$	Sm$_{0.2}$ MM$_{0.8}$ Co$_5$
Remanence B$_r$ in KG	10.0	16.3	8.0
Coercive force $_1$H$_c$ in KOe	38	16.6	16
Maximum energy product in M G Oe	24.6	26.0	16

magnetizing the magnets and measuring their magnetic properties is difficult. Methods of making the measurements using a superconducting solenoid have been described.[28]

The coercivity in kOe (kilo-Oersted), remanence in kG (kilo-Gauss) and maximum energy product in MGOe (Mega-Gauss-Oersted) are extremely high. In Table 23-2[29,30] values for commercial permanent magnets of the compositions SmCo$_5$, Sm$_{0.4}$Pr$_{0.6}$Co$_5$ (samarium-praseodymium-cobalt compound) and Sm$_{0.2}$MM$_{0.8}$Co$_5$ (compound of samarium, "mischmetal" and cobalt) are shown. The principal reason for replacing some of the samarium by other rare earth metals is to lower the cost. "Mischmetal" ("MM") is a cheaper cerium-rich rare earth alloy with 50 wt.% Ce, 27 wt.% La, 16 wt.% Nd and 5 wt.% Pr.

In the last few years a large amount of research and development effort has been expanded on the cobalt-rare earth permanent magnet materials. Besides the RE-Co$_5$ compositions, those based on the intermetallic compound RE$_2$Co$_{17}$ have been investigated. In spite of outstanding properties of the permanent magnet materials their commercial application has been limited primarily because of the very high cost of the raw materials.

V. Elongated Single Domain Permanent Magnets and Magnetic Core Materials for High Frequencies

The good permanent magnet properties of Alnico are due to the fact that its structure consists of magnetically aligned elongated single domain precipitated particles in a matrix which is weakly magnetic or non-magnetic. The idea of producing unsintered permanent magnets from single domain particles of a magnetic material aligned with, but magnetically insulated from each other, grew out of the advances in the domain theory of magnetization in the early 1940's. The early experimental work of Neel[31] in France was based upon the low temperature reduction of ferrous formate to which small amounts of calcium or magnesium formate had been added and

Table 23-3 Magnetic properties of permanent magnets from ESD particles

	Iron powder	Iron-cobalt powder
Remanence B_r in KG	7.9	9
Coercive force H_c in KOe	0.56	0.85
Maximum energy product in M G Oe	2.2	3.6

compacting the resulting very fine powder. These magnets were difficult to produce, were unstable and did not have sufficiently good magnetic properties for commercial production. A considerable improvement in magnetic properties was achieved with elongated powder particles having shape anisotropy. These are so-called "ESD" (elongated single domain) permanent magnets developed in the late 1950's.[32] The basic steps in the production of these magnets as described by Luborsky, Paine and Mendelson[32] are:

1. Highly elongated dendritic particles 10-20 nm in diameter of iron or an iron-cobalt alloy are electrolytically deposited from a sulfate electrolyte into a mercury cathode.

2. By a thermal treatment of the particles in the mercury suspension the dendritic branches of the particles are removed and the desired rod-like shape obtained.

3. By adding a metal such as tin as a mercury-tin amalgam to the dispersion, an alloy layer is formed on the surface of the particles which may act as the magnetically insulating layer between the particles upon which the quality of the compacted magnet depends.

4. The particles are magnetically aligned in a gradually increasing magnetic field with a maximum value of 4000 Oersteds and the powder particles compacted to a packing density of approximately 50%.

5. All traces of remaining mercury are removed by vacuum distillation.

6. The porous bar may be impregnated with an organic binder serving as an alternate matrix material in place of a metallic matrix. The binder is at least partially polymerized.

7. The mixture of aligned particles and matrix material is ground to a coarse powder.

8. The magnets are pressed to their final shape after the powder particles have been magnetically realigned.

The magnetic properties of commercial ESD iron and iron-cobalt permanent magnets are shown in Table 23-3. Considerably better properties can be obtained in the laboratory.

While the principal requirement for DC applications of soft ferromagnetic materials is a low hysteresis loss, eddy current losses are of primary importance for AC applications. Hysteresis losses increase with the first power of the frequency of the alternating current, eddy current losses with the square of the frequency. A principal method of decreasing eddy current losses is to increase the electrical resistance of the material. For power frequency applications, high resistance is obtained by dividing the iron-silicon cores of AC motors, generators or transformers into laminations, each separated from the next by a thin layer of electrically insulating material. There is, however, a practical limit to the reduction of losses by using laminations. For cores in applications for frequencies higher than power frequencies, cores from compressed insulated ferromagnetic powders are widely used. The history of these core materials has been described by Jones.[33] In many of the present applications of high frequency cores, insulated ferromagnetic oxide powders having high intrinsic electrical resistivities have replaced metal powder cores, but two of them, in which metal powders were or are used, will be briefly described.

The first of these are cores for radio frequency applications where the so-called "E" type of carbonyl iron powder, described in chapter 2, is used. The particles of this powder are spherical, 3 to 9 μm in diameter with an onionlike structure. They contain 1% O_2, 0.6% C and 0.5% N_2 and are extremely hard. The particles are coated with a thermosetting phenolic resin, small amounts of an insulating material are added and the mixture is pressed into the desired shape of the core. Because of the high intrinsic mechanical and magnetic hardness of the powder, strains introduced by the pressing operation, which are not relieved by the subsequent low temperature polymerizing heat treatment of the core, do not have much effect on the magnetic properties of the cores. Their permeability is on the order of 10, which is very low compared with material for DC magnetic applications, but much higher than the permeability of air or vacuum which is 1.

A second application using insulated alloy powder particles as the core material is the permalloy powder cores used in loading coils and inductors at audio frequencies.[34,35] The composition of the alloy is 81% nickel, 17% iron and 2% molybdenum. Powder of this composition is produced by the process described in chapter 2, in which the alloy is embrittled by the addition of a few thousandths of a percent of sulfur and then mechanically comminuted. A small amount of talc is added to the powder before it is heat treated between 630 and 760° C (1170 and 1400° F). The heat treated powder is coated with an inorganic insulating mixture of sodium silicate, magnesium oxide, colloidal clay and kaolin. Zinc stearate is added as a pressing lubricant and the mixture pressed into cores at the very high pressure of 1650 MPa (240,000 psi) into the shape of the desired core. In order to obtain the desired relatively high permeabilities of 190 to 220 the cores are heat treated at 630° C (1170° F). This heat treatment relieves the stresses introduced during pressing without impairing the insulating properties of the coating. The cores have acceptable core losses in the audio frequency range.

References

1. G. A. Bolze and J. M. Capus, Factors affecting magnetic properties of sintered iron, Mod. Dev. in Powder Metallurgy, vol. 11, ed by H. H. Hausner and P. W. Taubenblat, Princeton, p. 355-370, (1977).
2. ASTM standard A 596, Standard method of test for D.C. magnetic properties of material using ring test procedures and the ballistic method.
3. K. M. Moyer, The magnetic properties (DC) of atomized iron powder cores, Mod. Dev. in Powder Metallurgy, vol. 11, ed by H. H. Hausner and P. W. Taubenblat, Princeton, p. 371-384, (1977).
4. K. H. Moyer, P/M parts for magnetic applications, presented at P/M short course on "P/M magnetic materials and applications," Memphis, (Feb. 1979).
5. L. W. Baum, Theoretical and practical applications for the P/M production of magnetic parts, ibid.
6. O. W. Reen and R. J. Waltenbaugh, Properties and applications of ferrous magnetic P/M materials, ibid.
7. H. L. Sanderow, Soft magnetic materials—P/M applications, ibid.
8. E. V. Walker, D. K. Worn and R. E. S. Walters, Application of powder metallurgy to the production of high permeability magnetic alloy strip, "Symposium on Powder Metallurgy," Special Report No. 58, Iron and Steel Institute, London, p. 204-208, (1956).
9. Wiggin Nickel Alloys by Powder Metallurgy, Henry Wiggin and Co., Birmingham, England.
10. E. V. Walker and R. E. S. Walters. The production of grain-oriented 50-50 nickel-iron magnetic strip by cold rolling from sintered compacts, Powder Metallurgy, No. 4, p. 23-31, (1959).
11. T. Mishima, Ohm, vol. 19, p. 353, (1932). U.S. Patents 2,027,996 and 2,027,994, (1936).
12. R. Tebble and D. J. Craik, Magnetic Materials, N.Y., p. 430, (1969).
13. C. H. Howe, Sintered Alnico, in "Powder Metallurgy," ed by J. Wulff, Cleveland, p. 530-536, (1942).
14. Personal communication from R. Hendron, The Arnold Engineering Company, Marengo, Ill.
15. Oerstit Permanent Magnets, Deutsche Edelstahlwerke A.-G. Magnetfabrik Dortmund.
16. E. A. Nesbitt, H. J. Williams, J. H. Wernicke, and R. C. Sherwood, J. Appl. Phys., vol. 32S, p. 342-343, (1961).
17. E. A. Nesbitt, H. J. Williams, J. H. Wernicke, and R. C. Sherwood, J. Appl. Phys., vol. 33, p. 1674-1678, (1962).
18. J. H. Wernicke and S. Geller, Acta Cryst., vol. 12, p. 662-665, (1959).

19. E. Tatsumoto, F. Okamoto, H. Fujii, and C. Inore, J. de Physique, vol. 32, p. 550-551, (1971).
20. H. P. Kleis, A. Menth, and R. S. Perkins, Physica, vol. 80B, p. 153-163, (1975).
21. A. S. Ermolenko, IEEE Trans. Mag., vol. MAG 12, p. 992-996, (1976).
22. S. G. Sankar, V. U. S. Rao, E. Segal, W. E. Wallace, W. G. D. Frederick, and H. J. Garett, Phys. Rev., vol. 11, p. 435-439, (1975).
23. K. H. J. Buschow, W. Luiten, P. A. Naastepad, and F. F. Westendorp, Philips Tech., Rev., vol. 29, p. 336, (1968).
24. D. K. Das, IEEE Trans. Mag., vol. MAG5, p. 214-216, (1969).
25. M. G. Benz, and D. L. Martin, Cobalt-samarium permanent magnets, prepared by liquid phase sintering, Appl. Phys. Letters, vol. 17, p. 176-177, (1970).
26. M. G. Benz, and D. L. Martin, Mechanism of sintering in cobalt-rare earth permanent magnet alloys, J. Appl. Phys., vol. 43, p. 3165-3170, (1972).
27. G. H. Gessinger and E. DeLamotte, The sintering mechanism of samarium-cobalt alloys. Z. f. Metallkunde, vol. 64, p. 771-775, (1973).
28. D. L. Martin and M. G. Benz, Measurement of magnetic properties of cobalt-rare earth permanent magnets, IEEE Trans. Mag., vol. MAG7, p. 285-291, (1971).
29. D. L. Martin, J. T. Geertsen, R. P. Laforce, and A. C. Rockwood, Proc. 11th Rare Earth Res. Conf., Michigan, p. 342-352, (1974).
30. D. V. Ratman, and M. G. H. Wells, AIP Conf. Proc., vol. 18, p. 1154-1158, (1974).
31. L. Neel, Cahiers Phys., vol. 25, p. 1, 1944, U.S. Patent No. 2,651,105, (1953).
32. F. E. Luborsky, T. O. Paine, and L. T. Mendelson, Permanent magnets from elongated single domain particles, Powder Metallurgy, No. 4, p. 57-79, (1959).
33. W. D. Jones, Fundamental Principles of Powder Metallurgy, London, p. 680-710, (1960).
34. E. Schumacher, Magnetic powders and production of cores for induction coils, in "Powder Metallurgy," ed by J. Wulff, Cleveland, p. 166-172, (1942).
35. V. J. Pringel, Heat treatment of 2-81 molybdenum permalloy powder cores, The Western Electric Engineer, vol. 14, No. 2, p. 2-10, (1970).

Electrical Applications of Powder Metallurgy Materials

This chapter will be primarily concerned with materials for electrical contacts by powder metallurgy which include not only contacts as such, but also electrical resistance welding electrodes and metallic brushes for electric motors and generators. These metallic brushes replace those from graphite or carbon where higher electrical conductivity is required. In a final section, the contributions of powder metallurgy to superconducting materials are briefly discussed.

I. Electrical Contacts

Electrical contacts are as old as electrical technology, but a scientific understanding of the physics of these contacts is fairly recent. It is treated in the books by Holm[1] and by Jones.[2] The development of materials for electrical contacts has been largely empirical. In his book on electrical contact materials, Keil[3] proposed dividing contacts into five groups as a basis for the choice of contact materials and suggested materials for the first four of these groups. The groups and the materials suggested for them are:

     1. Contacts which switch practically without any electrical load, so that the current does not affect the contact surface even with frequent making and breaking. Materials: Ag, Au, Rh, Pt and their alloys

FIGURE 24-1 Assembly of composite contacts from metal powders varying in size and shape (Courtesy: Advanced Metallurgy, Inc.).

2. Contacts for low voltages and currents for which material transport from anode to cathode is observed under D.C. conditions. Materials: Cu, Ag, Au, Pd, Pt and their alloys.

3. Contacts for medium electrical loads for which in D.C. applications material transport from cathode to anode is observed. Materials: Cu, Ag, Pd, Pt, W and compound materials.

4. Contacts for the highest electrical loads, where burn-off occurs. Materials: Mo, W and compound materials.

5. Sliding contacts.

Compound materials are combinations of metals with second phases insoluble in the metal in the liquid. Most of them are produced by powder metallurgy. According to Keil's classification, it is these compound materials together with tungsten and molybdenum, which are primarily used for heavy duty contacts. They are the subject of a book by Schreiner.[4] Examples of electrical contacts produced by powder metallurgy are shown in Figure 24-1. The most important of the compound materials for electrical contacts are the following:

Silver-tungsten with 25 to 50 wt. % Ag, bal. W.
Silver-tungsten carbide with 35 to 65 wt. % Ag, bal. WC.
Copper-tungsten with 20 to 50 wt. % Cu, bal. W.
Silver-molybdenum with 30 to 50 wt. % Ag, bal. Mo.

Silver-cadmium oxide with 85 to 90 wt. % Ag, bal. CdO.
Silver-graphite with 90 to 99.75 wt. % Ag, bal. graphite.
Silver-nickel with 40 to 90 wt. % Ag, bal. Ni.

Some compound materials, less frequently used for electrical contacts, are copper-tungsten carbide, copper-graphite, copper-molybdenum, copper-lead, copper-molybdenum carbide, silver-molybdenum carbide, silver-molybdenum sulfide and silver-tin oxide. All of these materials consist of a combination of the metals Ag or Cu, which have high electrical conductivity, and a second component. Since the second components (W, WC, Mo, CdO, Graphite, Ni in Ag and W, WC and graphite in Cu) are completely insoluble in the base metal, they lower the electrical conductivity of the silver or copper much less than alloying elements which form solid solutions with silver or copper. This is important for their use as electrical contacts. The added components are chosen primarily because they increase the resistance of the contacts to burn-off and to sticking or welding, as is indicated in table 24-1.

Table 24-1[5]   Effect of added components upon the resistance of copper and silver to burn-off and to sticking and welding.

Copper → increase in burn-off resistance → Cu-W
                                            Cu-WC
Silver → increase in resistance to welding → Ag-CdO
                                             Ag-Graphite
Silver → increase in burn-off resistance → Ag-Ni → Ag-Mo → Ag-W
                                                            Ag-WC

A.   Production of Contact Materials[4]

Tungsten contacts. In chapter 14 the production of tungsten rods from powder by electrical resistance sintering, rolling and swaging is described. From the swaged round rods pure tungsten contacts up to 15 mm in diameter and 0.5 to 3 mm thick are cut.

Silver-graphite contacts. Silver-graphite compound materials are produced by mixing silver powder, produced by electrolysis or by precipitation (see chapter 2), with graphite powder, pressing at approximately 275 MPa (40,000 psi), sintering between 700 and 900°C (1300-1650°F) in an inert or reducing atmosphere and repressing at 600 to 900 MPa (87-130,000 psi). Particularly in materials with low graphite contents almost complete densification can be achieved in repressing.

For silver-graphite as well as for other silver base contact materials two methods of producing finished contacts are available. In the first method, fairly large blanks are produced by compacting, sintering and repressing, which are subsequently hot-rolled or sometimes cold-rolled into sheet or drawn into wire from which the contacts are punched or cut. With the other method, contacts are produced to finished shape on automatic compacting presses and again sintered and repressed. The powders used for these contact materials are generally not free flowing so that the powder mixtures have to be granulated before they can be fed into automatic compacting presses. Granulating may be done by pressing the powder lightly into compacts which can be readily broken up into coarse powder or by a thermal treatment.

After being finished to size, the contacts are attached to a substrate usually by brazing. Neither silver-graphite nor silver-cadmium oxide brazes well. Contacts from these compound materials may therefore be produced as two layer products with one layer consisting of the compound material, the other being pure silver. When the contacts are produced to finished size, the two layers may be produced in a single compact by using a specially designed automatic compacting press with two feed shoes. The die cavity is first filled with pure silver powder by one feed shoe, the lower punch is lowered to allow the second feed shoe to fill the die cavity with the compound powder and the two layer compact is pressed.

Silver-nickel compacts. These contacts may be produced from a mixture of silver and nickel powder. Electrolytic or precipitated silver powder and carbonyl nickel powder are used. In order to obtain a very fine distribution of the nickel in the silver matrix, a compound silver-nickel powder may be produced by precipitating an intimate mixture of the metals as carbonates with sodium carbonate from an aqueous solution of silver and nickel nitrate, reducing the carbonate mixture in hydrogen at 400°C (750°F) and washing the compound powder free of alkali. The powder mixture or the compound powder is pressed at relatively low pressure, e.g. 140 MPa (20,000 psi) in order to avoid trapping of gases during the subsequent sintering in the neighborhood of 800°C (1470°F) and repressed.

Silver-cadmium oxide contacts. Silver-cadmium oxide compact materials may be produced by powder metallurgy, but also by fusion metallurgy. In the latter case, a cast silver-cadmium alloy ingot is rolled into sheet or drawn into wire and the silver-cadmium oxide compound material produced by internal oxidation, i.e. by heating the alloy in an oxidizing atmosphere. Oxygen diffuses into the alloy, reacts with the cadmium solute in the alloy and produces cadmium oxide particles in situ. On the other hand, the compound material may be produced similarly to silver-graphite material by mixing silver powder and cadmium oxide powder, compacting, sintering in an oxidizing atmosphere and repressing. Both methods have advantages and disadvantages. The rate of internal oxidation of the alloy depends upon the rate of diffusion of oxygen into silver. The depth of the cadmium oxide layer is therefore proportional to the square root of the time of treatment. A subscale layer containing CdO 1 mm (0.04″) thick in an alloy of 90 Ag-10 Cd is formed during internal oxidation at 800°C (1470°F) in approximately 60 hours. The size of the oxide particles at the surface is extremely fine, but gradually coarsens with increasing distance from the surface. The compound material produced by internally oxidizing cast and worked ingots is completely dense. The particle size of the oxide in compound material prepared by mixing silver and cadmium oxide powders, compacting and sintering depends, of course, on the size of the original powders used. The cadmium oxide particles are uniform in size and distribution. The density depends upon the treatment parameters, particularly the repressing pressure.

Powder metallurgy and internal oxidation may be combined in producing the compound material. A silver-cadmium alloy powder is produced by atomizing the molten alloy. The alloy powder is internally oxidized, pressed, sintered and repressed into contacts.

Copper-tungsten, silver-tungsten and silver-tungsten carbide contacts. The method of production of these contact materials depends upon the ratio of high conductivity metal and refractory material. Tungsten-copper compound materials with 60% or less tungsten can be readily prepared by the classical method of mixing the powders, pressing, sintering, generally below the copper melting point, and repressing. These alloys can also be cold rolled, swaged or hot extruded. When a tungsten-copper compound material is formulated with small amounts of nickel, e.g. a 70 W, 25 Cu, 5 Ni material, it can be sintered to full density by liquid phase sintering as described in chapter 11. The addition of nickel lowers, however, the electrical conductivity of the material considerably. Materials with 60 to 80% tungsten are generally produced by infiltration either of loose tungsten powder or of a pressed and sintered tungsten powder compact. This method has been described in chapter 12. The fabrication of tungsten-silver, molybdenum-copper, molybdenum-silver and copper-tungsten carbide compound materials is similar to that of tungsten-copper materials.

B. Properties and Uses of Contact Materials[4]

Tungsten metal contacts are used in automobiles for horns, for voltage regulators and for spark plug electrodes.

Silver-nickel contacts. Adding nickel to silver lowers its electrical conductivity, from $63 \times 10^4 \, \text{Ohm}^{-1} \, \text{cm}^{-1}$ for pure silver to $38 \times 10^4 \, \text{Ohm}^{-1} \, \text{cm}^{-1}$ for an alloy with 40% Ni. Nickel increases the hardness with a 40 wt. % Ni alloy having a Vickers hardness of 72 in the annealed and 128 in the cold rolled condition. The burn-off of an 80 wt. % Ag - 20 wt. % Ni alloy is only one half that of pure silver. Because of the low material transport compared to pure silver, silver-nickel is used for switching direct current in air. Contacts with 40 wt. % nickel are used for the main contact in the load switches for electrical locomotives. Silver-nickel contacts are also used for temperature regulators in household applicances.

Tungsten-copper, molybdenum-copper, tungsten-silver and molybdenum-silver contacts. The conductivity of tungsten-copper contact materials increases with increasing copper content. The values lie between those calculated theoretically for parallel and series arrangements of the components in the microstructure. The hardness increases approximately linearly with increasing tungsten percentages by volume. The high resistance to burn-off by arcing has led to the use of these materials as arcing contacts. This is shown in Figure 24-2. The arcing tip does not carry

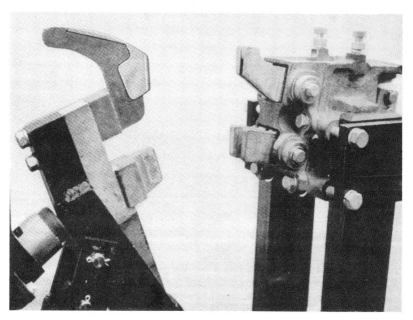

FIGURE 24-2 Circuit breaker; the upper contacts faced with tungsten-copper compound withstand the arc when the switch is opened.

the current, but only has to withstand the arc when the circuit is opened; the current may be carried by silver plated copper contacts. W-Ag contacts have a lower contact resistance than W-Cu contacts.

Silver-tungsten carbide contacts. The tungsten carbide-silver compound materials have outstanding resistance to mechanical loads and to burn-off by arcing. Their wear resistance increases with increasing WC content. Similar to tungsten-silver and tungsten-copper contacts they are used as arcing contacts, but WC-Ag contacts with 40 to 60% WC are also used for current carrying contacts in switches which require high resistance to burn-off and to welding with adequately low contact resistance.

Silver-graphite contacts. The electrical conductivity of Ag-graphite contacts with 5% graphite at approximately 40×10^4 Ohm^{-1} cm^{-1} is sufficiently high so that the currents carried during switching do not cause unduly high heating of the contact. The Vickers hardness decreases with increasing graphite content and reaches a value of 38 with 10% graphite. An outstanding property of silver-graphite contacts is their resistance to welding. They are therefore used for switches which must not weld even with high short circuit currents up to 10 KA as in elevators, protective switches and similar applications. Silver-graphite contacts are not highly resistant to burn-off, the less so the higher the graphite content. Their composition must be adjusted to give an adequate combination of resistance to welding and to burn-off.

II. Materials for Electrical Resistance Welding Electrodes

Resistance welding electrodes have three major functions:

> Conducting the welding current to the work pieces
> Transmitting to the work piece in the weld area the amount of force needed to produce a satisfactory weld
> Rapidly dissipating the heat from the weld zone

In applications, where high heat, long weld time, inadequate cooling or high pressure are involved, copper alloys produced by fusion metallurgy become inadequate and W-Cu compound materials of similar compositions and properties and produced by the same methods as those used for contacts are employed. These are the "group B" electrode materials, for which the Resistance Welding Manufacturers Association[6] has specified minimum properties, as shown in Table 24-2.

Classes 10, 11 and 12 are tungsten-copper compound materials. Class 11 electrodes contain 26 wt. % Cu, 74 wt. % W, class 10 is somewhat higher and class 12 somewhat lower in copper content. Class 13 is unalloyed tungsten,

Table 24-2   Minimum Properties for RWMA Group B (Refractory Metal Composition) Electrode Materials.

Class	Rockwell Hardness	Electrical Conductivity & IACS	Compressive psi	Strength MPa
10	B72	35	135,000	930
11	B94	28	160,000	1100
12	B98	27	170,000	1170
13	A69	30	200,000	1380
14	B85	30		

class 14 unalloyed molybdenum. Special electrode materials with higher strength have been developed from compound materials consisting of tungsten and a precipitation hardened copper alloy.[7] In addition to tungsten-copper compound material those from tungsten-silver and from tungsten carbide-copper have been used for resistance welding electrodes. Another material for electrical resistance welding electrodes based on powder metallurgy is a copper alloy dispersion strengthened with aluminum oxide. It has been discussed in chapter 21.

III.   Metal Graphite Brushes[8, 9]

Brushes are the components which transfer electrical current between a rotating element and a stationary element in electrical motors and generators. For high voltage, low current machines all carbon and all graphite brushes are used, but for low voltage, high current applications brushes with higher electrical conductivities are needed.

While graphite brushes carry at most 70 amp/in² (100 kamp/m²), those produced from a compound metal-graphite material are rated at 125 to 150 amp/in² (195 to 230 kamp/m²). Metal-graphite brushes may have compositions within the ranges:

    5-70% graphite
    0-10% Sn
    0-12% Zn
    0-10% Pb
    balance Cu.

They are produced by mixing the powders, pressing the mixture and sintering the compacts in a reducing atmosphere in a temperture range 200 to 300°C below the copper melting point (1083°C). Only the grades high in metal powder content, i.e. 80% or more metal, can be considered powder metallurgy products, in which extensive metal to metal contact is established during sintering. In order to heat treat the grades low in metal powder, additions of resins or oil have to be made. The decomposition of these organic materials during heat treatment is responsible for the mechanical bond in the finished products. The addition of tin, zinc or lead to the powder mixtures produces a liquid phase during sintering, but brushes containing these materials have lower electrical conductivities than the straight copper-graphite compositions. The graphite in the brushes gives the brushes the frictional characteristics needed in their operation.

Large metal-graphite brushes with sizes up to 1½ x 2 x 3½" are compacted as blanks which after sintering are ground to the desired size and shape. Small brushes, e.g. those for automotive starters, are compacted on automatic presses. Methods for incorporating flexible shunts, the "pigtails," in the metal graphite brushes during compacting have been developed. Strength and resistivity of metal carbon brushes depend upon their composition. Transverse rupture strength in the range of 2500 to 25,000 psi and resistivities from 4 to 500 Ohm-cm have been quoted. Figure 24-3 shows examples of metal-graphite brushes.

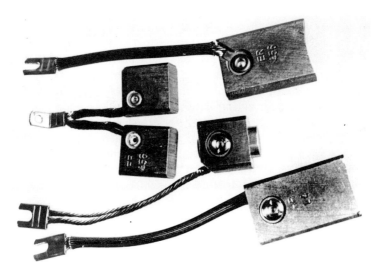

FIGURE 24-3 Photograph of industrial metal-graphite brushes (Courtesy: General Electric Co., Carbon Products Operation).

IV. Powder Metallurgy in the Fabrication of Superconducting Materials

Powder metallurgy is playing a role in the fabrication of superconducting materials of the Nb_3Me (Me = Ge, Ga, Al, Sn) type intermetallic compounds. A principal difficulty in fabricating these compounds is their extreme brittleness which makes it impossible to plastically deform them at a reasonable temperature. The original method of producing solenoids from Nb_3Sn, the "wind and react" method, consisted of packing an intimate mixture of niobium and tin powder into a niobium tube, cover the niobium tube with a monel metal jacket and draw it into a fine wire with diameters as small as 200 μm. The wire which had a core of the unreacted ductile powders was wound into a solenoid and reacted into the superconducting intermetallic compound by heating it for several hours at 950-1000°C. The principal disadvantage of this method is the necessity of winding the coil before carrying out the diffusion reaction.

In order to produce materials with stable superconducting properties, a number of conditions must be fulfilled, one of which is to have a maximum size of 10 μm for the superconducting component of the assembly. This is possible in the so-called "bronze process" which does not involve powder metallurgy.[10] Niobium filaments in a bronze matrix are heated to temperatures in the range of 650-750°C for up to 120 hours. The tin in the bronze diffuses to the surface of the niobium filaments and forms a Nb_3Sn layer, typically 2 μm thick between the Nb core and the bronze matrix.

In a variation of the "bronze process"[11] pure copper - niobium alloys with niobium concentrations above 15% are melted and chill cast resulting in a uniform dispersion of interconnected niobium particles in a copper matrix. The castings are drawn into fine wire, plated with Sn and the Sn diffused resulting in a multifilamentary high field superconductor. Instead of casting niobium-copper alloys, sintered compacts from a niobium-copper powder mixture have also been used for drawing into wire.

A third, strictly powder metallurgical approach to producing multi-filamentary Nb_3Sn wire was developed at the Lawrence Berkeley Laboratory of the University of California.[10] The process is shown schematically in Figure 24-4. Niobium powder in the range between 250-400 mesh is cold isostatically pressed at 30,000 psi (207 MPa) into rods, which are vacuum sintered for 12 minutes at 2250°C. The sintered rods are infiltrated in a tin bath for 30 seconds at 750°C. They are double clad with an inner jacket of niobium to prevent loss of tin and undesirable reactions and an outer jacket of monel or copper tubing to facilitate wire drawing. The composite rods are drawn into wire 0.010 to 0.015" (0.25 to 0.37 mm) thick and heat treated 2 minutes at 950°C to react the tin with the niobium

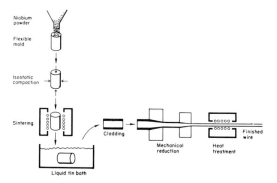

FIGURE 24-4 Flow chart of the infiltration process for producing multifilamentary superconducting wire.

matrix to form Nb_3Sn filaments with transverse dimensions of 1 to 2 μm. While the "bronze process" can be used only if Nb_3Sn (or V_3Sn) is the superconducting compound, the Lawrence Laboratory process can also be used with Nb_3Al, $Nb_3(Al,Si)$ and $Nb_3(Al,Ga)$ as the superconducting compounds.[12]

References

1. R. Holm, Electric Contact Handbook, Berlin, (1958).
2. F. Ll. Jones, The Physics of Electrical Contacts, Oxford, (1958).
3. A. Keil, Werkstoffe für elektrische Kontakte, Berlin, (1960).
4. H. Schreiner, Pulvermetallurgie elektrischer Kontakte, Berlin, (1964).
5. Reference 3, p. 267.
6. Metals Handbook, 8th edition, Vol. 6, American Soc. for Metals, Metals Park, Ohio, p. 408-409, (1971).
7. F. R. Hensel, E. I. Larsen and E. F. Swazy, Physical properties of metal compositions with a refractory metal base, ch. 42 in "Powder Metallurgy," ed. by J. Wulff, Cleveland, p, 483-492, (1942).
8. R. Kieffer and W. Hotop, Pulvermetallurgie und Sinterwerkstoffe, Berlin, p. 316-320, (1943).
9. R. R. Hoffman, Brushes and allied powder metal parts, Trans. AIME, Vol. 161, p. 550-554, (1945).
10. M. R. Pickus and J. L. Wang, Superconductors by powder metallurgy techniques, Mod. Dev. in Powder Metallurgy, Vol. 11, ed. by H. H. Hausner and P. W. Taubenblat, Princeton, p. 315-329, (1977).
11. Reports by S. Foner et. al. of the Francis Bitter National Magnet laboratory, Cambridge, Mass. at the TMS-AIME Fall Meeting in Milwaukee, Sept. (1979).
12. Report by J. L. F. Wang of Lawrence Berkeley Laboratory at the same meeting.

NAME INDEX

Hurst, H. L. — 467
Huseby, R. A. — 449, 465

I

Illis, A. — 57
Inman, M. C. — 519
Inore, C. — 548
International Nickel Company — 40,
41, 42, 57, 59, 98, 515, 516
Inukai, T. — 482
Irani, R. R. — 59, 98
Irmann, R. — 519
Ivensen, V. A. — 263, 267

J

Jackson, G. — 293, 306
Jackson, H. C. — 175
Jacobson, M. I. — 334
Jaffe, R. I. — 356
Jamison, R. S. — 443
Jangg, G. — 356, 357
Jeffes, J. H. E. — 209
Jenkins, A. — 493
Johnson, D. L. — 252, 266
Johnson, E. W. — 519
Jones, D. J. — 356
Jones, F. Ll, — 549, 559
Jones, P. K. — 482
Jones, R. L. — 519
Jones. W. D. — 4, 10, 14, 17, 103,
156, 175, 209, 212, 239, 313, 548
Joshi, A. — 356
Just, A. — 172

K

Kamm, R. — 111, 112, 116
Kang, T. K. — 306
Katushinskii, P. — 176
Kaufman, S. M. — 456, 460, 466, 467
476, 477, 482
Kawabata, Y. — 57
Kaysser, W. — 297, 306
Keil, A. — 549, 550, 559
Keith, G. H. — 335, 518
Kelly, A. — 519

Kempf, L. W. — 443
Kennedy, S. W. — 209
Keystone Carbon Company — 360, 363
Kieffer, R. — 177, 209, 313, 318, 356,
399, 559
Kienzl, F. — 356
Kimura, M. — 318
Kingery, W. D. — 266, 291, 295, 306
Kinsman, S. — 98
Klar, E. — 14, 56, 57, 334, 519
Kleis, H. P. — 548
Klimenko, P. A. — 176
Klinkenberg, I. J. — 373, 381
Klugston, E. J. — 57
Knapton, A. G. — 529
Knight, R. A. — 318
Koehring, R. P. — 465
Koerner, R. M. — 175
Kolaska, J. — 400
Kölbl, F. — 318
Kosco, J. C. — 318
Kotschy, J. — 467
Krall, F. — 209
Kramer, K. H. — 354, 357
Krantz, T. — 463, 466
Krupp, Fried. — 7, 383
Ku, R. C. — 342, 344, 356
Kubik, S. — 232, 240
Kuczynski, G. C. — 116, 249, 251, 252,
254, 259, 260, 266, 267, 278, 283,
284
Kuhn, H. A. — 139, 141, 450, 453, 454,
463, 465, 467
Kulkarni, K. M. — 519
Kunda, W. — 57
Kupfer, H. A. — 98
Kuzmick, J. F. — 225, 240
Kvist, S. A. — 441, 442, 443

L

Labbe, J. — 379, 381
Lacomb, P. — 56, 220, 232, 239, 251, 266
Lafferty, W. D. — 177
Laforce, R. P. — 548
Langer, C. — 57

Stross, F. H. — 98
Suh, S. K. — 465
Sumitomo Electric Company — 50
Suzuki, T. — 482
Svensson, L. E. — 442
Swazy, E. F. — 442, 559

T

Takahashi, N. — 399
Tallmadge, J. A. — 14, 56
Tamalet, M. — 477, 482
Tatsumoto, E. — 548
Taub, J. M. — 356
Taubenblat, P. W. — 319, 442, 443
Tebble, R. — 547
Teller, E. — 98
Tendolkar, G. S. — 14, 55
Tenzelius, J. — 441
Tequesta, J. — 519
Thellmann, E. L. — 57, 175, 357
Thermit, Inc. — 369
Thibodeau, R. D. — 177
Thölen, A. r. — 257, 267
Thornblad, O. — 283, 442
Thornburg, D. R. — 177
Thümmler, F. — 400
Tikkanen, M. H. — 262, 263, 267
Tiles, B. — 356
Titterington, R. — 57
Toeplitz, W. R. — 380
Tormyn — 31, 445
Torre, C. — 116
Towner, R. J. — 520
Tracey, V. A. — 176, 177, 318, 378, 381
Trent, E. M. — 57, 284, 399
Trzebiatowski, W. — 222, 229, 239, 240
Tsumuki, C. — 482
Tuffin, W. B. — 519
Tunderman, J. H. — 163, 176
Tuominen, S. M. — 334
Turk, R. — 379, 381

U

Udin, H. — 244, 245, 256, 265
Unckel, H. — 110, 111, 116

Universal Cyclops Corporation — 332
Ushitani, K. — 482

V

Vacek, V. J. — 281, 284
Vazquez, A. J. — 357
Vernia, P. — 457, 466
Vinogradov, G. a. — 163, 176
Volterra, R. — 240

W

Wagner, C. — 292, 306
Wagner, H. — 356
Wain, W. L. — 303, 306
Walker, E. V. — 547
Wallace, W. E. — 548
Waller, G. M. — 467
Walraedt, J. — 56
Waltenbaugh, R. J. — 547
Walters, R. E. S. — 547
Wang, C. T. — 355
Wang, J. L. F. — 559
Ward, N. L. — 481
Warner, J. S. — 57
Wartel, W. S. — 334, 519
Wartman, F. s. — 357
Watanabe, H. — 456, 466
Watanabe, K. — 303, 306
Watkinson, J. F. — 14, 56
Webber, H. M. — 209
Weber, C. E. — 493
Webster, D. — 518
Weertman, J. — 256, 267
Weiss, E. L. — 98
Wellman, S. K. — 484, 485, 486
Wellman, S. K., Inc. — 487, 488
Wellner, P. — 267
Wells, M. G. H. — 548
Wernicke, J. H. — 542, 547
Wert, C. — 356
West, G. C. — 98
Westendorp, F. F. — 548
Westerman, E. J. — 526, 529
White, J. — 293, 306
Wiberg, M. — 56

SUBJECT INDEX

279-280
loose powder sintering of sponge
layer — 310-311
copper-Ni-Sn, Cu-Be, Cu-Mn —
425-426
copper-tin
as bond in diamond tool
materials — 492
in filters — 310, 368-369, 373-374
in friction materials — 484-485
in selflubricating bearings —
361-363, 365-367
mechanism of sintering — 301-
305
copper-zinc
in brass structural parts —
427-429
in friction materials — 484, 485
copper powder metallurgy
powder production — 33-38
sintered powder compacts
dimensional changes — 213-216,
218-219, 221-223
microstructure — 227
copper-tin equilibrium diagram — 302
for use as infiltrant — 315-317,
435-437, 553
surface tension — 244-245
Core materials (magnetic) — 546-547
Core rod — 119, 129
material of construction — 141
Corrosion resistance
of filters — 368
of stainless steel structural parts —
429-430
Cost analysis — 469
for powder forgings — 480-481
for structural parts — 474-476
for wrought aluminum alloys,
conventional vs. P/M — 501-502
Coulter counter — 61-62, 85-86
Crack initiation and propagation
in cemented carbides — 395
Crank type presses — 132-133
Curved surfaces

compacts with — 131-132
Cyclic pressing
continuous compacting by — 171-172

D

DPG process — 18
DS - lead
dispersion strengthened lead alloy —
511-512
DS - nickel
dispersion strengthened nickel alloy
— 514-515
Darcy's law — 378
Deagglomeration of powder — 67, 84-85
Deflocculent — 312
Deicing
porous metal application — 368
Demagnetization curve — 532
Dendritic powder — 66, 69
Densification
densification parameter — 213
hot, by repressing — 447
in liquid phase sintering — 287, 290,
294-296
in sintering homogeneous metal
powder compacts — 212-224
Density distribution
in compacts — 99, 106-114
Dental restorations
with silver-tin amalgams — 527-529
Depth of fill — 115
Dewaxing — 101
dewaxing zone — 180, 189-192
of cemented carbide compacts — 387
Dewpoint of gases — 195, 202-203, 207
Diamond tool materials
metal bonded — 492-493
Die barrel — 119, 122, 131, 139-141
Die cavity — 117, 120, 121, 123
Die plate — 123, 126, 127
Die set — 123
Die table — 120
Die wall
friction — 110
lubrication — 101-102

147-148

Fountain pen tips
  by powder extrusion — 173

Fracture strain
  in cold forming P/M preforms — 464
  in hot forging P/M preforms —
    450-454

Friction, coefficent of
  in bearings — 367
  in friction materials — 484

Friction materials
  metallic — 484-489
    production — 484-487
  "organic" — 484
  "semimetallic" — 484

Fuel cell electrodes — 375, 377-378

Fuldens process
  for cutting tool materials — 33, 150,
    526-527

Furnaces, sintering — 179-193
  for powder rolled strip — 163-166

G

Gas adsorption method
  for specific surface measurement —
    62, 73-76, 91-92

Gatorizing
  process for superalloy components —
    503-504

Genus
  of particle aggregates — 242-243,
    261

Gibbs-Thomson equation — 247-248,
  292

Glass containers
  for hot isostatic pressing — 332-333

Glass filters — 368

Glidcop
  dispersion strengthened copper —
    513-514

Grainboundary diffusion
  mechanism in sintering — 249-253

Graingrowth
  during sintering — 227, 229-232,
    258-260

exaggerated in tungsten powder
  compacts — 229
in copper-tin compacts — 305
in doped tungsten — 341-344

Granular powder — 67, 70

Graphite
  as addition to iron for P/M steels —
    270-272, 405-416
  colloidal suspension as lubricant in
    hot forging — 459
  in friction materials — 485-487
  in metal-graphite brushes — 556-557
  in selflubricating bearings — 361-363

Graphite molds
  in loose powder sintering — 310

Green density — 218

Green strength — 20, 60, 97, 212

Grinding wheels
  with diamonds as abrasives — 492

Growth
  see Dimensional change during
    sintering

H

H-iron process — 23

Hall flow meter — 60-61, 93-96

Hammer forging
  of P/M preforms — 449, 458

Hard metal
  see cemented carbides

Hardness
  of cemented carbides — 389
  of sintered metal compacts — 238-239

Hardenability
  of sintered steels — 409-410, 439

Heat treating
  of structural parts — 402, 409-413,
    438-439

Heating elements
  for sintering furnaces — 184-185

Heavy alloy
  composition and properties — 348
  mechanism of sintering — 284-295
  presintering — 282

Height of rise

578

disadvantages of — 101-102
effect upon green density — 101-102
effect upon green strength — 101
volatilization of — 180
in selflubricating bearings — 364
starved lubrication — 362
paraffin in cemented carbides — 387

M

Machining
cost comparison with P/M parts —
470, 475-476
of structural parts — 402, 439-440
Macroporous layer
in porous electrodes — 377
Macrosegregation
avoidance by P/M processing — 496
Magnesium
alloy powder production — 46
dispersion strengthened alloys —
517-518
parts extrusion — 324-325
properties of extrusions — 498-499
Magnetic applications of powder
metallurgy — 531-548
core materials — 546-547
permanent magnets — 539-545
soft magnetic materials
pressed to shape — 533-538
wrought — 538-539
Magnetic properties of soft P/M
magnetic materials — 533-538
effect of density — 534-536
effect of particle size distribution —
536
effect of sintering temperature —
534-536
Main bearings, automotive — 522-525
Manganese
diffusion in iron — 271
manganese-molybdenum-steel for
hot forging — 455
Mannesmann process
for atomizing iron powder — 28
Martensitic stainless steel

(see also stainless steel)
properties of sintered compacts —
431
Master alloys
for hot forging compositions:
Ni, Mn or Mo with C; ferromolybdenum with C — 456
for structural parts:
Mn-Cr-Mo-C and Mn-V-Mo-C
— 415-416
for titanium alloys — 353
Mean free path
in cobalt binder of cemented
carbides — 394-395
Mechanical alloying
for producing dispersion strengthened
nickel, iron and aluminum-base
alloys — 515-517
Mechanical presses
for hot forging — 458
for structural parts — 117, 132-139
Mechanism of sintering
by interdiffusion — 269-284
of single phase metal powders —
241-267
with liquid phase — 285-307
Melt-drop technique of atomizing — 16
Meshbelt furnaces — 182-185
Metal-graphite brushes — 556-557
Metal-oxide equilibria — 199-203
Metal Powder Industries Federation
specifications for bearings — 365
specifications for structural parts
(standard 35) — 421-426
standard 2 — 60
3 — 93
4 — 93
5 — 76-77
6 — 60
10 — 80
15 — 97
28 — 94
32 — 87
45 — 94
46 — 94

NOTES

NOTES

NOTES

NOTES

NOTES

30C-UL-5/80 Production, Century Graphics Inc.-Printing, Universal Lithographers Inc.